Lab 8 10/20

Ex 4 — 16/20
Ex 13 = 14/20
Ex 14 = 20/20
Ex 15 = 20/20
Ex 16 = 18/20

Applications and Investigations in Earth Science

Applications and Investigations in Earth Science

SEVENTH EDITION

Edward J. Tarbuck

Frederick K. Lutgens

Kenneth G. Pinzke

Illustrations by Dennis Tasa

Boston   Columbus   Indianapolis   New York   San Francisco   Upper Saddle River
Amsterdam   Cape Town   Dubai   London   Madrid   Milan   Munich   Paris   Montréal   Toronto
Delhi   Mexico City   São Paulo   Sydney   Hong Kong   Seoul   Singapore   Taipei   Tokyo

Acquisitions Editor, Geology: *Andrew Dunaway*
Marketing Manager: *Maureen McLaughlin*
Editorial Director: *Frank Ruggirello*
Assistant Editor: *Sean Hale*
Editorial Assistant: *Michelle White*
Marketing Assistant: *Nicola Houston*
Managing Editor, Geosciences and Chemistry:
 Gina M. Cheselka
Project Manager, Science: *Edward Thomas*
Full Service/Composition: *Element, LLC.*
Production Editor, Full Service: *Heidi Allgair/Element*

Art Director: *Derek Bacchus*
Cover and Interior Design: *Derek Bacchus*
Senior Technical Art Specialist: *Connie Long*
Art Studio: *Spatial Graphics*
Photo Research Manager: *Maya Melenchuk*
Photo Researcher: *Kristin Piljay*
Senior Manufacturing and Operations Manager: *Nick Sklitsis*
Operations Supervisor: *Maura Zaldivar*
Cover Photo: *Kayaker takes an early morning tour on Lake Moraine, Alberta, Canada;* ©Charlie Munsey/Corbis.

©2012, 2009, 2006, 2003, 2000, 1997, 1994 by Pearson Education, Inc.
Pearson Prentice Hall
Pearson Education, Inc.
Upper Saddle River, New Jersey 07458

10 9 8 7 6 5 4 3 2 1

Library of Congress Cataloging-in-Publication Data

Tarbuck, Edward J.
  Applications and investigations in Earth science / Edward J. Tarbuck, Frederick K.
Lutgens, Dennis Tasa. — 7th ed.
    p. cm.
  Includes bibliographical references and index.
  ISBN-13: 978-0-321-68955-9 (alk. paper)
  ISBN-10: 0-321-68955-0 (alk. paper)
  1. Earth sciences—Laboratory manuals.   I. Lutgens, Frederick K.   II. Tasa, Dennis.
III. Title.
  QE44.T37 2011
  550.78—dc22   2011000586

ISBN-10:   0-321-68955-0
ISBN-13: 978-0-321-68955-9

Prentice Hall
is an imprint of

www.pearsonhighered.com

Dedication

The course of human life is much like the dynamic planet we occupy: Unexpected *events* occur, resulting in unexpected *changes*. The unexpected and untimely death of our longtime friend and colleague Ken Pinzke (1945–2009) resulted in a thorough examination of *Applications and Investigations in Earth Science*. The success of the first six editions of *Applications* may have ensured a successful seventh edition, but a complete review and assessment of the needs of faculty and students seemed the most appropriate way to honor Ken's more than forty years of tireless efforts educating students in the Earth sciences. It is in his memory that we dedicate this edition.

Contents

Preface

Like previous editions, the *Seventh Edition* is intended to be a supplemental tool for achieving an understanding of the basic principles of geology, meteorology, oceanography, and astronomy. While enrolled in what may be their first, and possibly only, Earth science course, students will greatly benefit from putting the material *presented* in the classroom to *work* in the laboratory. Learning, we believe, becomes significant and long-lasting when accomplished through discovery. While some students may be inspired to pursue careers in the Earth sciences based on something they encounter during this course, most will not. Regardless of the path on which students find themselves, the goal of this edition is to expand and diversify the knowledge base of every student as it relates to the ever-changing, complex, and fascinating planet we call home.

New to this Edition

Faculty who have used previous editions of *Applications and Investigations* may recognize little more than the sequential order of the exercises of this edition. Our goal for this edition was to minimize the need for faculty instruction in the lab, freeing them to use their valuable time to interact with students. The most significant changes to this edition are described in the following sections.

Exercises That Are More Self-Contained

Significant effort has been put into making the exercises less reliant on traditional text material and/or direct faculty instruction. In some exercises, this meant providing additional background material within the exercises themselves. In others, questions that relied heavily on outside material have been modified or replaced. We are confident that these changes, some major and some minor, make these exercises more useful and meaningful for both students and instructors.

A More Inquiry-Based Lab Experience

Whenever possible, the exercises have been modified in order to provide a "hands-on" approach to learning. For example, rather than providing rules for reading contour lines, we ask students to explore the nature of contour lines through a series of activities and questions. Furthermore, we have endeavored to engage students in the process of gathering and analyzing scientific data to enhance their critical reasoning skills.

Improved Visual Components

In an effort to respond to the changing nature of our students, who are accustomed to stunning visual images provided by the many technological tools at their disposal, we have completely transformed the visual elements of this manual. In this edition, over 80 new photos support the updated manuscript and questions. In addition, the vast majority of the illustrations have been redrawn to enhance their visual clarity and appeal.

Content Revised to Improve Clarity

Our many years of classrooms instruction have made us keenly aware of the frustration that students (and hence, by default, their instructors) face when instructions and questions are ambiguous. Likewise, we recognize that instructors are genuinely interested in making learning experiences meaningful for their students. With those things in mind, we ensured that the exercises were reviewed by a support team with educational backgrounds other than Earth science—a reflection, essentially, of the majority of the students who utilize this manual. Following these reviews, the material was extensively revised to provide a manuscript, questions, and supporting illustrations that more effectively convey the course material.

Layout

In the Seventh Edition, we focused on revising the design and layout of this manual to increase ease of use and promote student learning.

Safety

As students work in laboratory settings, safety is a critical element. With this in mind, Caution boxes are highlighted throughout the text that detail the proper handling of materials and the care needed to effectively complete these labs.

Finally, we sincerely hope that this Seventh Edition enhances the planning and implementation of instructional goals of all faculty—those who have used our materials for many years as well as new faces who bring fresh ideas and perspectives to the varied classrooms of the 21st century.

Acknowledgements

Writing a laboratory manual requires the talents and cooperation of many people. We value the support of Teresa Tarbuck and Mark Watry of Spring Hill College, and Amy Sword, a student at Kansas State University, whose evaluation of the Sixth Edition greatly enhanced this Seventh Edition. They helped make the exercises more current, readable, and engaging.

Working with Dennis Tasa, who creates the manual's outstanding illustrations, is always enjoyable and rewarding. We value his outstanding artistic talent, imagination, and unconditional patience with extensive revisions. Dennis and his excellent staff have made preparing this edition a pleasure.

Great thanks also go to those colleagues who prepared in-depth reviews. Their critical comments and thoughtful input helped guide and strengthen our efforts. Special thanks to:

Mark Borderlon, Irvine Valley College

Brett Burkett, Collin County Community College

Lisa Chaddock, Cuyamaca College

Dora Devery, Alvin Community College

Janis Forehand, Tarrant County Junior College

Carlos Garza, Clark Atlanta University

Brian Kirchner, Henry Ford Community College

Karen Koy, Missouri Western State University

Rob Martin, Florida State College at Jacksonville

Neil Mulchan, Broward College

Steven Newton, Laney College

As always, we want to acknowledge the team of professionals at Pearson Prentice Hall. We sincerely appreciate the company's enduring support of excellence, innovation, and commitment to producing outstanding texts and laboratory manuals. Special thanks to our editor, Andy Dunaway, and to our conscientious project manager, Sean Hale, for another job well done. The production team, led by Ed Thomas at Pearson and Heidi Allgair at Element LLC, did an excellent job of maintaining the quality and efficiency of this project. All are true professionals with whom we are very fortunate to be associated.

Ed Tarbuck
Fred Lutgens

PART 1
Geology

Photo by SCPhotos/Alamy

The Study of Minerals

Objectives

Completion of this exercise will prepare you to:

A. Describe the physical properties commonly used to identify minerals.

B. Identify minerals using a mineral identification key.

C. Visually identify some common rock-forming minerals.

D. List the uses of several economic minerals.

Materials

mineral samples   dilute hydrochloric acid
streak plate   magnet
glass plate   hand lens

Introduction

For Earth science professionals and students learning about our planet, identifying minerals using relatively simple techniques is an important skill. Knowledge of common minerals and their properties is basic to an understanding of rocks. This exercise introduces the common physical and chemical properties of minerals and how these properties are used to identify minerals.

What Is a Mineral?

Geologists define a **mineral** as *any naturally occurring inorganic solid that possesses an orderly crystalline structure and can be represented by a chemical formula.* Thus, Earth materials that are classified as minerals exhibit all of the following characteristics:

- **Naturally occurring** Minerals form by natural, geologic processes. Synthetic materials, meaning those produced in a laboratory or by human intervention, are not considered minerals.

- **Solid substance** Only crystalline substances that are solid at temperatures encountered at Earth's surface are considered minerals.

- **Orderly crystalline structure** Minerals are crystalline substances, which means their atoms are arranged in an orderly, repetitive manner.

- **Generally inorganic** Inorganic crystalline solids, such as ordinary table salt (halite), that are found naturally in the ground are considered minerals, whereas organic compounds, are generally not. Sugar, a crystalline solid like salt that comes from sugarcane or sugar beets, is a common example of an organic compound that is not considered a mineral.

- **Can be represented by a chemical formula** The common mineral quartz, for example, has the formula SiO_2, which indicates that quartz consists of silicon (Si) and oxygen (O) atoms in a ratio of one-to-two. This proportion of silicon to oxygen is true for any sample of quartz, regardless of its origin. However, the compositions of some minerals vary *within specific, well-defined limits.*

1. Use the geologic definition of a mineral to determine which of the items listed in Figure 1.1 are minerals and which are not minerals. Check the appropriate box.

Yes	No	Mineral	Yes	No	Mineral
	✓	Rock candy	✓	✓	Obsidian
✓		Quartz		✓	Cubic zirconia
	✓	Motor oil		✓	Hydrogen
✓		Emerald		✓	Rain water
	✓	Vitamin D	✓		Halite

Figure 1.1   Which of these items are minerals?

Physical Properties of Minerals

Minerals have definite crystalline structures and chemical compositions that give them unique sets of physical and chemical properties shared by all specimens of that mineral, regardless of when or where they form. For example, if you compare two samples of the mineral quartz, they will be equally hard and equally dense, and they will break in the same manner. However, the physical properties of individual samples may vary within specific limits due to ionic substitutions, inclusions of foreign elements (impurities), and defects in the crystal structure.

Diagnostic properties are particularly useful in mineral identification. The mineral magnetite, for example will attract a magnet, while the mineral halite has a salty taste. Because so few minerals share these properties, magnetism and salty taste are considered diagnostic properties of magnetite and halite, respectively.

Optical Properties

Luster The appearance of light reflected from the surface of a mineral is known as **luster**. Minerals that have the appearance of metals, regardless of color, are said to have a **metallic luster**. Some metallic minerals, such as native copper and galena, develop a dull coating or tarnish when exposed to the atmosphere. Because they are not as shiny as samples with freshly broken surfaces, these samples exhibit a *submetallic luster*.

Most minerals have a **nonmetallic luster** and are described as *vitreous (glassy)*, *dull* or *earthy* (a dull appearance like soil), or *pearly* (such as a pearl or the inside of a clamshell). Still others exhibit lusters that are *silky* (like silk or satin cloth) or *greasy* (as though coated in oil).

2. Examine the luster of the minerals in Figure 1.2. Place the letter A, B, C, D, or E in the space provided that corresponds to the luster exhibited. Letters may be used more than once. **A.** Metallic luster, **B.** Nonmetallic luster—glassy, **C.** Nonmetallic luster—dull, **D.** Nonmetallic luster—silky, **E.** Nonmetallic luster—greasy.

3. Examine the mineral specimens provided by your instructor and separate them into two groups—metallic and nonmetallic.

Metallic: *Magnetite, Pyrite, Hornblende*

Nonmetallic: *Biotite, Plagioclase Feldspar, Halite*

The Ability to Transmit Light Minerals are able to transmit light to different degrees. A mineral is described as **opaque** when no light is transmitted; **translucent** when light, but not an image, is transmitted; and **transparent** when both light and an image are visible through the sample.

B Quartz   A Galena

C Limonite   B Gypsum (selenite)

D Talc   A Native copper

Figure 1.2 Photos illustrating various types of mineral luster.

Color Although **color** is generally the most conspicuous characteristic of a mineral, it is infrequently considered a diagnostic property.

4. Based on the samples of fluorite and quartz in Figure 1.3, why isn't color a diagnostic property of these two minerals?

They vary in colors so using color is not a reliable way to identify them.

Streak The *color* of a mineral in powdered form, called **streak**, is often useful in identification. A mineral's streak is obtained by rubbing it across a *streak plate* (a piece of unglazed porcelain) and observing the color of the mark it leaves. Although the color of a particular mineral may vary from sample to sample, its streak is usually consistent in color.

Streak can also help distinguish between minerals with metallic luster and those with nonmetallic luster. Metallic minerals generally have a dense, dark streak

A. Fluorite   **B. Quartz**

Figure 1.3 Color variation exhibited by **A.** fluorite and **B.** quartz.

Color: brassy
Streak: dark gray

Figure 1.4 Streak test.

Figure 1.6 Using streak to assist in describing luster.

Hematite var. specular
Fluorite
Feldspar var Amazonite

	COLOR OF SPECIMEN	STREAK
Specimen A:	Gray & sparkly	Bronze
Specimen B:	Light Green	White
Specimen C:	Green/Blue (Teal)	White

(Figure 1.4), whereas minerals with nonmetallic luster typically have a light-colored streak.

Not all minerals produce a streak when rubbed across a streak plate. For example, the mineral quartz is harder than a streak plate. Therefore, it produces no streak using this method.

5. Figure 1.5 shows two specimens of the mineral hematite and their corresponding streaks. For both samples, describe the color of the specimen and the streak.

	COLOR OF SPECIMEN	STREAK
Specimen A:	Red-Orange	Red-Orange
Specimen B:	Dark w/ Red	Red-orange

6. Select three of the mineral specimens provided by your instructor. Do they exhibit a streak? If so, is the streak the same color as the mineral specimen?

Figure 1.5 Hematite, an ore of iron, is found in both nonmetallic and metallic forms.

7. To some observers, the mineral shown in Figure 1.6 exhibits a metallic luster, while others describe its luster as nonmetallic. Based on the streak of this sample, how would you describe its luster?

Luster: Non-metallic because it has a light streak, whereas metallics have a darker streak.

Crystal Shape, or Habit

Recall that all minerals are crystalline, and when they form in unrestricted environments, they develop **crystals** that exhibit geometric shapes. For example, well-developed quartz crystals are hexagonal with pyramid-shaped ends, and garnet crystals are 12-sided (Figure 1.7). In addition, some crystals tend to grow and form characteristic shapes or patterns called **crystal shape**, or **habit**. Commonly used terms to describe various crystal habits include *bladed* (flat, elongated strips), *fibrous* (hair-like), *tabular* (tablet shaped), *granular* (aggregates of small crystals), *blocky* (square), and *banded* (layered).

Although crystal shape, or habit, is a diagnostic property for some specimens, many of the mineral samples you will encounter consist of crystals that are too minute to be seen with the unaided eye or are intergrown such that their shapes are not obvious.

A. Quartz **B. Garnet**

Figure 1.7 Characteristic crystal forms of **A.** quartz and **B.** garnet.

Specimen A   Specimen B

Specimen C   Specimen D

Figure 1.8 Crystal shapes and habits.

8. Select one of the following terms to describe the crystal shape, or habit, of each specimen shown in Figure 1.8: cubic crystals, hexagonal crystals, fibrous habit, banded habit, blocky habit, bladed habit, tabular habit.

Specimen A: <u>Fibrous</u>

Specimen B: <u>Bladed Habit</u>

Specimen C: <u>Banded Habit</u>

Specimen D: <u>Cubic Crystals</u>

SKIP ➤ Use a *contact goniometer*, illustrated in Figure 1.9, to measure the angle between adjacent faces on the quartz crystals on display in the lab.

a. Are the angles about the same for each quartz specimen, or do they vary from one sample to another? _____

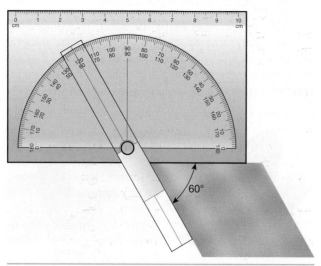

Figure 1.9 Contact goniometer.

Figure 1.10 Sheets of elastic minerals, like muscovite, can be bent but will snap back when the stress is released.

b. Write a generalization that describes how the angle between crystal faces relates to the shape of the sample. _____

Mineral Strength

How easily minerals break or deform under stress is determined by the type and strength of the chemical bonds that hold the crystals together.

Tenacity The term **tenacity** describes a mineral's resistance to breaking or deforming. Minerals that are ionically bonded, such as fluorite and halite, tend to be *brittle* and shatter into small pieces when struck. By contrast, minerals with metallic bonds, such as native copper, are *malleable*, or easily hammered into different shapes. Minerals, including gypsum and talc, that can be cut into thin shavings are described as *sectile*. Still others, notably the micas, are *elastic* and will bend and snap back to their original shape after the stress is released (Figure 1.10).

Hardness One of the most useful diagnostic properties is **hardness**, a measure of the resistance of a mineral to abrasion or scratching. This property is determined by rubbing a mineral of unknown hardness against one of known hardness or vice versa. A numerical value of hardness can be obtained by using the Mohs scale of hardness, which consists of 10 minerals arranged in order from 1 (softest) to 10 (hardest), as shown in Figure 1.11. It should be noted that the Mohs scale is a relative ranking; it does not imply that mineral number 2, gypsum, is twice as hard as mineral 1, talc.

10. The minerals shown in Figure 1.12 are fluorite and topaz that have been tested for hardness. Use the Mohs scale in Figure 1.11 to identify which is fluorite and which is topaz.

MINERAL NAME

Specimen A: <u>Fluorite</u>

Specimen B: <u>Topaz</u>

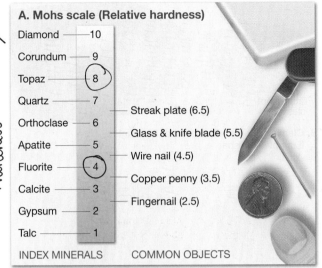

A. Mohs scale (Relative hardness)

INDEX MINERALS		COMMON OBJECTS
Diamond	10	
Corundum	9	
Topaz	⑧	
Quartz	7	
Orthoclase	6	Streak plate (6.5)
Apatite	5	Glass & knife blade (5.5)
Fluorite	④	Wire nail (4.5)
Calcite	3	Copper penny (3.5)
Gypsum	2	Fingernail (2.5)
Talc	1	

Hardness → (handwritten, left margin)

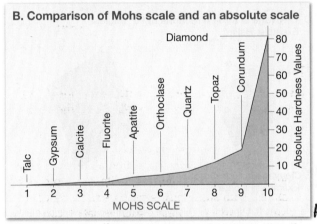

B. Comparison of Mohs scale and an absolute scale

(Graph: MOHS SCALE on x-axis, 1–10; Absolute Hardness Values on y-axis, 10–80. Minerals labeled: Talc, Gypsum, Calcite, Fluorite, Apatite, Orthoclase, Quartz, Topaz, Corundum, Diamond.)

Figure 1.11 Hardness scales. **A.** Mohs scale of hardness, with the hardness of some common objects. **B.** Relationship between the Mohs relative hardness scale and an absolute hardness scale.

In the laboratory, common objects are often used to determine the hardness of a mineral. These objects include a human fingernail, which has a hardness of about 2.5, a copper penny (3.5), and a piece of glass (5.5). The mineral gypsum, which has a hardness of 2,

Less Dense (handwritten) *Harder* (handwritten)

Scratch

A.   **B.**

Figure 1.12 Hardness test.

Table 1.1 Hardness Guide

HARDNESS	DESCRIPTION
Less than 2.5	A mineral that can be scratched by your fingernail (hardness = 2.5)
2.5 to 5.5	A mineral that cannot be scratched by your fingernail (hardness = 2.5) and cannot scratch glass (hardness = 5.5)
Greater than 5.5	A mineral that scratches glass (hardness = 5.5)

can be easily scratched with a fingernail. On the other hand, the mineral calcite, which has a hardness of 3, will scratch a fingernail but will not scratch glass. Quartz, one of the hardest common minerals, will easily scratch glass. Diamonds, hardest of all, scratch anything, including other diamonds. Table 1.1 serves as a guide for describing the hardness of minerals.

11. Select three mineral specimens from the set provided by your instructor. Determine the hardness of each mineral, using Table 1.1 as a guide.

HARDNESS

Rose Quartz — Specimen A: ~6

Talc — Specimen B: <2.5 (1)

Anthracite Coal — Specimen C: ~3

Cleavage In the crystalline structure of many minerals, some chemical bonds are weaker than others. When minerals are stressed, they tend to break (cleave) along these planes of weak bonding, a property called **cleavage**. When broken, minerals that exhibit cleavage produce smooth, flat surfaces, called **cleavage planes**, or **cleavage surfaces**.

Cleavage is described by (1) noting the number of **directions of cleavage**, which is the number of different sets of cleavage planes that form on the surfaces of a mineral when it cleaves, and (2) determining the **angle(s)** at which the directions of cleavage meet (Figure 1.13). Each cleavage surface of a mineral that has a different orientation is counted as a different direction of cleavage. However, when cleavage planes are parallel, they are counted only *once*, as one direction of cleavage. Minerals may have one, two, three, four, or more directions of cleavage (Figure 1.13).

When minerals such as muscovite, calcite, halite, and fluorite are broken, they display cleavage surfaces that are easily detected. However, other minerals exhibit cleavage planes that consist of multiple offset surfaces that are not as obvious. A reliable way to determine whether a specimen exhibits cleavage is to

A. Cleavage in one direction.
Example: Muscovite

B. Cleavage in two directions at 90° angles.
Example: Feldspar

Fracture not cleavage

C. Cleavage in two directions not at 90° angles.
Example: Hornblende

Fracture not cleavage

D. Cleavage in three directions at 90° angles.
Example: Halite

E. Cleavage in three directions not at 90° angles.
Example: Calcite

F. Cleavage in four directions.
Example: Fluorite

Figure 1.13 Common cleavage directions of minerals. **A.** Basal cleavage produces flat sheets. **B.** This type of prismatic cleavage produces an elongated form with a rectangular cross section. **C.** This type of prismatic cleavage produces an elongated form with a parallelogram cross section. **D.** Cubic cleavage produces cubes or parts of cubes. **E.** Rhombohedral cleavage produces rhombohedrons. **F.** Octahedral cleavage produces octahedrons.

rotate it in bright light and look for flat surfaces that reflect light.

Do not confuse cleavage with crystal shape. When a mineral exhibits cleavage, it will break into pieces that all have the same geometry. By contrast, the smooth-sided quartz crystals shown in Figure 1.7 illustrate crystal shape rather than cleavage. If broken, quartz crystals fracture into shapes that do not resemble one another or the original crystals.

12. Describe the cleavage of the mineral shown in Figure 1.14A. *(Muscovite)*
    Cleavage in one direction

13. Refer to the photograph in Figure 1.14B, which shows a mineral that has several smooth, flat cleavage surfaces, to complete the following.

    a. How many cleavage planes or surfaces are present on the specimen?
    Number of cleavage planes: _____

A. Muscovite

B. Calcite

Figure 1.14 Identifying cleavage.

b. How many directions of cleavage are present on the specimen?

Number of directions of cleavage: _3_

c. The cleavage directions meet at (90-degree angles *or* angles other than 90 degrees).

Degrees of cleavage angles: Angles other than 90° (75°)

14. Select one mineral specimen supplied by your instructor that exhibits more than one direction of cleavage. How many directions of cleavage does it have? What are the angles of its cleavage?

Halite 4 Planes

Number of directions of cleavage: _3_

Degrees of cleavage angles: _90°_

Fracture   Minerals that do not exhibit cleavage when broken are said to **fracture** (Figure 1.15). Fractures are described using terms such as *irregular*, *splintery*, or *conchoidal* (smooth, curved surfaces resembling broken glass). Some minerals may cleave in one or two directions and fracture in another.

Density and Specific Gravity

Density is defined as mass per unit volume and is expressed in grams per cubic centimeter (g/cm^3). Mineralogists also use a related measure called *specific gravity* to describe the density of minerals. **Specific gravity (SG)** is a number representing the ratio of a mineral's weight to the weight of an equal volume of water. Water has a specific gravity of 1.

Most common rock-forming minerals have specific gravities of between 2 and 3. For example, quartz has a specific gravity of 2.7. By contrast, some metallic minerals such as pyrite, native copper, and magnetite are more than twice as dense as quartz and thus are considered to have a high specific gravity. Galena, an ore of lead, is even more dense, with a specific gravity of about 7.5.

With a little practice, you can estimate the specific gravity of a mineral by hefting it in your hand. Ask yourself whether the mineral feels about as "heavy" as

Figure 1.16   A variety of magnetite, lodestone, is like a magnet and will attract iron objects.

similar-sized rocks you have handled. If the answer is "yes," the specific gravity of the sample is likely between 2.5 and 3. (*Note:* Exercise 23, "The Metric System, Measurements, and Scientific Inquiry," contains a simple experiment for estimating the specific gravity of a solid.)

SKIP

15. Hefting each specimen supplied by your instructor. Identify the minerals from this group that exhibit high specific gravities.

a. How many minerals have a high specific gravity? _____

b. Of those with a high SG, did most of them have a metallic or nonmetallic luster? _____

Other Properties of Minerals

Magnetism   Magnetism is characteristic of minerals, such as magnetite, that have a high iron content and are attracted by a magnet. One variety of magnetite called *lodestone* is magnetic and will pick up small objects such as pins and paper clips (Figure 1.16).

Taste   The mineral halite has a "salty" taste.

> **CAUTION:** Do not taste any minerals or any other materials unless you know it is *absolutely* safe to do so.

Odor   A few minerals have distinctive odors. For example, minerals that are compounds of sulfur smell like rotten eggs when rubbed vigorously on a streak plate.

Feel   The mineral talc often feels "soapy," and the mineral graphite has a "greasy" feel.

Striations   Striations are closely spaced, fine lines on the crystal faces of some minerals. Certain plagioclase

A. Irregular fracture   **B. Conchoidal fracture**

Figure 1.15   Minerals that do not exhibit cleavage are said to fracture.

Figure 1.17 These parallel lines, called *striations*, are a distinguishing characteristic of the plagioclase feldspars.

Figure 1.18 Calcite reacting with dilute hydrochloric acid. (Photo by Chip Clark)

feldspar minerals exhibit striations on one cleavage surface (Figure 1.17).

Reaction to Dilute Hydrochloric Acid A very small drop of dilute hydrochloric acid, when placed on the surface of certain minerals, will cause them to "fizz" (effervesce) as carbon dioxide is released (Figure 1.18). The acid test is used to identify the *carbonate minerals*, especially the mineral calcite ($CaCo_3$), the most common carbonate mineral.

> **CAUTION:** Hydrochloric acid can discolor, decompose, and disintegrate mineral and rock samples. Use the acid only after you have received specific instructions on its use from your instructor. Never taste minerals that have had acid placed on them.

Identification of Minerals

Now that you are familiar with the physical properties of minerals, you are ready to identify the minerals supplied by your instructor. To complete this activity, you need the Mineral Data Sheet (Figure 1.19) and the Mineral Identification Key (Figure 1.20).

The mineral identification key divides minerals into three primary categories: (1) those with metallic luster, (2) those with nonmetallic luster that are dark colored, (3) and those with nonmetallic luster that are light colored. Hardness is used as a secondary identifying factor. As you complete this exercise, remember that the objective is to learn the *procedure* for identifying minerals through *observation* and *data collection* rather than simply to name the minerals.

16. Identify the specimens supplied by your instructor, using the following steps:

    Step 1: Leaving enough space for each mineral, number a piece of paper (up to the number of samples you've been assigned) and place your specimens on the paper.

    Step 2: Select a specimen and determine its physical properties by using the tools provided (glass plate, streak plate, magnet, etc.).

    Step 3: List the properties of that specimen on the Mineral Data Sheet (Figure 1.19).

    Step 4: Use the Mineral Identification Key (Figure 1.20) as a resource to identify the specimen.

    Repeat steps 2 through 4 until all samples have been identified.

In the following section you will examine some common rock-forming minerals and selected economic minerals. This will provide you with the information needed to complete the last column of the mineral data sheet (Figure 1.19).

MINERAL DATA SHEET #1

Sample number	Luster	Hardness	Color	Streak	Fracture or Cleavage (number of directions and angles)	Other Properties	Name	Economic Use or Rock-forming
48	Metallic	6.5-7.5	Mostly Dark	Brown	Fracture	Magnetic	Magnetite	Ore of Iron
20	Metallic	6	Darkish with some Gold	Darkish Yellow	Fracture	Rough Feeling	Pyrite	Rock Forming
25	Non Metallic	6.5	Black/Dark Red	Deep Red	Fracture	X	Hematite	Ore of Iron, Pigment
27	Metallic	1-2	Silvery Gray	Gray	Fracture	Feels Greasy	Graphite	Pencil Lead, Lubricant
21	Metallic	6-7	Black	No Streak	Cleavage 2 Directions (Not 90°)	Rubs off Easy	Hornblende	Rock Forming
43	Metallic	5.5-6	Gray Green	White	Fracture	Bad Odor	Augite	Rock Forming
17	Non-Metallic	7	Light Pink	White	Fracture	Feels Soapy (?)	Rose Quartz	Primary ingredient in glass
45	Non-Metallic	6.5-7	Yellow Green	White	Fracture	Bad Odor	Olivine	Rock Forming
15	Metallic	3.5	Dark with Orange	Bronze	Fracture	X	Sphalerite	Major ore of Zinc
32	Non-Metallic	2.5	Yellow	White (Kind of) Yellow	Fracture	Feels Soapy	Sulfur	Rock Forming

Refer to the section on "Mineral Groups" to complete this part

Figure 1.19 Mineral data sheet.

MINERAL DATA SHEET #2

Sample number	Luster	Hardness	Color	Streak	Fracture or Cleavage (number of directions and angles)	Other Properties	Name	Economic Use or Rock-forming
29	Metallic	2.5	Black/Gold	Gold	Cleavage one direction (not 90°)	Some Striations Brittle	Biotite	Rock-Forming
35	Non Metallic	6-7	Pink/Gray	White	Cleavage one direction (not 90°)	Pearly	Potassium Feldspar	Rock-Forming
34	Non Metallic	6-7	Gray	White	Fracture	Pearly	Plagioclase Feldspar	Rock-Forming
7	Non Metallic	7	White	White	Fracture	Pearly	Quartz	Primary Ingredient in Glass
3	Non Metallic	2.5-3	Blue Gray	White	Fracture	Glassy	Calcite	cement, soil conditioning
12	Non Metallic	2.5	Clear	White	Cleavage 3 Directions (90°)	Glassy	Halite	Table Salt Road Salt
4	Non Metallic	4	Gray/White	White	Cleavage one direction (not 90°)		Fluorite	used in steel manufacturing Toothpaste
10	Metallic	2.5	Clear/Tan	White	Cleavage one direction (not 90°)	Brittle	Muscovite	Insulator in Electrical Applications
5A	Non Metallic	2	Clear	White	Cleavage two direction (not 90°)	Glassy	Selenite Gypsum	Wallboard Plaster
1	Non Metallic	1	Blue Green	White	Fracture	Feels Soapy	Talc	Paint Cosmetics

Refer to the section on "Mineral Groups" to complete this part

Figure 1.19   Mineral data sheet (continued)

MINERAL DATA SHEET #3

Sample number	Luster	Hardness	Color	Streak	Fracture or Cleavage (number of directions and angles)	Other Properties	Name	Economic Use or Rock-forming

Refer to the section on "Mineral Groups" to complete this part

Figure 1.19 Mineral data sheet (*continued*)

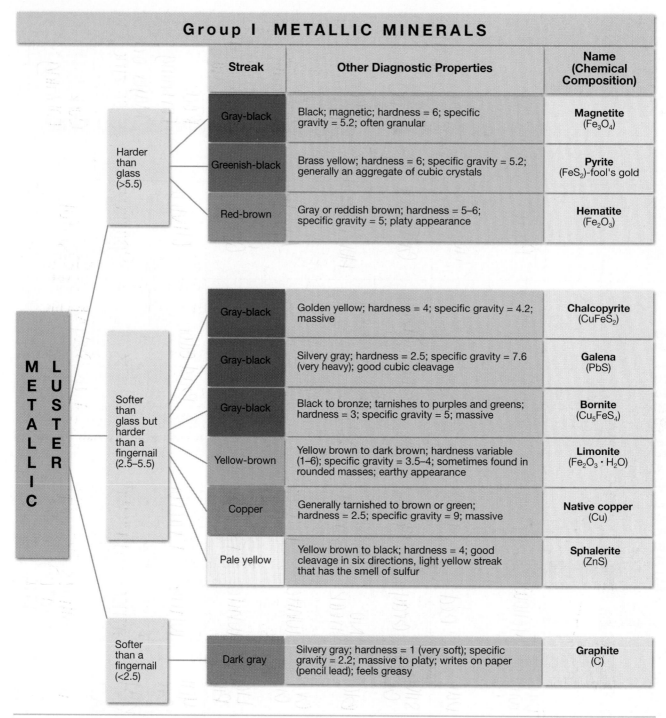

Figure 1.20   Mineral identification key.

Mineral Groups

More than 4000 minerals have been named, and several new ones are identified each year. Fortunately for students who are beginning to study minerals, no more than a few dozen are abundant! Collectively, these few make up most of the rocks of Earth's crust and, as such, are referred to as the **rock-forming minerals**.

Although less abundant, many other minerals are used extensively in the manufacture of products and are called **economic minerals**. However, rock-forming minerals and economic minerals are not mutually exclusive groups. When found in large deposits, some rock-forming minerals are economically significant. For example, the mineral calcite, the primary component of the sedimentary rock limestone, has many uses, including the production of concrete.

Group II NONMETALLIC MINERALS / DARK COLOR

	Cleavage	Other Diagnostic Properties	Name (Chemical Composition)
Harder than glass (>5.5)	Cleavage Present	Greenish black to black; hardness = 5–6; specific gravity = 3.4; fair cleavage, two directions at nearly 90 degrees	**Augite** (Ca, Mg, Fe, Al silicate)
		Black to greenish black; hardness = 5–6; specific gravity = 3.2; fair cleavage, two directions at nearly 60 degrees and 120 degrees	**Hornblende** (Ca, Na, Mg, Fe, OH, Al silicate)
		White to dark gray; hardness = 6; specific gravity = 2.6; two directions of cleavage at nearly right angles; striations on some faces	**Plagioclase feldspar** (Na, Ca, AlSi$_3$O$_8$)
	Cleavage poor or absent	Red to reddish brown; hardness = 6.5–7.5; conchoidal fracture; glassy luster	**Garnet** (Fe, Mg, Ca, Al silicate)
		Gray to brown; hardness = 9; specific gravity = 4; hexagonal crystals common	**Corundum** (Al$_2$O$_3$)
		Dark brown to black; hardness = 7; conchoidal fracture; glassy luster	**Smoky quartz** (SiO$_2$)
		Olive green; hardness = 6.5–7; small glassy grains	**Olivine** (Mg, Fe)$_2$SiO$_4$
Softer than glass but harder than a fingernail (2.5–5.5)	Cleavage present	Yellow brown to black; hardness = 4; good cleavage in six directions, light yellow streak that has the smell of sulfur	**Sphalerite** (ZnS)
		Dark brown to black; hardness = 2.5–3, excellent cleavage in one direction; elastic in thin sheets; black mica	**Biotite mica** (K, Mg, Fe, OH, Al silicate)
	Cleavage absent	Generally tarnished to brown or green; hardness = 2.5; specific gravity = 9; massive	**Native copper** (Cu)
Softer than a fingernail (<2.5)	Cleavage poor or absent	Reddish brown; hardness = 1–5; specific gravity = 4–5; red streak; earthy appearance	**Hematite** (Fe$_2$O$_3$)
		Yellow brown; hardness = 1–3; specific gravity = 3.5; earthy appearance; powders easily; not a true mineral	**Limonite** (Fe$_2$O$_3$ · H$_2$O)

Figure 1.20 Mineral identification key (*continued*)

Important Rock-Forming Minerals

Feldspar Group Feldspar is the most abundant mineral group and is found in many igneous, sedimentary, and metamorphic rocks (Figure 1.21). One group of feldspar minerals contains potassium ions in its crystalline structure and is referred to as *potassium feldspar*. The other group, called *plagioclase feldspar*, contains calcium and/or sodium ions (Figure 1.21). All feldspar minerals have two directions of cleavage that meet at 90-degree angles and are relatively hard (6 on the

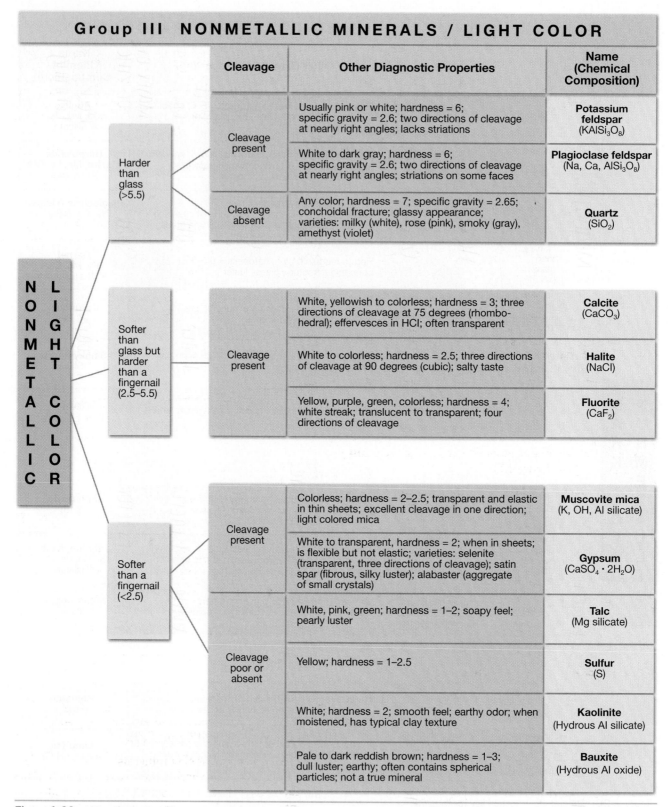

Group III NONMETALLIC MINERALS / LIGHT COLOR

	Cleavage	Other Diagnostic Properties	Name (Chemical Composition)
Harder than glass (>5.5)	Cleavage present	Usually pink or white; hardness = 6; specific gravity = 2.6; two directions of cleavage at nearly right angles; lacks striations	**Potassium feldspar** ($KAlSi_3O_8$)
		White to dark gray; hardness = 6; specific gravity = 2.6; two directions of cleavage at nearly right angles; striations on some faces	**Plagioclase feldspar** (Na, Ca, $AlSi_3O_8$)
	Cleavage absent	Any color; hardness = 7; specific gravity = 2.65; conchoidal fracture; glassy appearance; varieties: milky (white), rose (pink), smoky (gray), amethyst (violet)	**Quartz** (SiO_2)
Softer than glass but harder than a fingernail (2.5–5.5)	Cleavage present	White, yellowish to colorless; hardness = 3; three directions of cleavage at 75 degrees (rhombohedral); effervesces in HCl; often transparent	**Calcite** ($CaCO_3$)
		White to colorless; hardness = 2.5; three directions of cleavage at 90 degrees (cubic); salty taste	**Halite** (NaCl)
		Yellow, purple, green, colorless; hardness = 4; white streak; translucent to transparent; four directions of cleavage	**Fluorite** (CaF_2)
Softer than a fingernail (<2.5)	Cleavage present	Colorless; hardness = 2–2.5; transparent and elastic in thin sheets; excellent cleavage in one direction; light colored mica	**Muscovite mica** (K, OH, Al silicate)
		White to transparent, hardness = 2; when in sheets; is flexible but not elastic; varieties: selenite (transparent, three directions of cleavage); satin spar (fibrous, silky luster); alabaster (aggregate of small crystals)	**Gypsum** ($CaSO_4 \cdot 2H_2O$)
	Cleavage poor or absent	White, pink, green; hardness = 1–2; soapy feel; pearly luster	**Talc** (Mg silicate)
		Yellow; hardness = 1–2.5	**Sulfur** (S)
		White; hardness = 2; smooth feel; earthy odor; when moistened, has typical clay texture	**Kaolinite** (Hydrous Al silicate)
		Pale to dark reddish brown; hardness = 1–3; dull luster; earthy; often contains spherical particles; not a true mineral	**Bauxite** (Hydrous Al oxide)

Figure 1.20 Mineral identification key (*continued*)

Mohs scale). The only reliable way to physically distinguish the feldspars is to look for striations that are present on some cleavage surfaces of plagioclase feldspar (see Figure 1.17) but do not appear in potassium feldspar.

Quartz Quartz is a major constituent of many igneous, sedimentary, and metamorphic rocks. Quartz is found in a wide variety of colors (caused by impurities), is quite hard (7 on the Mohs scale), and exhibits conchoidal fracture when broken (Figure 1.22). Pure

A. Potassium feldspar crystal

B. Potassium feldspar with well developed cleavage

C. Sodium-rich plagioclase feldspar

D. Plagioclase feldspar (labradorite)

Figure 1.21 Feldspar group. **A.** Characteristic crystal form of plagioclase feldspar. **B.** Like this sample, most pink feldspar belongs to the potassium feldspar subgroup. **C.** Most sodium-rich plagioclase feldspar is light colored and has a porcelain luster. **D.** Calcium-rich plagioclase feldspar tends to be gray, blue-gray, or black in color. Labradorite, the variety shown here, exhibits iridescence.

quartz is clear, and if allowed to grow without interference, it will develop hexagonal crystals with pyramid-shaped ends (see Figure 1.7).

Mica *Muscovite* and *biotite* are the two most abundant members of the mica family. Both have excellent

A. Smoky quartz

B. Rose quartz

C. Milky quartz

D. Jasper

Figure 1.22 Quartz is one of the most common minerals and has many varieties. **A.** Smoky quartz is commonly found in coarse-grained igneous rocks. **B.** Rose quartz owes its color to small amounts of titanium. **C.** Milky quartz often occurs in veins that occasionally contain gold. **D.** Jasper is a variety of quartz composed of minute crystals.

A. Muscovite

B. Biotite

Figure 1.23 Two common micas: **A.** muscovite and **B.** biotite.

Kaolinite

Figure 1.24 Kaolinite, a common clay mineral.

Olivine

Figure 1.25 Olivine.

cleavage in one direction and are relatively soft (2.5 to 3 on the Mohs scale) (Figure 1.23).

Clay Minerals Most clay minerals originate as products of chemical weathering and make up much of the surface material we call soil. Clay minerals also account for nearly half of the volume of sedimentary rocks (Figure 1.24).

Olivine Olivine is an important group of minerals that are major constituents of dark-colored igneous rocks and make up much of Earth's upper mantle. Olivine is black to olive green in color, has a glassy luster, and exhibits conchoidal fracture (Figure 1.25).

A. Augite   **B. Hornblende**

Figure 1.26 These dark-colored silicate minerals are common constituents of igneous rocks: **A.** augite and **B.** hornblende.

Pyroxene Group The *pyroxenes* are a group of silicate minerals that are important components of dark-colored igneous rocks. The most common member, *augite*, is a black, opaque mineral with two directions of cleavage that meet at nearly 90-degree angles (Figure 1.26A).

Amphibole Group *Hornblende* is the most common member of the amphibole group and is usually dark green to black in color (Figure 1.26B). Except for its cleavage angles, which are about 60 degrees and 120 degrees, it is very similar in appearance to augite. Found in igneous rocks, hornblende makes up the dark portion of otherwise light-colored rocks.

Calcite *Calcite*, a very abundant mineral, is the primary constituent in the sedimentary rock limestone and the metamorphic rock marble. A relatively soft mineral (3 on the Mohs scale), calcite has three directions of cleavage that meet at 75-degree angles (see Figure 1.14B).

17. When feldspar minerals are found in igneous rocks, they tend to occur as elongated, rectangular crystals. By contrast, quartz (most commonly the smoky and milky varieties) usually occurs as irregular or rounded grains that have a glassy texture. Which of the crystals (A, B, C, or D) in the igneous rocks shown in Figure 1.27 are feldspar crystals and which are quartz?

    Feldspar: ___C, D___

    Quartz: ___A, B___

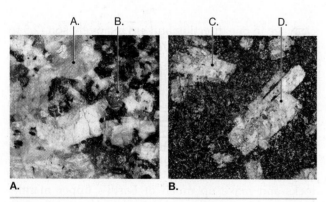

Figure 1.27 Identifying crystals of feldspar and quartz in coarse-grained igneous rocks.

Table 1.2 Economic Minerals

MINERAL	INDUSTRIAL AND COMMERCIAL USES
Calcite	Cement; soil conditioning
Chalcopyrite	Major ore of copper
Corundum	Gemstones, sandpaper
Diamond	Gemstones, drill bits
Fluorite	Used in steel manufacturing, toothpaste
Galena	Major ore of lead
Graphite	Pencil lead; lubricant
Gypsum	Wallboard; plaster
Halite	Table salt, road salt
Hematite	Ore of iron; pigment
Kaolinite	Ceramics; porcelain
Magnetite	Ore of iron
Muscovite	Insulator in electrical applications
Quartz	Primary ingredient in glass
Sphalerite	Major ore of zinc
Sulfur	Sulfa drugs; sulfuric acid
Sylvite	Potassium fertilizers
Talc	Paint, cosmetics

Economic Minerals

Many of the minerals selected for this exercise are metallic minerals that are mined to support our modern society. In addition, nonmetallic minerals such as fluorite, halite, and gypsum have economic value. Table 1.2 provides a list of some economic minerals and their industrial and commercial uses.

18. Complete the last column in the mineral data sheet (Figure 1.19) by indicating "rock-forming" or by listing the economic use of the samples used in this exercise (see Table 1.2).

Companion Website

The companion website provides numerous opportunities to explore and reinforce the topics of this lab exercise. To access this useful tool, follow these steps:

1. Go to www.mygeoscienceplace.com.
2. Click on "Books Available" at the top of the page.
3. Click on the cover of *Applications and Investigations in Earth Science, 7e.*
4. Select the chapter you want to access. Options are listed in the left column ("Introduction," "Web-based Activities," "Related Websites," and "Field Trips").

The Study of Minerals

Name _Michelle Ferro_   **Course/Section** _____

Date _____   **Due Date** _____

1. Name the physical property (hardness, color, streak, etc.) described by each of the following statements.

DESCRIPTION	PHYSICAL PROPERTY
Breaks along smooth planes:	_____
Scratches glass:	_____
Shines like a metal:	_____
Red powder:	_____
Looks like broken glass:	_____

2. What term is used to describe the shape of a mineral that has three directions of cleavage that intersect at 90 degrees?

   _____

3. Describe the cleavage of the minerals listed below.

MINERAL	CLEAVAGE
Muscovite:	_____
Calcite:	_____
Halite:	_____

4. Describe the cleavage of the feldspar minerals (potassium feldspar and plagioclase feldspar).

   _____

   Number of directions of cleavage: _____

   Degrees of cleavage angles: _____

5. What physical property most distinguishes biotite mica from muscovite mica?

   _____

6. Name a mineral that exhibits the physical properties listed below. (Use the photos in this exercise, if needed.)

PROPERTY	MINERAL
One direction of cleavage:	_____
Striations:	_____
Multiple colors:	_____
Cubic cleavage:	_____
Nonmetallic, vitreous Luster:	_____
Fracture:	_____
Metallic luster:	_____

7. Figure 1.28 illustrates the common crystal form of the mineral fluorite and the characteristic shape of a cleaved sample of fluorite. Identify each specimen (A *or* B) next to its appropriate description below.

   Crystal form of fluorite: _____

   Cleavage of fluorite: _____

A.   **B.**

Figure 1.28 Comparing crystal shape and cleavage.

8. Refer to the photo in Figure 1.29 to complete the following:

   a. Describe the crystal form of this specimen.

   _____

   b. What term is used to describe the lines on this sample?

   _____

   c. Based on what you can discern from this photo, use the mineral identification key (Figure 1.20) to name this mineral.

   _____

9. A photo of *agate*, a variety of quartz composed of minute crystals, is provided in Figure 1.30. Based on this image, describe the habit of this sample.

   _____

10. If a mineral can be scratched by a penny but not by a human fingernail, what is its hardness on the Mohs scale?

    Hardness: _____

11. What term is used to describe the tenacity of muscovite?

    Tenacity of muscovite: _____

12. Use the mineral identification key (Figure 1.20) to identify a mineral that is nonmetallic, colored, lacks cleavage, and is green in color.

    Mineral name: _____

Figure 1.29 Identifying mineral properties through observation.

Figure 1.30 What habit does this sample display?

13. For each mineral listed below, list at least one diagnostic property. (Refer to your mineral data sheet in Figure 1.19, if necessary.)

MINERAL NAME	DIAGNOSTIC PROPERTY
Halite:	_____
Galena:	_____
Magnetite:	_____
Muscovite:	_____
Hematite:	_____
Fluorite:	_____
Talc:	_____
Graphite:	_____
Calcite:	_____

14. List the two most common rock-forming mineral groups.

    _____

    _____

15. Provide an economic use for each mineral listed below:

    Galena: _____

    Hematite: _____

    Graphite: _____

    Sphalerite: _____

    Gypsum: _____

    Calcite: _____

Rocks and the Rock Cycle

Objectives

Completion of this exercise will prepare you to:

A. Determine whether rocks are igneous, sedimentary, or metamorphic.

B. Describe what the texture of a rock indicates about its history.

C. Name the dominant mineral(s) found in common igneous, sedimentary, and metamorphic rocks.

D. Identify the environments in which sediments are deposited.

E. Use a classification key to identify rocks.

F. Recognize and name some common rocks.

Materials

metric ruler	igneous rocks
glass plate	sedimentary rocks
iron nail	metamorphic rocks
dilute hydrochloric acid	hand lens

Introduction

Most **rocks** are aggregates (mixtures) of mineral grains or fragments (gravel, sand, and silt) of preexisting rocks. However, there are some important exceptions including *obsidian*, which is a rock made of volcanic glass (a non-crystalline substance), and *coal*, which is a rock made of decayed plant material (an organic substance).

Rocks are classified into three groups—igneous, sedimentary, and metamorphic—based on the processes by which they were formed. One very useful device for understanding rock types and the geologic processes that transform one rock type into another is the **rock cycle** (Figure 2.1). Examine the rock cycle as you read the following descriptions of each rock group and pay particular attention to the processes that formed them.

Igneous rocks are the solidified products of once-molten material that was created by melting in the upper mantle or crust. Geologists call molten rock **magma** when it is found at depth and **lava** when it erupts at Earth's surface. The distinguishing feature of most igneous rocks is the interlocking arrangement of their mineral crystals that develops as the molten material cools and solidifies. *Intrusive* igneous rocks form below Earth's surface from magma, and *extrusive* igneous rocks form at the surface from lava. Volcanism produces a variety of distinctive rock structures (Figure 2.2).

Sedimentary rocks form at or near Earth's surface from the products of *weathering*. This material, called **sediment**, is transported by erosional agents (water, wind, or ice) as solid particles or ions in solution to their site of deposition. The process of **lithification** (meaning "to turn to stone") transforms the sediment into hard rock. Sedimentary rocks cover much of Earth's surface and may contain organic matter (oil, gas, and coal) and fossils. The layering that develops when sediment is deposited is the most recognizable feature of sedimentary rocks. These layers, called **strata**, or **beds**, usually accumulate in nearly horizontal sheets that can be as thin as a piece of paper or tens of meters thick (Figure 2.3).

Metamorphic rocks are produced from preexisting igneous, sedimentary, or other metamorphic rocks that have been subjected to conditions within Earth that are significantly different from those under which the rock originally formed. *Metamorphism*, the process that causes the transformation, generally occurs at depths where both the temperatures and pressures are significantly higher than at Earth's surface.

Rock Textures and Compositions

The task of distinguishing the three rock groups and naming individual samples relies heavily on the ability to recognize their *textures* and *compositions.*

Texture refers to the shape, arrangement, and size of mineral grains in a rock. The shape and arrangement

ROCK CYCLE

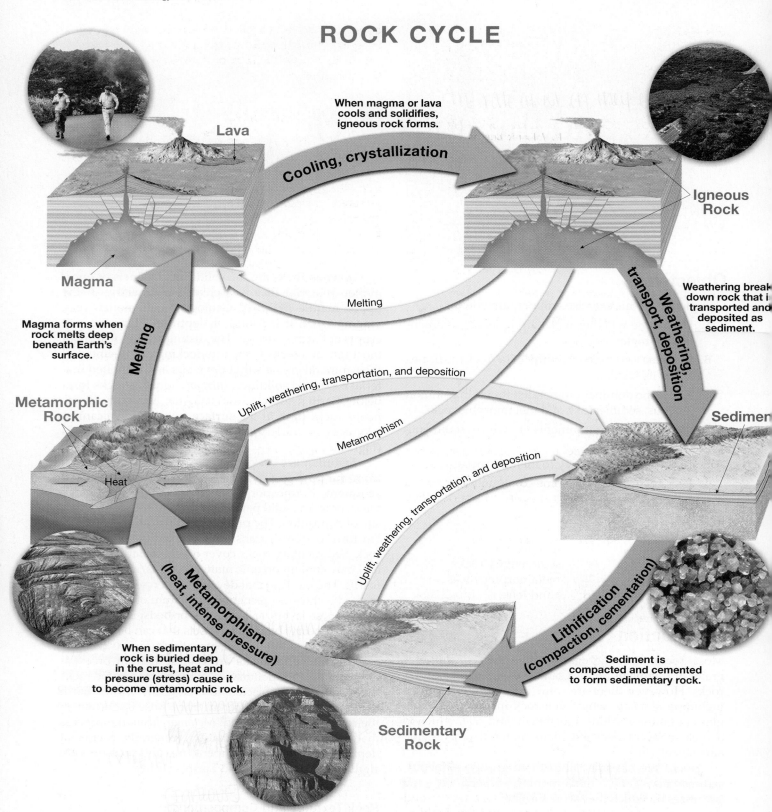

When magma or lava cools and solidifies, igneous rock forms.

Cooling, crystallization

Lava

Magma

Igneous Rock

Melting

Weathering, transport, deposition

Weathering breaks down rock that is transported and deposited as sediment.

Magma forms when rock melts deep beneath Earth's surface.

Melting

Uplift, weathering, transportation, and deposition

Metamorphism

Metamorphic Rock

Heat

Sediment

Uplift, weathering, transportation, and deposition

Metamorphism (heat, intense pressure)

Lithification (compaction, cementation)

When sedimentary rock is buried deep in the crust, heat and pressure (stress) cause it to become metamorphic rock.

Sediment is compacted and cemented to form sedimentary rock.

Sedimentary Rock

Figure 2.1 Over long spans, rocks are constantly forming, changing, and reforming. The rock cycle helps us understand the origin of the three basic rock groups. Arrows represent processes that link each group to the others.

A. Lava flows **B. Volcanic bombs**

C. Vesicular texture **D. Columnar jointing**

Figure 2.2 Volcanism produces a variety of distinctive rock structures. **A.** Lava flow that consists of rough jagged blocks sits atop a flow that exhibits a smooth ropy surface. **B.** A volcanic bomb produced when molten lava was hurled from a volcano. **C.** Vesicles are somewhat spherical voids left by gas bubbles that form as lava solidifies. **D.** Columns in a lava flow that formed by contraction and fracture as the lava cooled.

minerals can be identified by a particular diagnostic property. For example, rocks composed of the mineral calcite fizz when a drop of dilute hydrochloric acid (HCl) is applied.

Rock Classification

One of the first steps in rock identification is to determine the group to which a sample belongs. Every rock has characteristics that aid in determining whether it is igneous, sedimentary, or metamorphic. Igneous rocks generally consist of randomly oriented, interlocking crystals of two or more different minerals and tend to be relatively hard. A few igneous rocks exhibit easily recognized properties such as having the appearance of dark broken glass or exhibiting many bubble-shaped holes.

Most sedimentary rocks, on the other hand, are composed of weathered rock debris (silt, sand, pebbles, and gravel) that is often easily identified in a hand sample. Other sedimentary rocks are recognized because they contain shells or plant debris that has been compacted or cemented into a solid mass. Some sedimentary rocks form when material dissolved in water precipitates to generate a rock composed of intergrown crystals—distinguishable from crystalline igneous rocks because they tend to be composed of easily identifiable minerals—calcite, halite, or gypsum.

Metamorphic rocks are produced from preexisting rocks by heat and pressure, and they are sometimes difficult to distinguish from their parent rock. Some

of mineral grains provide clues to whether a rock is igneous, sedimentary, or metamorphic.

Composition refers to the abundance and type of minerals found in a rock. Large mineral grains can often be identified by sight or by their physical properties, while a hand lens or microscope may be needed to identify small mineral grains. In addition, a few

Figure 2.3 Sedimentary layers, called strata, or beds.

Figure 2.4 Highly deformed metamorphic rock.

metamorphic rocks, however, exhibit a characteristic parallel alignment of flat crystals or elongated crystals that can be diagnostic. Others show clear evidence of having been deformed by folding (Figure 2.4).

The origin of a particular rock may also be revealed by the presence of certain minerals within it. For example, halite and gypsum form when saltwater evaporates and, thus, are found almost exclusively in sedimentary rocks. When mica crystals are abundant, it is a good indication that a rock is metamorphic. The presence of quartz, however, is of no value in determining rock type because it is common in all three rock groups.

Examine the close-up images of the rock samples in Figure 2.5 to answer the following questions.

1. Which of the samples (A–H) are composed of weathered rock material or shells (fossils) that have been lithified? (sedimentary)

   Sample(s): B, C, D, G

2. Which of the samples you listed in question 1 is/are composed mainly of rounded rock fragments? (Metamorphic)

   Sample(s): ~~B C D G~~ B and D, G

3. Which of the samples you listed in question 1 is/are composed mainly of angular rock fragments? (Igneous)

   Sample(s): ~~A B H~~ C

4. Which of the samples you listed in question 2 is/are composed of sand-size particles?

   Sample(s): B

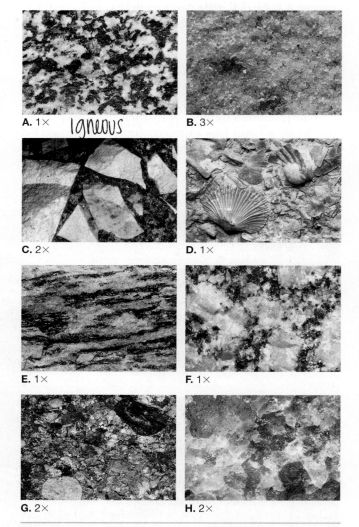

A. 1× Igneous   B. 3×

C. 2×   D. 1×

E. 1×   F. 1×

G. 2×   H. 2×

Figure 2.5 Comparison of igneous, sedimentary, and metamorphic rocks.

5. Which of the samples (A–H) are crystalline—composed of interlocking crystals? (*Igneous*)

   Sample(s): ___A, F, and H___

6. Which of the samples you listed in question 5 consists of randomly oriented interlocking crystals of two or more different minerals?

   Sample(s): ___A, F, and H___

7. The samples you listed in question 6 belong to which rock group?

   Rock type: ___Igneous___

8. Which of the samples (A–H) has/have the mineral crystals aligned, or arranged so they are oriented in the same direction? (*Metamorphic*)

   Sample(s): ___E___

9. Which of the samples you listed in question 5 consist of interlocking crystals of what appears to be the same mineral?

   Sample(s): ___H___

10. Classify the rocks in the samples (A–H in) as either igneous, sedimentary, or metamorphic. (Hint: Sample H has a salty taste.)

    Igneous: ___A, F, and H___

    Sedimentary: ___B, C, D, and G___

    Metamorphic: ___E___

Igneous Rocks

Igneous rocks are generated by cooling and crystallization of molten rock. The interlocking network of silicate minerals that develop as the molten material cools gives most igneous rocks their distinctive crystalline appearance.

Textures of Igneous Rocks

The rate at which magma cools determines the size of the interlocking crystals found in igneous rocks. The slower the cooling rate, the larger the mineral crystals.

Coarse-Grained (Phaneritic) Texture   When a large mass of magma solidifies at depth, it cools slowly and forms igneous rocks that exhibit a **coarse-grained texture**. These rocks have intergrown crystals that are roughly equal in size and large enough that the individual minerals within can be identified with the un-aided eye. A hand lens or binocular microscope can greatly assist in mineral identification.

Fine-Grained (Aphanitic) Texture   Igneous rocks that form when molten material cools rapidly at the surface or as small masses within the upper crust exhibit a **fine-grained texture**. Fine-grained igneous rocks are composed of individual crystals that are too small to be identified without strong magnification.

Porphyritic Texture   A **porphyritic texture** results when molten rock cools in two different environments. The resulting rock consists of larger crystals embedded in a matrix of smaller crystals. The larger crystals are termed *phenocrysts*, and the smaller, surrounding crystals are called *groundmass*, or *matrix*.

Glassy Texture   During explosive volcanic eruptions, molten rock is ejected into the atmosphere, where it is quenched very quickly. When the material solidifies before the atoms arrange themselves into an orderly crystalline structure, the rocks exhibit a **glassy texture** that resembles manufactured glass.

Fragmental, or Pyroclastic Texture   Volcanoes sometimes blast fine ash, molten blobs, and/or angular blocks torn from the walls of the vent into the air during an eruption. Igneous rocks composed of these rock fragments have a **fragmental, or pyroclastic texture**.

Vesicular Texture   Common features of some fine-grained extrusive igneous rocks are the voids left by gas bubbles that escape as lava solidifies. These somewhat spherical openings are called *vesicles*, and the rocks that contain them have a **vesicular texture**.

Use the close-up photos of igneous rocks in Figure 2.6 to answer the following questions.

11. Which sample(s) (A–H) exhibit a porphyritic texture?

    Sample(s): ___B and G___

12. Of the sample(s) you listed in question 11, what terms are used to denote the larger crystals and the surrounding smaller crystals? Label the larger and smaller crystals on the photo(s).

    Larger crystals: ___you can see with unaided eye___

    Smaller crystals: ___Need "aid" in order to see___

13. Which sample(s) (A–H) exhibit(s) a coarse-grained texture?

    Sample(s): ___A___

14. Complete the description of the environment in which a *coarse-grained igneous rock* forms by choosing the appropriate terms. (at great depth *or* on the surface) (slowly *or* rapidly) (extrusive *or* intrusive).

    When molten rock solidifies ___at great depth___, it cools ___slowly___, and produces an ___intrusive___ igneous rock.

Quartz Feldspar

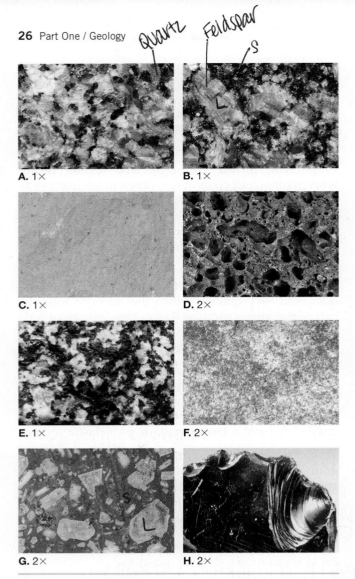

A. 1× B. 1×

C. 1× D. 2×

E. 1× F. 2×

G. 2× H. 2×

Figure 2.6   Identifying igneous textures.

15. Which of the sample(s) (A–H in Figure 2.6) exhibit(s) a fine-grained texture?

    Sample(s): _C and F_

16. Complete the description of the environment in which a *fine-grained igneous rock* forms, by choosing the appropriate terms. (at great depth *or* on the surface) (slowly *or* rapidly) (extrusive *or* intrusive).

    When molten rock solidifies _at the surface_, it cools _rapidly_, and produces an _extrusive_ igneous rock.

17. Which of the sample(s) (A–H in Figure 2.6) exhibit(s) a vesicular texture?

    Sample(s): _D_

18. Which sample(s) (A–H in Figure 2.6) exhibit(s) a glassy texture?

    Sample(s): _H_

19. Although Samples A and C (Figure 2.6) appear different, they have very similar mineral compositions. Briefly explain what accounts for their different appearances.

    The rate at which they coould and the depth because one is course-grained and the other is fine-grained.

Composition of Igneous Rocks

Despite their significant compositional diversity, igneous rocks (and the magmas from which they form) can be divided into four groups, based on the proportions of light and dark silicate minerals.

Felsic (*granitic*) igneous rocks are composed mainly of the light-colored minerals quartz and potassium feldspars, with lesser amounts of plagioclase feldspar. Recall that feldspar crystals can be identified by their rectangular shapes, flat surfaces, and tendency to be pink, white, or dark gray in color. Quartz grains, on the other hand, are glassy and somewhat rounded and tend to be light gray. Dark-colored minerals account for no more than 15 percent of the minerals in rocks in this group.

Intermediate (*andesitic*) rocks are mixtures of both light-colored minerals (mainly plagioclase feldspar) and dark-colored minerals (mainly amphibole). Dark minerals comprise between 15 percent and 45 percent of these rocks.

Mafic (*basaltic*) rocks contain abundant dark-colored minerals (mainly pyroxene and olivine) that account for between 45 percent and 85 percent of their composition. Plagioclase feldspar makes up the bulk of the remainder.

Ultramafic rocks are composed almost entirely of the dark silicate minerals pyroxene and olivine and are seldom observed at Earth's surface. However, the ultramafic rock peridotite is a major constituent of Earth's upper mantle.

20. On the photo of Sample A in Figure 2.6, label a quartz crystal.

21. On the photo of Sample B in Figure 2.6, label a feldspar crystal.

22. What mineral appears to be dominant in Sample A (Figure 2.6)?

    Dominant mineral: _Quartz_

An important skill used in identifying igneous rocks is being able to classify them based on mineral composition—felsic, intermediate, mafic, or ultramafic. However, it is not possible to identify minerals in fine-grained igneous rocks without sophisticated equipment. Further, the dark silicate minerals are often difficult to differentiate—even in coarse-grained rocks. Consequently, geologists working in the field estimate mineralogy using the **color index**, a value based on the percentage of dark silicate minerals (see top of Figure 2.7). Rocks with a low color index are felsic in composition, whereas those with the highest color index are ultramafic.

It is important to note that although the dark silicate minerals are often black in color, the mineral pyroxene can be dark green, and olivine is often olive green in color. Therefore, rocks with a greenish color have a very high color index and are usually ultramafic in composition.

23. Use Samples A–F in Figure 2.8 as well as the Color Index located at the top of Figure 2.7 to complete the following statements:
    a. Sample(s) __A, F__ has/have a felsic (granitic) composition.
    b. Sample(s) __C, D__ has/have an intermediate (andesitic) composition.
    c. Sample(s) __B__ has/have a mafic (basaltic) composition.
    d. Sample(s) __E__ has/have an ultramafic composition.

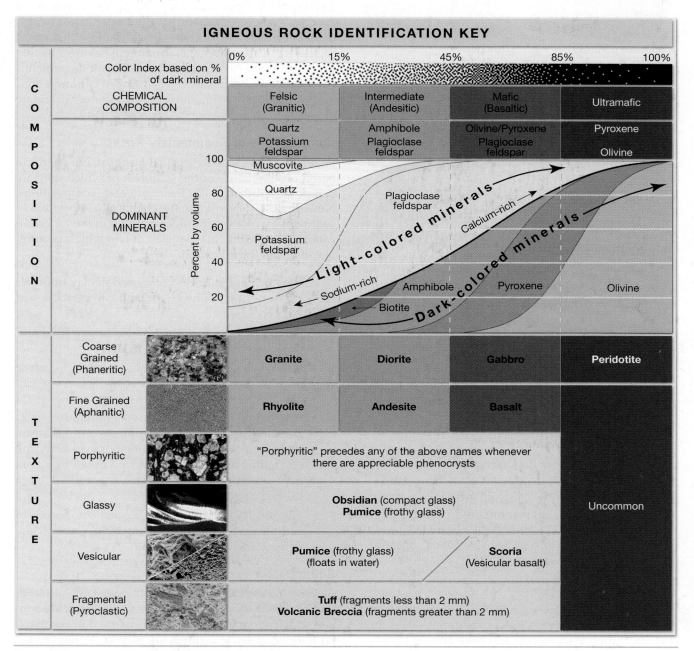

Figure 2.7 Igneous rock identification key.

A. Rhyolite   B. Gabbro

C. Andesite porphyry   D. Diorite

E. Peridotite   F. Granite

Figure 2.8 Hand samples that illustrate igneous compositions.

Igneous Rock Identification

To determine the name of an igneous rock, use the *Igneous Rock Identification Key* found in Figure 2.7 and follow these steps:

Step 1: Identify the texture of the rock.

Step 2: Determine the mineral composition of the rock by:

    a. identifying the abundance of major minerals or

    b. estimating the mineral composition using the color index

Step 3: Use the *Igneous Rock Identification Key* to name the rock.

24. Place each of the igneous rocks supplied by your instructor on a numbered piece of paper. Then complete the *Igneous Rock Chart* (Figure 2.9) and name each specimen.

Sedimentary Rocks

Weathering breaks down rock into finer material that is transported and deposited as sediment. Over long periods of time, compaction and cementation turn this sediment into sedimentary rock.

Materials that accumulate as sediment have two principal sources. First, sediments may originate as solid particles from weathered rocks. These particles are called *detritus*, and the rocks that they form are called *detrital sedimentary rocks*. The second major source of sediment is soluble material produced by chemical weathering.

Detrital sedimentary rocks consist of mineral grains or rock fragments derived from mechanical and chemical weathering that are transported and deposited as solid particles. Clay minerals are the most abundant solid products of chemical weathering. Quartz is abundant in detrital rocks because it is extremely durable.

Chemical sedimentary rocks are products of mineral matter that was dissolved in water and later precipitated. Precipitation may occur as a result of processes such as evaporation or temperature change, or as a result of life processes, such as those that result in the formation of shells. Sediment formed by life processes has a *biochemical* origin. Limestone, which is composed of calcite ($CaCO_2$) is the most common mineral in chemical sedimentary rocks and may originate either from chemical or biological processes.

Compositions of Sedimentary Rocks

Sedimentary rocks are classified as detrital or chemical based on the nature of the sediment of which they are composed, rather than their mineral compositions. Nevertheless, determining the chemical composition of sedimentary rocks is an important step in their identification. Most sedimentary rocks contain a high percentage of one of the following:

- **Clay minerals** Rocks composed of clay minerals are very fine grained, soft, and can be scratched with an iron nail.
- **Quartz** Rocks composed of quartz are hard and can scratch glass.
- **Calcite** Rocks composed of calcite effervesce when a drop of dilute HCl is applied. They can also be easily scratched with an iron nail.
- **Evaporite minerals** These sedimentary rocks contain salts, usually halite or gypsum, that are deposited when saltwater evaporates. They are crystalline and can be scratched by an iron nail; gypsum is soft enough to be scratched by a human fingernail.
- **Altered plant fragments** Rocks composed of carbon and other organic material have a low density and are easily broken.

Sedimentary Textures

Sedimentary rocks exhibit two basic textures—*clastic* and *nonclastic*. Rocks that display a **clastic texture** consist of discrete particles that are cemented or compacted together. Clastic textures are further divided based on particle size and shape.

IGNEOUS ROCK CHART

Sample Number	Texture	Color (light- intermediate- dark)	Dominant Minerals	Rock Name
58	coarse-grained	Mostly light colored	Potassium Feldspar Quartz, Muscovite, Plagioclase Feldspar	Granite
05	fine-grained	Dark	Pyroxene Plagioclase Feldspar Calcium-Rich	Basalt
60	coarse-grained	Intermediate	Amphibole Plagioclase Feldspar	Diorite
03	fine-grained	Light	Quartz Potassium Feldspar	Rhyolite
60	Glassy	Dark	Pyroxene Olivine	Obsidian

Figure 2.9   Igneous rock chart.

Rocks that have a **nonclastic texture** are often crystalline and, consist of minerals that form patterns of interlocking crystals. Although crystalline sedimentary rocks can be similar in appearance to some igneous rocks, they tend to consist of minerals—most often calcite, halite, or gypsum—that are easily distinguished from the silicate minerals found in igneous rocks.

Examining Sedimentary Rocks

Carefully examine the common sedimentary rocks shown in Figure 2.10. Use these photos and the preceding discussion to answer the following questions.

25. What characteristic can be used to distinguish Sample A (conglomerate) from Sample B (breccia)?
    Fragments (Angular/Rounded)

26. List the name(s) of the chemical sedimentary rocks that clearly exhibit a clastic texture.
    K and H

27. Samples C, O, P, and Q are all primarily composed of quartz. What property of quartz would assist you in identifying these samples?
    They can scratch glass

CLASSIFICATION OF SEDIMENTARY ROCKS

Detrital Sedimentary Rocks

A. Conglomerate
(rounded fragments)

B. Breccia
(angular fragments)

C. Sandstone
(usually quartz)

D. Arkose
(feldspar, quartz)

E. Siltstone
(quartz, clay minerals)

F. Shale or Mudstone
(clay minerals)

Chemical and Biochemical Sedimentary Rocks

G. Crystalline limestone
(calcite)

H. Microcrystalline limestone
(calcite)

I. Fossiliferous limestone
(calcite)

J. Coquina
(calcite)

K. Chalk
(calcite)

L. Travertine
(calcite)

M. Rock salt
(halite)

N. Rock gypsum
(gypsum)

O. Chert
(microcrystalline quartz)

P. Flint
(microcrystalline quartz)

Q. Agate
(microcrystalline quartz)

R. Bituminous coal
(carbon)

Figure 2.10 Hand samples of common sedimentary rocks.

28. Samples G–L all contain calcite. What property of calcite would assist you in identifying these samples?

   Easily scratched with iron nail
   Effervesce when diluted HCl is applied

29. Compare and contrast Sample I (fossiliferous limestone) and Sample J (coquina).

   They are both made of calcite but one has larger fragments

30. Sample G (rock salt) in Figure 2.10 and Sample A in Figure 2.6 are both crystalline. How do these samples differ in appearance?

   Their color and A has more distinct particles.

31. What characteristic do Sample Q (agate) and Sample L (travertine) share?

   They both have a nonclastic texture

32. Which of these terms most accurately describes the texture of Sample N—*clastic* or *nonclastic*?

   Nonclastic

33. Sample H is a microcrystalline rock composed of calcite, and Sample O is a microcrystalline rock composed of quartz. What test(s) could you use to figure out which is which?

   Scratch them with nail and glass

34. What mineral does Sample D contain that gives it a pinkish color?

   Quartz

Identifying Sedimentary Rocks

The *Sedimentary Rock Identification Key* in Figure 2.11 divides sedimentary rocks into two groups—detrital and chemical. The primary subdivisions for detrital

SEDIMENTARY ROCK IDENTIFICATION KEY

Detrital Sedimentary Rocks			Chemical Sedimentary Rocks		
Clastic Texture (particle size)	**Distinctive Properties**	**Rock Name**	**Composition**	**Distinctive Properties**	**Rock Name**
Gravel Coarse (over 2 mm)	Rounded rock or mineral fragments, typically poorly sorted	Conglomerate	Calcite, CaCO₃ (effervesces in HCl)	Fine to coarse crystalline	Crystalline Limestone
	Angular rock or mineral fragments, typically poorly sorted	Breccia		No visible grains, may exhibit conchoidal fracture	Micrite
Sand Medium (1/16 to 2 mm)	Quartz grains, typically rounded, well sorted	Quartz Sandstone (Sandstone)		Visible shells and shell fragments loosely cemented	Coquina
	At least 25% feldspar, typically poorly sorted, angular fragments	Arkose (Sandstone)		Various size shells and shell fragments, well cemented	Fossiliferous Limestone
	Mixture of sand and mud, typically poorly sorted	Graywacke (Sandstone)		Microscopic shells and clay, soft	Chalk
Silt/mud Fine (1/16 to 1/256 mm)	Mostly silt-size quartz and clay, blocky, gritty	Siltstone (Mudstone)		Faint layering, may contain cavities or pores	Travertine
Clay/mud Very fine (less than 1/256 mm)	Mostly clay, splits into layers, may contain fossils	Shale (Mudstone)	Quartz, SiO₂	Microcrystalline, may exhibit conchoidal fracture, will scratch glass	Chert (light colored)
	Mostly clay, crumbles easily	Claystone (Mudstone)			Flint (dark colored)
					Agate (banded)
			Gypsum CaSO₄•2H₂O	Fine to coarse crystalline, soft	Rock Gypsum
			Halite, NaCl	Fine to coarse crystalline, tastes salty	Rock Salt
			Altered plant fragments	Black brittle organic rock, may be layered	Bituminous Coal

Where Limestone spans the Calcite composition column's rock names: Crystalline Limestone, Micrite, Coquina, Fossiliferous Limestone, Chalk, Travertine.

Figure 2.11 Sedimentary rock identification key.

rocks are based upon grain size, whereas composition is used to subdivide the chemical sedimentary rocks.

35. Place each of the sedimentary rocks supplied by your instructor on a numbered piece of paper. Then complete the *Sedimentary Rock Chart* in Figure 2.12 for each rock. Use the Sedimentary Rock Identification Key (Figure 2.11) to determine each specimen's name.

Sedimentary Environments

Sedimentary rocks are extremely important in the study of Earth history. The texture and mineral composition of

sedimentary rocks often suggest something about the location (environment) where the sediment accumulated. For example, rock salt forms in warm, shallow seas, where the loss of water by evaporation exceeds the annual rainfall. In addition, fossils and sedimentary structures provide important clues about a rock's history. Think of each sedimentary rock as representing a "place" on Earth where sediment was deposited and later lithified.

Figure 2.13 illustrates and describes a few modern environments (places) where sediment accumulates.

SEDIMENTARY ROCK CHART

Specimen Number	Texture and Grain Size	Composition or Sediment Name	Detrital or Chemical (Biochemical)	Rock Name
84	Sand Texture Medium Grains	Quartz	Detrital	Sandstone
88	Fine to coarse Crystalline	Calcite, CaCO$_3$	chemical	Limestone- Both kinds
82	Gravel Texture coarse Grains	Rounded Rock or Mineral Fragments	Detrital	conglomerate
80	Clay/Mud Texture very Fine Grains	Mostly clay	Detrital	Shale
93	Black Brittle	Altered Plant Fragments [carbon]	chemical	coal
49	Fine Grained Soft Texture	calcite	chemical	chalk

Figure 2.12 Sedimentary rock chart.

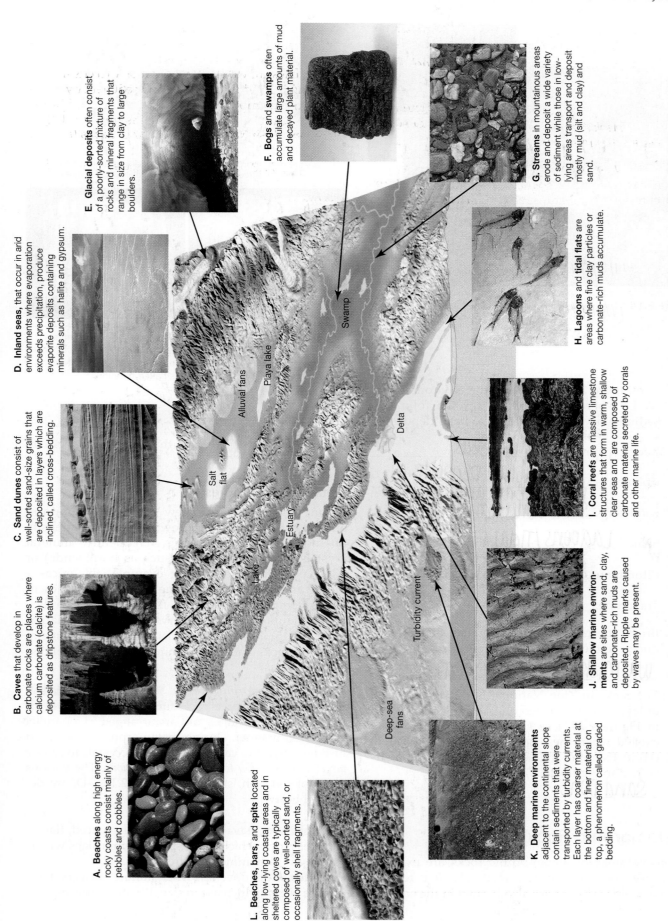

B. Caves that develop in carbonate rocks are places where calcium carbonate (calcite) is deposited as dripstone features.

C. Sand dunes consist of well-sorted sand-size grains that are deposited in layers which are inclined, called cross-bedding.

D. Inland seas, that occur in arid environments where evaporation exceeds precipitation, produce evaporite deposits containing minerals such as halite and gypsum.

E. Glacial deposits often consist of a poorly-sorted mixture of rocks and mineral fragments that range in size from clay to large boulders.

F. Bogs and **swamps** often accumulate large amounts of mud and decayed plant material.

G. Streams in mountainous areas erode and deposit a wide variety of sediment while those in low-lying areas transport and deposit mostly mud (silt and clay) and sand.

H. Lagoons and **tidal flats** are areas where fine clay particles or carbonate-rich muds accumulate.

I. Coral reefs are massive limestone structures that form in warm, shallow clear seas and are composed of carbonate material secreted by corals and other marine life.

J. Shallow marine environments are sites where sand, clay, and carbonate-rich muds are deposited. Ripple marks caused by waves may be present.

K. Deep marine environments adjacent to the continental slope contain sediments that were transported by turbidity currents. Each layer has coarser material at the bottom and finer material on top, a phenomenon called graded bedding.

A. Beaches along high energy rocky coasts consist mainly of pebbles and cobbles.

L. Beaches, bars, and **spits** located along low-lying coastal areas and in sheltered coves are typically composed of well-sorted sand, or occasionally shell fragments.

Swamp, Alluvial fans, Playa lake, Salt flat, Lake, Estuary, Delta, Turbidity current, Deep-sea fans

Figure 2.13 Sedimentary environments. (Photo C by John S. Shelton; Photo G by Michael Collier; Photo K by Marli Miller; Photo L by David R. Frazier Photolibrary, Inc./Alamy

Figure 2.14 The orange and yellow cliffs of Utah's Zion National Park.

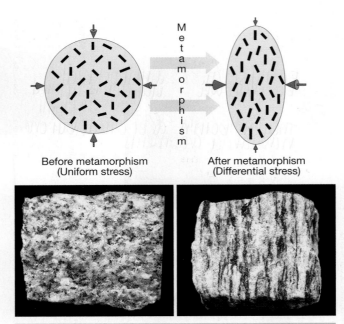

Before metamorphism (Uniform stress)

After metamorphism (Differential stress)

M e t a m o r p h i s m

Figure 2.15 Under directed pressure, linear or platy minerals, such as the micas, become reoriented or recrystallized so that their surfaces are aligned at right angles to the stress. The resulting planar orientation of mineral grains is called foliation.

36. Use Figure 2.13 to identify the environment(s) (A–L) where the sediment for the following sedimentary rocks could have been deposited.

    Rock gypsum: _Inland Seas_

    Conglomerate: _Glacial Deposits_

    Sandstone: _Beaches, Bars, and spits Sand Dunes and (J)_

    Shale: _Lagoons/Tidal Flats (H, J, and K)_

    Bituminous coal: _Bogs/Swamps_

    Travertine: _Coral Reefs (B) caves_

37. Briefly describe the environment that is conducive to the formation of coral reefs.

    Warm, shallow clear seas

38. The rocks in Zion National Park, Utah, consist of layers of well-sorted quartz sandstone shown in Figure 2.14. Describe the environment that existed when these sediments were deposited. (Hint: Examine the sedimentary environments shown in Figure 2.13.)

    Sand Dune

Metamorphic Rocks

Extensive areas of metamorphic rocks are exposed on every continent in the relatively flat regions known as *shields*. They are also located in the cores of mountains and buried beneath sedimentary rocks on the continents.

Metamorphic Textures

During metamorphism, new minerals often form and/or existing minerals grow larger as the intensity of metamorphism increases. Frequently, mineral crystals that are elongated (such as hornblende) or have a sheet structure (for example, the micas—biotite and muscovite) become oriented perpendicular to compressional forces. The resulting parallel, linear alignment of mineral crystals is called **foliation** (Figure 2.15). Foliation is associated with many metamorphic rocks and gives them a layered or banded appearance.

Foliated Metamorphic Rocks The mineral crystals in foliated metamorphic rocks are either elongated or have thin, platy shapes and are arranged in a parallel or layered manner. During metamorphism, increased heat and pressure can cause mineral crystals to *become larger* and *foliation to become more obvious.* The various types of foliation, shown in Figure 2.16A–D, are described below:

Slaty or rock cleavage refers to closely spaced, flat surfaces along which rocks split into thin slabs when struck with a hammer (Figure 2.16A).
Phyllite texture develops when minute mica crystals in slate begin to increase in size. Phyllite surfaces

FOLIATED TEXTURES

A. Slaty or rock cleavage
Slate

B. Phyllite texture
Phyllite

C. Schisosity
Mica Schist

D. Gneissic texture
Gneiss

NONFOLIATED TEXTURES

E. Coarse grained
Marble

F. Fine grained
Quartzite

Figure 2.16 Metamorphic textures.

have a shiny, somewhat metallic sheen and often have a wavy surface (Figure 2.16B).

Schisosity can be identified by a scaly layering of glittery, platy minerals (mainly micas) that are often found in association with deformed quartz and feldspar grains (Figure 2.16C).

Gneissic texture forms during high-grade metamorphism when ion migration results in the segregation of light and dark minerals (Figure 2.16D).

Nonfoliated Metamorphic Rocks Nonfoliated metamorphic rocks consist of intergrown crystals of various size and are most often identified by determining their mineral composition (Figure 2.16E,F). The minerals that comprise them are most often quartz or calcite. Therefore, the hardness test and/or the acid test can be used.

Other Distinguishing Features Some metamorphic rocks have stretched or deformed pebbles or fossils

(Figure 2.17A). Others contain unusually large crystals called *porphyroblasts* that are surrounded by a fine-grained matrix of other minerals (Figure 2.17B). These large crystals tend to be minerals that are mainly associated with metamorphic rocks, including garnet, staurolite, and andalusite. Another distinguishing characteristic of some highly deformed metamorphic rocks is that their foliation becomes contorted, as shown in Figure 2.17C.

Metamorphic Rock Identification

Metamorphic rocks are classified according to their texture and mineral composition. To identify metamorphic rocks, use the *Metamorphic Rock Identification Key* (Figure 2.18) and follow these steps:

Step 1: Determine whether the rock is foliated or nonfoliated.

Step 2: Determine the size of the mineral grains.

Step 3: If possible, determine the rock's mineral composition and note any other distinctive features.

Step 4: Use the information collected and the Metamorphic Rock Identification Key to determine the name of the rock. The names of the medium- and coarse-grained rocks are often modified with the mineral composition placed in front of the name (for example, *mica schist*).

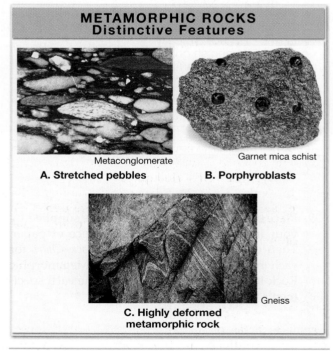

METAMORPHIC ROCKS
Distinctive Features

A. Stretched pebbles
Metaconglomerate

B. Porphyroblasts
Garnet mica schist

C. Highly deformed metamorphic rock
Gneiss

Figure 2.17 Distinctive features of metamorphic rocks.

METAMORPHIC ROCK IDENTIFICATION KEY

Distinctive Properties	Texture	Grain Size	Rock Name	Parent Rock
Excellent rock cleavage, smooth dull surfaces	Foliated	Very fine	Slate	Shale, or siltstone
Breaks along wavy surfaces, glossy sheen	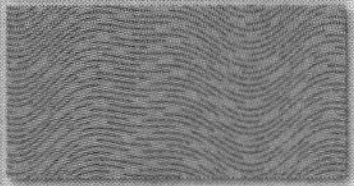	Fine	Phyllite	Shale, slate, or siltstone
Micas dominate, breaks along scaly foliation	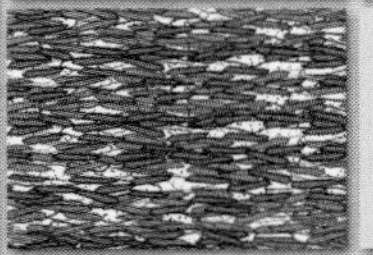	Medium to Coarse	Schist Mica schist Chlorite schist Talc schist Garnet mica schist	Shale, slate, phyllite, or siltstone
Compositional banding due to segregation of dark and light minerals	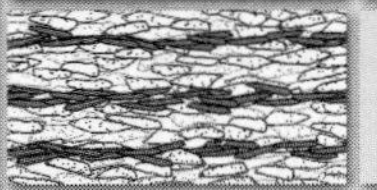	Medium to Coarse	Gneiss	Shale, schist, granite, or volcanic rocks
Banded, deformed rock with zones of light and dark colored crystalline minerals		Medium to Coarse	Migmatite	Shale, granite, or volcanic rocks
Interlocking calcite or dolomite crystals nearly the same size, soft, reacts to HCl	Nonfoliated	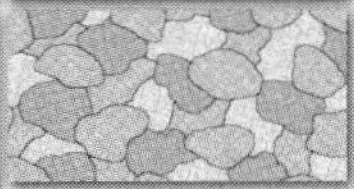 Medium to coarse	Marble	Limestone, dolostone
Fused quartz grains, massive, very hard		Medium to coarse	Quartzite	Quartz sandstone
Round or stretched pebbles that have a preferred orientation	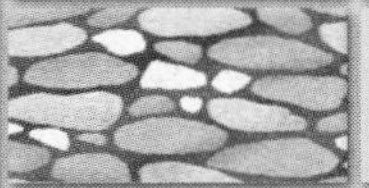	Coarse-grained	Metaconglomerate	Quartz-rich conglomerate
Shiny black rock that may exhibit conchoidal fracture		Fine	Anthracite	Bituminous coal
Usually, dark massive rock with dull luster		Fine	Hornfels	Any rock type
Very fine grained, typically dull with a greenish color, may contain asbestos fibers		Fine	Serpentinite	Basalt or ultramafic rocks
Broken fragments in a haphazard arrangement		Medium to very coarse	Fault breccia	Any rock type

(Increasing Metamorphism indicated for Slate through Migmatite)

Figure 2.18 Metamorphic rock identification key.

39. Place each of the metamorphic rocks supplied by your instructor on a numbered piece of paper. Then complete the *Metamorphic Rock Chart* for each rock (Figure 2.19). Use the Metamorphic Rock Identification Key to determine each specimen's name (Figure 2.18).

METAMORPHIC ROCK CHART

Sample Number	Foliated or Nonfoliated	Grain Size	Mineral Composition (if identifiable)	Rock Name
75	Foliated	Very Fine	Shale or Siltstone	Slate
78	Nonfoliated	Medium to Coarse	Limestone Dolostone	Marble
70	Foliated	Medium to Coarse	Shale, Schist, granite or volcanic rocks	Gneiss
76	Foliated	Medium to Coarse	Shale, Slate, phyllite or Siltstone	Schist

Figure 2.19 Metamorphic rock chart.

Companion Website

The companion website provides numerous opportunities to explore and reinforce the topics of this lab exercise. To access this useful tool, follow these steps:

1. Go to www.mygeoscienceplace.com.
2. Click on "Books Available" at the top of the page.
3. Click on the cover of *Applications and Investigations in Earth Science, 7e.*
4. Select the chapter you want to access. Options are listed in the left column ("Introduction," "Web-based Activities," "Related Websites," and "Field Trips").

Notes and calculations.

Rocks and the Rock Cycle

Name _____   **Course/Section** _____

Date _____   **Due Date** _____

1. The rock samples you encountered while completing this exercise are called *hand samples* because you can pick them up in your hand and study them. You can also easily examine hand samples with a hand lens, test them for hardness and reactivity to HCl, and feel and heft them. However, some characteristics of each rock group are so diagnostic that you can observe them at a distance and classify the rocks as igneous, sedimentary, or metamorphic. Examine the rock outcrops shown in Figure 2.20 and indicate whether each is igneous, sedimentary, or metamorphic.

   Outcrop A: _____

   Outcrop B: _____

   Outcrop C: _____

2. Match each of the metamorphic rocks listed below with one possible parent rock from the following list: bituminous coal, shale, limestone, slate, granite, sandstone.

METAMORPHIC ROCK	NAME OF PARENT ROCK
Marble:	_____
Slate:	_____
Phyllite:	_____
Gneiss:	_____
Quartzite:	_____
Anthracite:	_____

3. Match each term or characteristic with the appropriate rock group.

   A. Foliated texture   **B.** Clastic texture

   C. Gneissic texture   **D.** Chemical rocks

   E. Porphyritic texture   **F.** Lithification

Outcrop A

Outcrop B

Outcrop C

Figure 2.20 Photos to accompany question 1.

G. Alignment of mineral grains

H. Silt-size particles

I. Vesicular texture

J. Strata or beds

K. Feslic composition

L. Slaty texture

M. Glassy texture

N. Evaporite deposits

O. Detrital rocks

Igneous rocks: _____

Sedimentary rocks: _____

Metamorphic rocks: _____

4. Name the foliated metamorphic rocks shown in Figure 2.21 and list their names in order from lowest to highest metamorphic grade.

5. Name each of the rocks shown in Figure 2.22 and name the rock group to which each belongs.

(Name) _____

(Name) _____

(Name) _____

(Name) _____

Lowest _____

_____

_____

Highest _____

Figure 2.21 Photos to accompany question 4.

A. _____ (Name)

B. _____ (Name)

_____ (Rock Group)

_____ (Rock Group)

C. _____ (Name)

D. _____ (Name)

_____ (Rock Group)

_____ (Rock Group)

E. _____ (Name)

F. _____ (Name)

_____ (Rock Group)

_____ (Rock Group)

G. _____ (Name)

H. _____ (Name)

Figure 2.22 Photos to accompany question 5.

Aerial Photographs, Satellite Images, and Topographic Maps

Objectives

Completion of this exercise will prepare you to:

A. Use aerial photos and satellite images to investigate Earth's surface.

B. Explain what a topographic map is and how it is used to study landforms.

C. Use map scales to determine distances.

D. Determine the latitude and longitude of a location on a topographic map.

E. Use the Public Land Survey system to locate features.

F. Use topographic maps to determine elevation, relief, and the shapes of various landforms.

G. Construct a simple contour map and a topographic profile.

Materials

ruler	hand lens
stereoscope	topographic maps

Aerial Photographs and Satellite Images

Aerial photographs and satellite images are important tools for studying Earth's surface. When overlapping aerial photographs are combined, they produce **stereograms** that can be viewed in three dimensions. To view a stereogram, a **stereoscope** is placed directly over the line that separates the two images (Figure 3.1). As you look through a stereoscope, you may have to adjust it slightly until the image appears in three dimensions. The topography you see through the stereoscope will be vertically exaggerated—making slopes and other features appear steeper than they actually are.

Obtain a stereoscope from your instructor, unfold it, and center it over the line that separates the two aerial photographs of the volcanic cone in Figure 3.2. View this stereogram to complete the following.

1. Identify and label the crater at the summit of the volcano.

2. Outline and label the lava flow, located directly north of the volcano.

3. What is the white, curved feature that extends from the base of the cone to its summit?

_____

4. Mark the highest point on the volcano with an X.

Since the early 1970s, hundreds of satellites have been launched that systematically collect images of Earth's surface using a variety of remote sensing techniques. Computer enhancement of these images

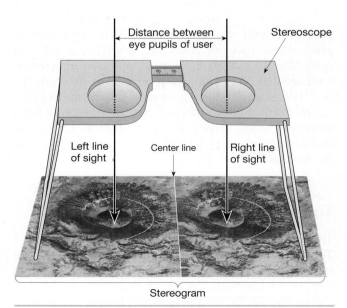

Figure 3.1 Aligning a stereoscope to view a stereogram.

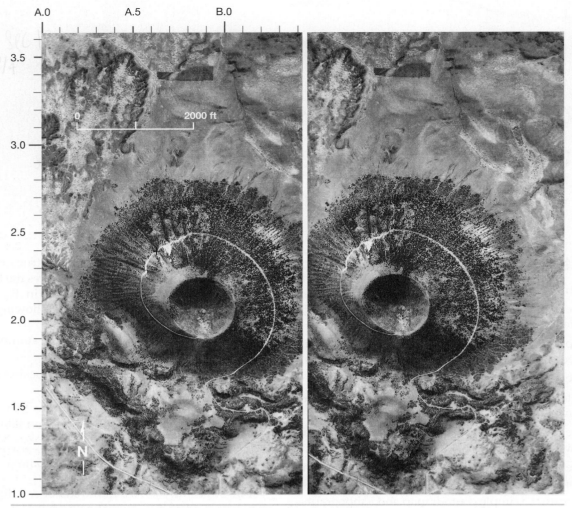

Figure 3.2 Stereogram of Mt. Capulin, a volcanic cinder cone located in northeastern New Mexico. The stereogram consists of two overlapping aerial photographs taken from an altitude of approximately 16,000 feet. To view the three-dimensional image of the volcano, center a stereoscope over the line that separates the two photographs. Then, while looking through the stereoscope, adjust the stereoscope until the image appears in three dimensions. (Courtesy of U.S. Geological Survey)

provides Earth scientists with data to study current phenomena such as volcanoes, hurricanes, and oil spills (Figure 3.3A,B). Satellite images may also reveal features that were previously unknown or unidentified. For example, heavily eroded impact craters that struck Earth millions of years ago have recently been identified using satellites images (Figure 3.3C). Further investigations of satellite imagery can be found at the website listed at the end of this exercise.

Topographic Maps

Topography means "the shape of the land." **Topographic maps**, also called **quadrangles**, are two-dimensional representations of Earth's three-dimensional surface. Each map is a scale model that shows the width, length, and the height of the land above a reference plane—generally, average sea level.

Topographic maps have been produced by the U.S. Geological Survey (USGS) since the late 1800s and follow a standard format. In addition to using standard colors and symbols, each contains information about the location of the mapped area, the date the map was drawn and/or revised, a map scale, a north arrow, and the names of adjoining maps.

Examining Topographic Maps

Obtain a copy of a topographic map from your instructor and examine it. Use this map to answer the questions that follow.
PLEASE DO NOT WRITE OR MARK ON THE MAP.

5. A topographic map is assigned a name that is found in the upper-right corner of the map.

What is the name of your map?

Map name: _____

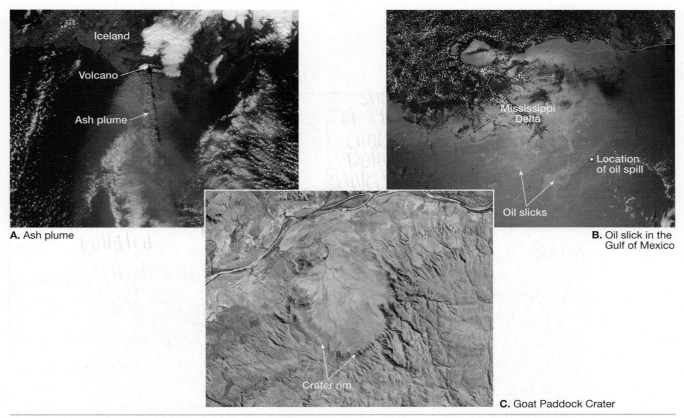

Figure 3.3 Satellite images are invaluable to our study of Earth. **A.** Satellite image of an ash plume produced by Iceland's Eyjafjallajökul Volcano, April 2010. **B.** Image of oil slick produced by a leak on the Deepwater Horizon platform located south of the Mississippi Delta. **C.** Goat Paddock Crater is an impact crater, located in northwestern Australia, which is largely covered by sediment deposited during the Pleistocene Epoch. (Photos courtesy of NASA)

6. Notice the small reference map and compass arrow in the lower margin of the map. In what part of the state (northern, southwestern, etc.) is the area located?

   Part of state: _____

7. What is the name of the map that adjoins the western edge of your map?

   Adjoining map: _____

8. When was the area surveyed? When was the map published?

   Surveyed: _____ Published: _____

9. Refer to the inside cover of this manual and examine the standard U.S. Geological Survey topographic map symbols. Using the standard map symbols as a guide, locate examples of various types of roads, buildings, woodland areas, and streams on the topographic map supplied by your instructor.

10. In general, what colors are used for the following types of features?

    Highways and roads: _____

    Buildings: _____

    Urban areas: _____

    Wooded areas: _____

    Water features: _____

Map Scales

You are probably familiar with scale model airplanes or cars that are miniature representations of the actual objects. Maps are similar in that they are "scale models" of Earth's surface. Each map has a **map scale** that expresses the relationship between distance on the map and the actual distance on Earth's surface. Different map scales depict an area on Earth with varying degrees of detail. On a topographic map, scale is usually indicated in the lower margin and is expressed in two format, fractional scales and graphic scales.

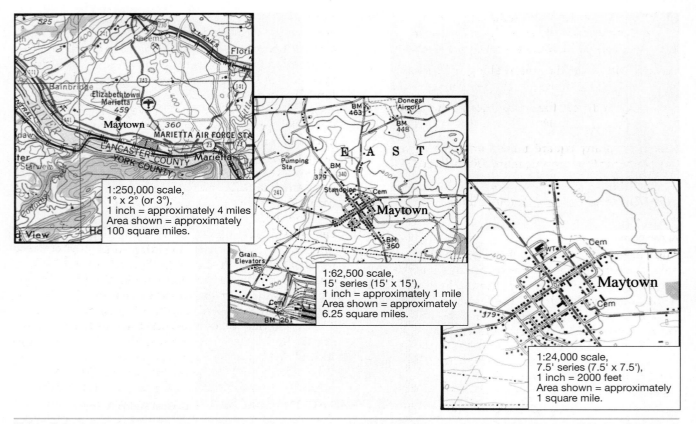

Figure 3.4 Portions of three topographic maps of the same area, showing the effect that different map scales have on the detail illustrated.

Fractional scales indicate how much the portion of Earth's surface represented on the map has been reduced from its actual size. For example, a map with a scale of 1/24,000 (or 1:24,000) tells the viewer that a distance of 1 unit on the map represents a distance of 24,000 of the *same* units on Earth (e.g., 1 inch equals 24,000 inches). The U.S. Geological Survey publishes maps of various scales with expansive as well as highly detailed coverage.

11. Depending on the map scale, *1 inch* on a topographic map represents various distances on Earth. Convert the following scales and round to the nearest mile. (Hint: 5280 feet = 1 mile.)

SCALE	1 INCH ON THE MAP REPRESENTS
1:62,500	_____ inches, or _____ miles on Earth
1:250,000	_____ inches, or _____ miles on Earth

12. Examine Figure 3.4 and complete the map description by choosing the appropriate terms: (smaller *or* larger) (more *or* less).

Maps with small fractional scales (e.g., 1:250,000) cover a _____ area and provide _____ detail. Maps with large fractional scales (e.g., 1:24,000) cover a _____ area and provide _____ detail.

A **graphic scale**, or **bar scale**, is a bar that is divided into segments that show the relationship between distance on the map and actual distance on Earth (Figure 3.5). Bar scales showing miles, feet, and kilometers are generally provided on most topographic maps. The left side of the bar is often divided into fractions to allow for more accurate measurements. A graphic scale is more useful than a fractional scale for measuring distances between points and can be used to make your own "map ruler" by transferring the bar scale to a piece of paper.

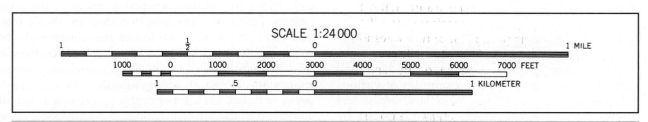

Figure 3.5 Typical graphic, or bar, scale.

13. Use the graphic scale on the topographic map provided by your instructor to construct a "map ruler" in miles and measure the following distances.

    a. Width of the map along the south edge = _____ miles

    b. Length of the map along the east edge = _____ miles

14. How many square miles (area) are represented on your topographic map? (Recall: area = width × length.)

    Map area = _____ square miles.

Location

One function of a topographic map is to determine the precise location of a feature on Earth's surface. Two frequently used methods for designating location are (1) **latitude** and **longitude**, used mainly to determine the location of a small area, and (2) the **Public Land Survey System (PLS)** system, which is generally used to define a precise land area in order to establish ownership. Both methods of location can be used to provide valuable information to engineers, surveyors, real estate agents, and others. A third method, the **Universal Transverse Mercator (UTM)** grid, is presented at the website listed at the end of this exercise. The **Global Positioning System (GPS)**, a widely used navigational tool, is based on the Universal Transverse Mercator coordinate system.

Latitude and Longitude Topographic maps are bounded by parallels of latitude on the north and south and meridians of longitude on the east and west. The lines of latitude and longitude that identify the boundaries of a quadrangle are labeled at the four corners of the map in degrees (°), minutes ('), and seconds (") and are indicated at intervals along the margins. Maps that cover 15 minutes of latitude and 15 minutes of longitude are called *15-minute series topographic maps*. A $7\frac{1}{2}$-minute series topographic map covers $7\frac{1}{2}$-minutes of latitude and $7\frac{1}{2}$-minutes of longitude. (*Note:* There are 60 minutes of arc in 1 degree and 60 seconds of arc in 1 minute of arc. Therefore, $\frac{1}{2}$ minute is the same as 30 seconds.) A more complete examination of latitude and longitude can be found in Exercise 22, "Location and Distance on Earth."

Public Land Survey System The Public Land Survey (PLS) uses a grid system to precisely describe the location of a parcel of land in most of the states located west of the Appalachian Mountains (Figure 3.6). The PLS began shortly after the Revolutionary War, when the federal government became responsible for large land areas beyond the 13 original colonies and acquired other land areas, such as the Louisiana Purchase. This arduous task involved teams of surveyors—working mostly on foot—that divided almost 1.5 billion acres into parcels of land called *congressional townships* and *sections*. This survey system provided a means for the government to sell land in order to raise public funds. As in the past, today the Public Land Survey System is used to describe property being transferred from one party to another. Several states, including Texas, Louisiana, New Mexico, Maine, Ohio, and Hawaii, use systems that incorporate land descriptions, such as Spanish land grants, that existed prior to their becoming states.

The PLS system begins at a point where an east–west line, called a **base line**, meets a north–south line, called a **principal meridian**, and provides the basis of the grid, as shown in Figure 3.6A. Horizontal lines at 6-mile intervals that parallel the base line establish east–west tracts, called **townships**. Each township is numbered north and south from the base line. The first horizontal 6-mile-wide tract north of the base line is designated Township One North (T1N), the second T2N, and so on. Vertical lines at 6-mile intervals that parallel the principal meridian define north–south tracts, called **ranges**. Each range is numbered east and west of the principal meridian. The first vertical 6-mile-wide tract west of the principal meridian is designated Range One West (R1W), the second R2W, and so on. On a topographic map, the townships and ranges covered by the map are printed in red along the margins.

The intersection of a township and a range defines a 6-mile-by-6-mile rectangle, called a **congressional township**, which is identified by referring to its township and range numbers. For example, in Figure 3.6A, the shaded congressional township would be identified as T1N, R4W.

Each congressional township is divided into 36 1-mile-square parcels, called **sections**—each with 640 acres. Sections are numbered beginning with number 1 in the northeast corner of the congressional township and ending with number 36 in the southeast corner, as shown in Figure 3.6B. The shaded section of land in Figure 3.6B would be designated Section 11, T1N, R4W. On a topographic map, the sections are outlined and their numbers are printed in red.

To describe smaller areas, sections may be subdivided into halves, quarters, or quarters of a quarter (Figure 3.6C). Each subdivision is identified by its compass position. For example, the 40-acre area designated with the letter X in Figure 3.6C would be described as the SW $\frac{1}{4}$ (southwest $\frac{1}{4}$) of the SE $\frac{1}{4}$ (southeast $\frac{1}{4}$) of Section 11. Hence, the complete locational description of the area marked with an X would be SW $\frac{1}{4}$, SE $\frac{1}{4}$ Sec. 11, T1N, R4W.

T13N R10W Sec 12
SE ¼ SE ¼

$$\begin{array}{r} 1000 \\ -\ 120 \\ \hline 880 \end{array}$$

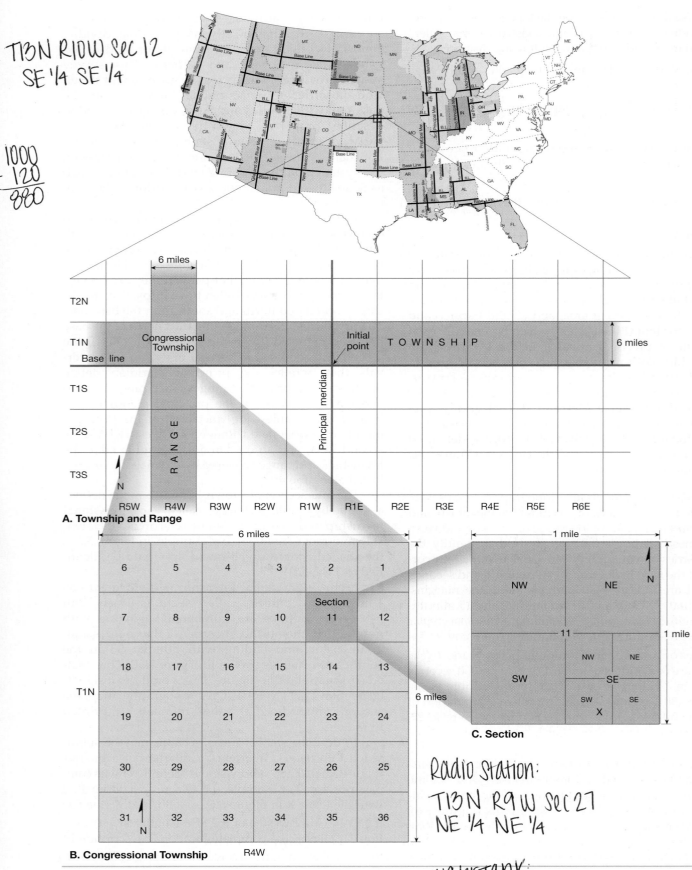

A. Township and Range

6 miles

6	5	4	3	2	1
7	8	9	10	Section 11	12
18	17	16	15	14	13
19	20	21	22	23	24
30	29	28	27	26	25
31	32	33	34	35	36

T1N

B. Congressional Township   R4W

C. Section

Radio station:
T13N R9W Sec 27
NE ¼ NE ¼

Water tank:
T12N R10W Sec 1
NE ¼ ~~NE~~ ¼
SE

Figure 3.6   The Public Land Survey System (PLS).

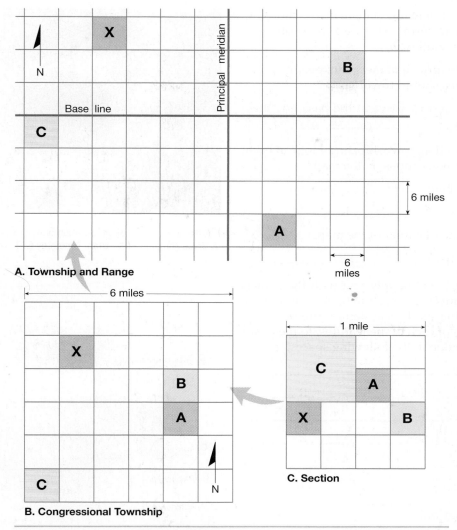

Figure 3.7 Hypothetical Public Land Survey System map, showing the locations of various parcels of land.

Figure 3.7 illustrates a hypothetical area that has been surveyed using the Public Land Survey system. Figure 3.7A is a township and range diagram, Figure 3.7B represents a congressional township within the township and range system, and Figure 3.7C is a section of the congressional township. Use Figure 3.7 to complete the following questions.

15. Use the PLS system to label the townships along the west edge and ranges along the bottom of Part A.

16. Use the PLS system to label each of the sections in the congressional township shown in Part B.

17. In the space provided below, use the PLS system to describe Plots A–C. Plot X has been completed as an example.

Plot X: $\underline{\text{NW}}$ $\frac{1}{4}$, $\underline{\text{SW}}$ $\frac{1}{4}$, Sec. $\underline{\text{8}}$, T $\underline{\text{3N}}$, R $\underline{\text{4W}}$

Plot A: ____ $\frac{1}{4}$, ____ $\frac{1}{4}$, Sec. ____ , T ____ , R ____

Plot B: ____ $\frac{1}{4}$, ____ $\frac{1}{4}$, Sec. ____ , T ____ , R ____

Plot C: ____ $\frac{1}{4}$, Sec. ____ , T ____ , R ____

18. Locate each of the areas described below on diagrams A, B, and C in Figure 3.7 by placing the appropriate letter (D or E) on each of the three diagrams.

D: SW $\frac{1}{4}$, SW $\frac{1}{4}$, Sec. 2, T3N, R3E

E: SE $\frac{1}{4}$, NE $\frac{1}{4}$, Sec. 34, T4S, R5W

Use the topographic map supplied by your instructor to complete the following.
PLEASE DO NOT WRITE OR MARK ON THE MAP.

19. List the townships and ranges shown on the map.

Townships: _____

Ranges: _____

20. Complete the following statement by choosing the appropriate term: (eastward *or* westward) (northward *or* southward).

    To arrive at the principal meridian, people living within this area should travel

    _____ , and to arrive at the base line they should travel _____ .

21. What is the section, township, and range at each of the following locations on the map?

    Center of the map:

    Sec. _____ , T _____ , R _____

    Extreme northeast corner of the map:

    Sec. _____ , T _____ , R _____

22. Your instructor will supply you with the names of three features (school, church, etc.) located on the map. Using the PLS system, write the complete location of each of the features, to the nearest $\frac{1}{4}$ of a $\frac{1}{4}$ section in the following spaces.

    Feature name: _____

    Location: _____ $\frac{1}{4}$, Sec. _____ , T _____ , R _____

    Feature name: _____

    Location: _____ $\frac{1}{4}$, Sec. _____ , T _____ , R _____

    Feature name: _____

    Location: _____ $\frac{1}{4}$, Sec. _____ , T _____ , R _____

Determining Elevations

Depicting the height or elevation of the land and illustrating the shape of landforms are the most important uses of topographic maps. Land elevations were originally established by land surveys that determined the elevations of some selected locations, called **bench marks (BM)**, which are highly accurate. These elevations are marked with brass plates at various locations around the United States (Figure 3.8). Surveyors work outward from these locations to establish elevations of numerous other sites.

Examining Contour Lines

Topographic maps, miniature models of Earth's surface, make use of contour lines. **Contour lines** connect all points on a map that have the same elevation above sea level. To visualize this concept, imagine a small volcanic island, as shown in Figure 3.9. Where sea level intersects the land at the shoreline represents the 0-foot contour line. The intersection of imaginary planes 50 feet and 100 feet above the ocean represent the 50-foot and 100-foot contour lines, respectively.

To effectively use topographic maps, familiarity with the rules for reading contour lines is necessary. Use Figure 3.10, a simplified contour map of a small

Figure 3.8 Photo of an elevation bench mark (BM) at a location in Bryce National Park, Utah. (Photo by E. J. Tarbuck)

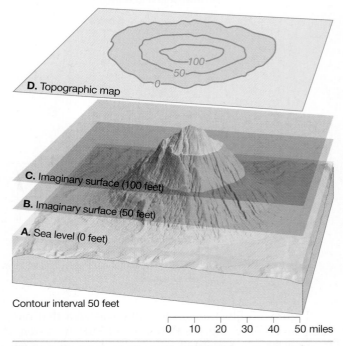

Contour interval 50 feet

```
0   10   20   30   40   50 miles
```

Figure 3.9 Schematic illustration showing how topographic maps are constructed. A contour line is drawn where an imaginary horizontal plane intersects the land surface. Where the ocean surface (plane A) intersects the land, it forms the 0-foot contour line. Plane B is 50 feet above sea level, and it intersects the land to form the 50-foot contour line. Plane C is 100 feet above the ocean, and it intersects the land to form the 100-foot contour line. D is the topographic map that results when the contour lines that mark where the imaginary planes intersect the surface of the volcanic island are drawn on a map.

volcanic cone and nearby landforms, to complete the following.

23. The difference in elevation between adjacent contour lines is called the **contour interval**. Look on the bottom of this map to determine the contour interval.

    Contour interval: _____

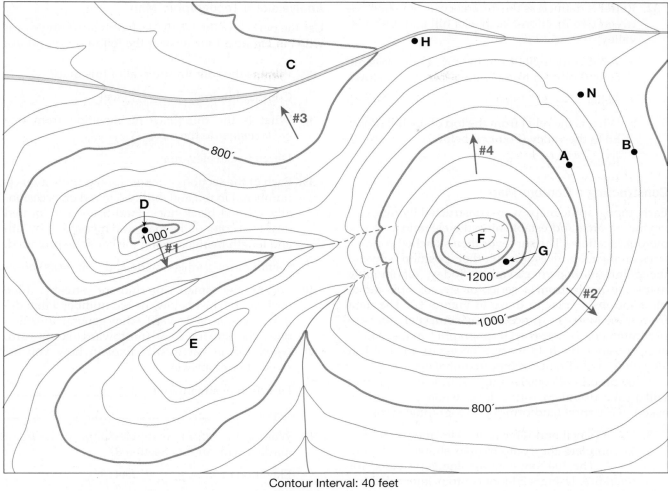

Contour Interval: 40 feet

Figure 3.10 Contour map to be used to answer questions 23–32.

24. What is the difference in elevation between points A and B?

   Difference in elevation: _____ feet

25. Notice that every fifth contour line, called an **index contour**, is printed as a bold brown line, and the elevation of that line is provided (in feet). List the elevations for each index contour shown on this map.

   _____ feet, _____ feet, _____ feet

26. Contour lines do not cross except to show an overhanging cliff—indicated as a "hidden contour" with a dashed line. Briefly explain why contour lines do not typically cross.

   _____

27. Steep slopes are shown by closely spaced contours. Which of the four slopes shown with red arrows labeled (1–4) is the steepest? Which is the least steep?

   Steepest: _____ Least steep: _____

28. Hills are represented by a series of roughly circular closed contours. Which of the landforms labeled B–E are hills?

   Hill landforms: _____

29. Depressions (basins without outlets) are shown by closed contours with hachures (short lines) that point downslope. Which of the landforms labeled D–G are closed depressions?

   Closed depressions: _____

30. When contour lines cross streams or dry stream channels, they form a V *that points upstream.* Draw arrows next to the three streams (shown in blue) to show the direction in which each is flowing.

31. Estimating the elevations of places not located on a contour line is done by extrapolation. For example, a point halfway between the 500- and 600-foot contour line would have an elevation of approximately 550 feet. Which of the following elevations is the best estimation for Point N: (830 feet, 870 feet, *or* 890 feet)?

   Elevation of Point N: _____ feet

32. **Relief** is defined as the difference in elevation be-tween two locations, such as a hill and a nearby valley.

    a. What is the relief from the top of the volcanic cone (Point G) to the valley (point H) below?

    Relief: _____ feet

    b. What is the relief from the top of the hill (Point D) to the valley (Point H) below?

    Relief: _____ feet

Constructing a Contour Map

Early topographic maps were constructed by first sur-veying an area and establishing the elevations at nu-merous locations. The surveyor then sketched contour lines on the map by estimating their position between the points of known elevation. Today, topographic maps are made from a series of stereoscopic aerial photographs similar to those in Figure 3.2. The data from these images is processed by computers that are programmed to construct topographic maps. How-ever, considerable field work is necessary to establish elevations and eliminate computational errors.

The process of constructing a simple contour map will prove useful in future labs, when you will be asked to interpret landforms on topographic maps.

33. Use a pencil and refer to the elevation provided to complete the contour map shown in Figure 3.11A. The 100-foot contour line is provided for reference. Using a 20-foot contour interval, draw a contour line for each 20-foot change in eleva-tion below and above 100 feet (e.g., 60 feet, 80 feet, 120 feet). You will have to estimate the ele-vations between the points. Label each contour line with its elevation.

34. The land shown on the contour map you con-structed generally slopes toward the (north *or* south).

    Land slopes toward the _____.

35. Show the directions in which the streams are flowing by drawing arrows on the map.

Drawing a Topographic Profile

Topographic maps depict Earth's surface as it would ap-pear when viewed from above. In some circumstances, a topographic profile (side view) provides a more valuable representation of an area. To change an over-head map view into a profile, follow the steps illustrat-ed in Figure 3.12.

36. Use the *profile graph* in Figure 3.11B to construct a west–east profile along the line A–A' on the con-tour map you completed in Figure 3.11A. Follow the guidelines for preparing a topographic pro-file in Figure 3.12.

Analysis of a Topographic Map

Use the portion of the Leadville, North quadrangle pro-vided in Figure 3.13 to answer the following questions.

37. What is the contour interval of this map?

    Contour interval: _____ feet

38. What is the difference in elevation from one *guide contour* to the next?

    Difference in elevation: _____ feet

39. Each of the sections on most topographic maps is numbered in red and outlined in red or, occasion-ally, dashed black lines. Find Section 9, located near the center of the map, and measure its width and length in miles, using the bar scale provided.

    Sections are __ mile(s) wide and __ miles(s) deep.

40. Locate the small intermittent stream (blue dashed line) just below the red number 9 that de-notes Section 9. Toward what general direction does the stream flow (northward, southward, westward, *or* eastward)? Explain how you ar-rived at your answer.

    Direction of stream flow: _____

    Explain: _____

41. What is the approximate elevation of the point marked with an X in Section 8?

    Elevation of X: _____ feet

42. What is the approximate relief between the loca-tion marked X and Turquoise Lake?

    Relief: _____ feet

43. Which of the following phrases best describes the topography of Tennessee Park (steep, rugged terrain; hilly terrain *or* gradually sloping to flat terrain)?

    The topography is _____.

44. The area on both sides of Tennessee Creek in Sec-tion 9 is marked with blue symbols that do not rep-resent streams. What do these symbols indicate? (See the inside cover of the lab book, if necessary.)

    The blue symbols represent _____.

45. Describe the change in elevation you would encounter as you walked from the boat ramp (Point A) on the northeast shore of Turquoise Lake directly northward along the dashed line to Point B. (Use terms such as steep, gradual, moderate, uphill, downhill, and relatively flat.)

    _____

    _____

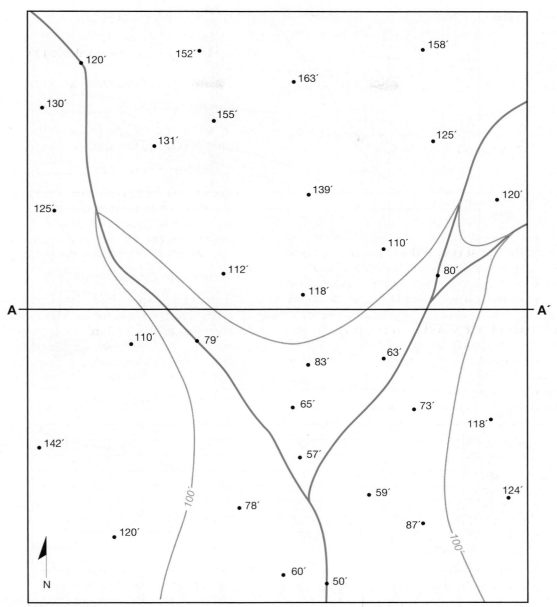

A. Contour Map

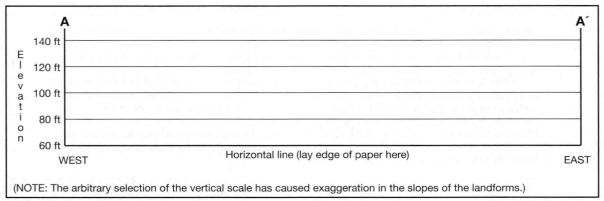

(NOTE: The arbitrary selection of the vertical scale has caused exaggeration in the slopes of the landforms.)

B. Profile

Figure 3.11 Construction of a contour map and profile. **A.** Complete this contour map. **B.** Graph to be used to construct contour map profile.

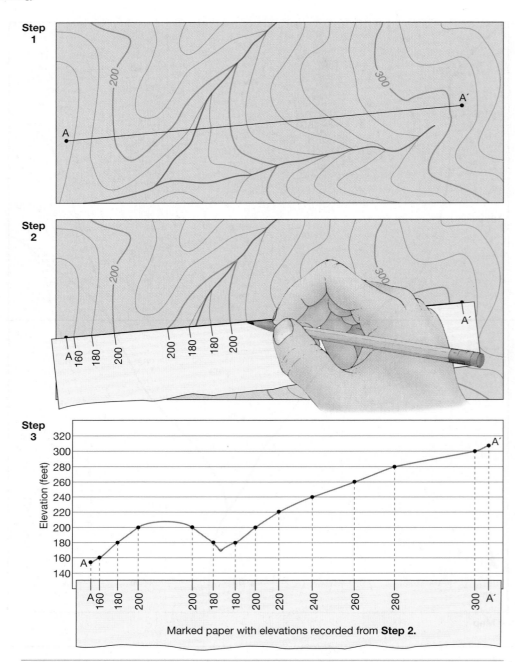

Figure 3.12 Construction of a topographic profile. **Step 1.** On the topographic map, draw a line along which the profile is to be constructed. Label the line A–A´.

As shown in **Step 2**, lay a piece of paper along line A–A´. Mark each place where a contour line intersects the edge of the paper and note the elevation of the contour line.

In **Step 3**, on a separate piece of paper, draw a horizontal line slightly longer than your profile line, A–A´. Select a vertical scale for your profile that begins slightly below the lowest elevation along the profile and extends slightly beyond the highest elevation. Mark this scale on either side of the horizontal line. Lay the marked paper edge (from Step 2) along the horizontal line. Wherever you have marked a contour line on the edge of the paper, place a dot directly above the mark at an elevation on the vertical scale equal to that of the contour line. Connect the dots on the profile with a smooth line to see the finished product.

Figure 3.13 Leadville North quadrangle.

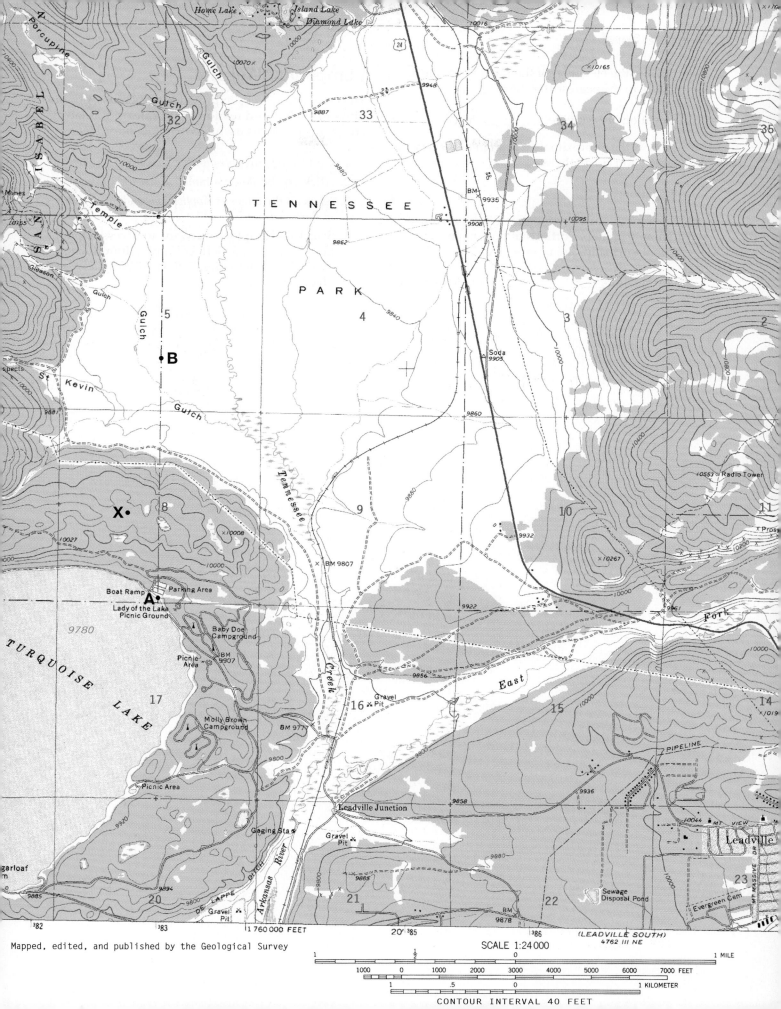

Mapped, edited, and published by the Geological Survey

SCALE 1:24000

(LEADVILLE SOUTH)
4762 III NE

CONTOUR INTERVAL 40 FEET

46. What is the elevation of the town Leadville, Colorado, to the nearest 1000 feet?

    Elevation of Leadville: _____ feet

47. What is the approximate elevation of the town in which you currently live?

    Elevation: _____

48. Examine the numbers of each section shown in red on this map. Portions of how many congressional townships are represented?

    Portions of _____ congressional townships

Companion Website

The companion website provides numerous opportunities to explore and reinforce the topics of this lab exercise. To access this useful tool, follow these steps:

1. Go to www.mygeoscienceplace.com.
2. Click on "Books Available" at the top of the page.
3. Click on the cover of *Applications and Investigations in Earth Science, 7e.*
4. Select the chapter you want to access. Options are listed in the left column ("Introduction," "Web-based Activities," "Related Websites," and "Field Trips").

Aerial Photographs, Satellite Images, and Topographic Maps

Name _____ **Course/Section** _____

Date _____ **Due Date** _____

Use Figure 3.14, which shows both a perspective view and a contour map of a hypothetical area, to answer questions 1–10.

1. What contour interval was used on this map?

   Contour interval: _____ feet

A. Perspective aerial view

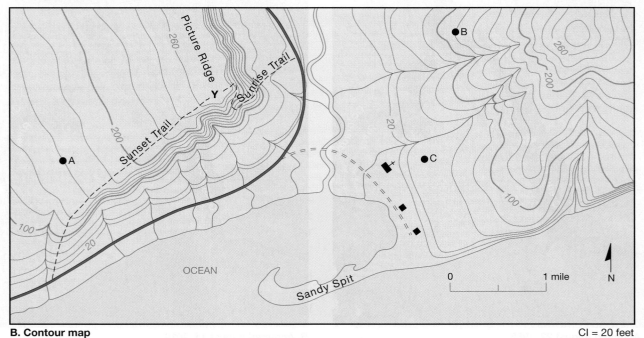

B. Contour map CI = 20 feet

Figure 3.14 Two views of a hypothetical coastal area. (After U.S. Geological Survey)

Hydrologic Cycle

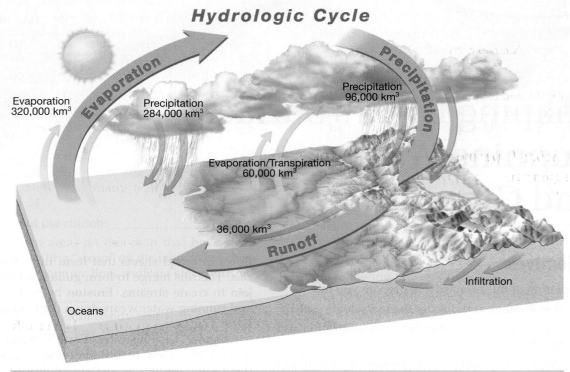

Evaporation

Evaporation
320,000 km³

Precipitation
284,000 km³

Precipitation

Precipitation
96,000 km³

Evaporation/Transpiration
60,000 km³

36,000 km³

Runoff

Infiltration

Oceans

Figure 4.1 Earth's water balance, a quantitative view of the hydrologic cycle.

$$\frac{36,000}{96,000} = 0.375$$

4. Worldwide, about how much of the precipitation that falls on the land becomes runoff (35, 55, *or* 75 percent)?

About _____35_____ percent becomes runoff.

5. Much of the water that falls on the land does not immediately return to the ocean via runoff. Instead, it is temporarily stored in reservoirs such as lakes. In some mountainous and polar regions, what features serve as reservoirs to temporarily store water?

glaciers

Water is temporarily stored ___in lakes___.

6. Label the drawing in Figure 4.2 with the letters that correspond to the following terms:

A. runoff

B. infiltration

C. groundwater

D. evaporation

E. precipitation

F. reservoir

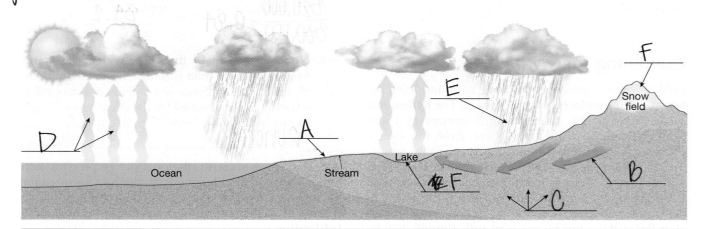

D

A

E

F

Snow
field

Ocean

Stream

Lake

F

C

B

Figure 4.2 Illustration (cross section) of the hydrologic cycle.

Infiltration and Runoff

When it rains, most of the water that reaches the land surface will infiltrate or run off. The balance between infiltration and runoff is influenced by factors such as the *permeability of the surface material, slope of the land, intensity of the rainfall,* and *type and amount of vegetation.* When the ground becomes saturated—that is, when it contains all the water it can hold—runoff occurs on the surface.

Permeability Experiment

This experiment is designed to help you understand how the **permeability** (that is, the ability of a material to transmit fluid) of various Earth materials affects how rapidly rainwater can be absorbed into the ground. Examine the equipment setup in Figure 4.3 and complete the following experiment.

Step 1: Obtain the following equipment and materials from your instructor:

graduated cylinder

beaker

small funnel

pieces of cotton

coarse sand

fine sand

soil

Step 2: Place a small wad of cotton in the neck of the funnel.

Step 3: Fill the funnel above the cotton about two-thirds full with coarse sand.

Step 4: With the bottom of the funnel placed in the beaker, pour in 50 ml of water and measure

Table 4.1 Data Table for Permeability Experiment

	LENGTH OF TIME TO DRAIN 50 ML OF WATER THROUGH FUNNEL	MILLILITERS OF WATER DRAINED INTO BEAKER
Coarse sand	_____ seconds	_____ ml
Fine sand	_____ seconds	_____ ml
Soil	_____ seconds	_____ ml

the time it takes until the water stops draining. Record the time in Table 4.1.

Step 5: Using the graduated cylinder, measure the amount of water (in milliliters) that has drained into the beaker and record the measurement in Table 4.1.

Step 6: Empty and clean the graduated cylinder, funnel, and beaker.

Step 7: Repeat the experiment two additional times, using fine sand and then soil. Record the results of each experiment at the appropriate place in Table 4.1. (*Note:* In each case, fill the funnel with the material to the *same level* that was used for the coarse sand and use the *same size* piece of cotton.)

Step 8: Clean the glassware and return it to your instructor, along with any unused sand and soil.

Questions 7–9 refer to the permeability experiment.

7. Of the three materials you tested (coarse sand, fine sand, and soil), which has the greatest permeability?

_____ has the greatest permeability.

8. Suggest a reason why different amounts of water were recovered in the beaker for each material that was tested.

_____

9. Write a brief statement summarizing the results of your permeability experiment.

_____

_____

_____

10. Describe how each of the following conditions influences infiltration and runoff.

Highly permeable material: _____

_____

Steep slope: _____

_____

Gentle rainfall: _____

_____

Dense vegetation: _____

_____

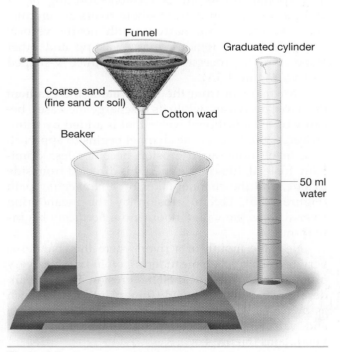

Funnel

Graduated cylinder

Coarse sand (fine sand or soil)

Cotton wad

Beaker

50 ml water

Figure 4.3 Equipment setup for permeability experiment.

Infiltration and Runoff in Urban Areas

In urban areas, much of the surface is covered with buildings, streets, and parking lots—all of which alter the amount of water that runs off compared to the amount that soaks in.

Figure 4.4 shows two hypothetical **hydrographs** (plots of streamflow, or runoff, over time) for an area before and after urbanization. Runoff is measured by stream **discharge**, which is the volume of water flowing past a given point per unit of time, usually measured in cubic feet per second. Use Figure 4.4 to complete the following.

11. Does urbanization (increase *or* decrease) the peak, or maximum, streamflow?

    Urbanization ___*increases*___ streamflow.

12. What is the effect of urbanization on the lag time between the time of the rainfall and the time of peak stream discharge?

    It decreases the lag time

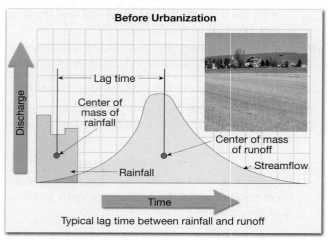

Before Urbanization

Discharge

Lag time

Center of mass of rainfall

Center of mass of runoff

Rainfall

Streamflow

Time

Typical lag time between rainfall and runoff

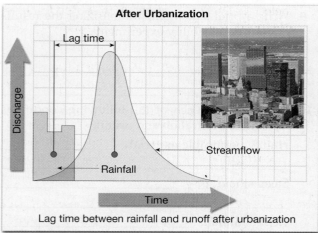

After Urbanization

Discharge

Lag time

Streamflow

Rainfall

Time

Lag time between rainfall and runoff after urbanization

Figure 4.4 The effect of urbanization on streamflow before urbanization (top) and after urbanization (bottom).

13. Does total runoff occur over a (longer *or* shorter) period of time in an area that has been urbanized?

    Runoff occurs over a ___*shorter*___ time after urbanization.

14. Based on what you have learned from the hydrographs, explain why urban areas often experience flash-flooding during intense rainfalls.

    Because they already have high streamflow so adding in more causes issues.

Running Water

Of all the agents that shape Earth's surface, running water is the most important. Rivers and streams are responsible for producing a vast array of erosional and depositional landforms in both humid and arid regions. As illustrated in Figure 4.5, many of these features are associated with the *headwaters* of a river, while others are found near the *mouth*.

An important factor that governs the flow of a river is its *base level*. The **base level** is the lowest point to which a river or stream may erode. The *ultimate base level* is *sea level*. However, lakes, resistant rocks, and main rivers often act as *temporary*, or *local*, *base levels* that control the erosional and depositional activities of a stream for a period of time.

The *head*, or source area, of a river is usually well above base level (Figure 4.5A). At the headwaters, rivers typically have steep gradients, or slopes, and downcutting prevails. As these headwater streams deepen their valleys, they may encounter rocks that are especially resistant to erosion—forming *rapids* and *waterfalls*. In arid areas where rivers are eroding bedrock, valleys are narrow, with nearly vertical walls. In humid regions, mass wasting and other slope erosion processes tend to produce V-shaped valleys (Figure 4.5A).

Downstream from the headwaters, the gradient of a river decreases, while its discharge increases, because of the additional water that is added by tributaries. As the level of the channel begins to approach base level, downward erosion becomes less dominant. Instead, the river's energy is directed from side to side, and the channel takes on a *meandering* path (Figure 4.5B). Lateral erosion by the meandering river widens the valley floor, and a *floodplain* begins to form.

Near the mouth of a river where the channel is very near base level, meandering often becomes very pronounced. Widespread lateral erosion by the meandering river produces a wide floodplain. Features such as *oxbow lakes*, *natural levees*, *backswamps* or *marshes*, and *yazoo tributaries* commonly develop on broad floodplains (Figure 4.5C).

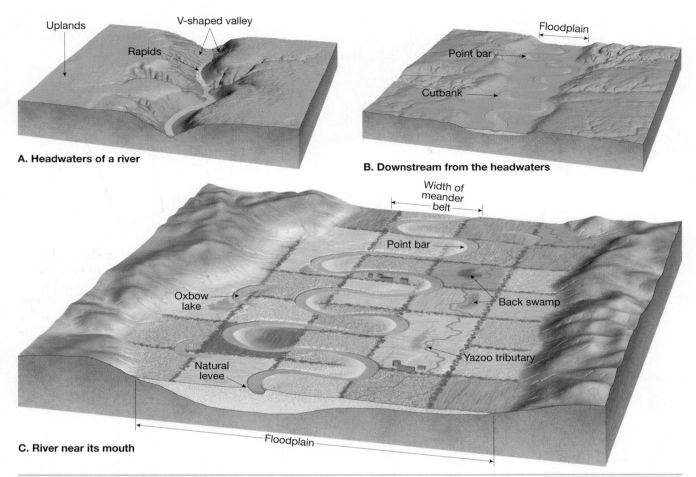

A. Headwaters of a river

B. Downstream from the headwaters

C. River near its mouth

Figure 4.5 Common features of stream valleys. **A.** Near its headwaters, a stream is well above base level, its gradient is steep, and downcutting is the dominant activity. **B.** In the middle reaches of a stream, the channel often exhibits a meandering pattern, directing energy from side to side. **C.** Near its mouth, a stream frequently flows on a wide floodplain and erodes its valley walls only in a few places.

Use the Portage, Montana, topographic map (Figure 4.6) and the stereogram (Figure 4.7) of a portion of the same region to complete the following.

15. Compare the stereogram in Figure 4.7 to the map in Figure 4.6. Then, on the topographic map, outline the area shown in the photo.

16. Describe the topography of Section 14, located on the west side of this map.

    It looks to be a very flat open area. (upland?)

17. Label the areas that topographically resemble Section 14 on the topographic map as "upland."

18. Describe the topography in the lower half of Section 17, located directly east of Section 14.

    steep _There is water there because the river passes through here_

19. Section 17 contains a portion of the valley occupied by the Missouri River. Approximately what percentage of the area shown on the map is stream valley (similar to the lower half of Section 17) and what percentage is upland?

Stream valley: __20__ % _or 25_

Upland: __80__ % _75_

20. Which of the following best describes the shape of the Missouri River Valley along the line labeled D–D′ (wide valley with a floodplain *or* steep-sided V-shaped valley with no floodplain)?

    The valley is a _steep-sided V-shaped with no flood plain_

21. Calculate the gradient, or slope, of this portion of the Missouri River by following the steps below.

Step 1: Locate a contour line that crosses the river. *Note:* A *guide contour* crosses the river in Section 26 (west side of map). Follow the guide contour to the right to establish its elevation.

Elevation: __3200__ feet _3,100_

Step 2: Locate a second guide contour crosses the river. *Note:* Look in Section 11, between the letters *V* and *E* in the word *RIVER*. What is its elevation?

Elevation: __2900__ feet _2800_

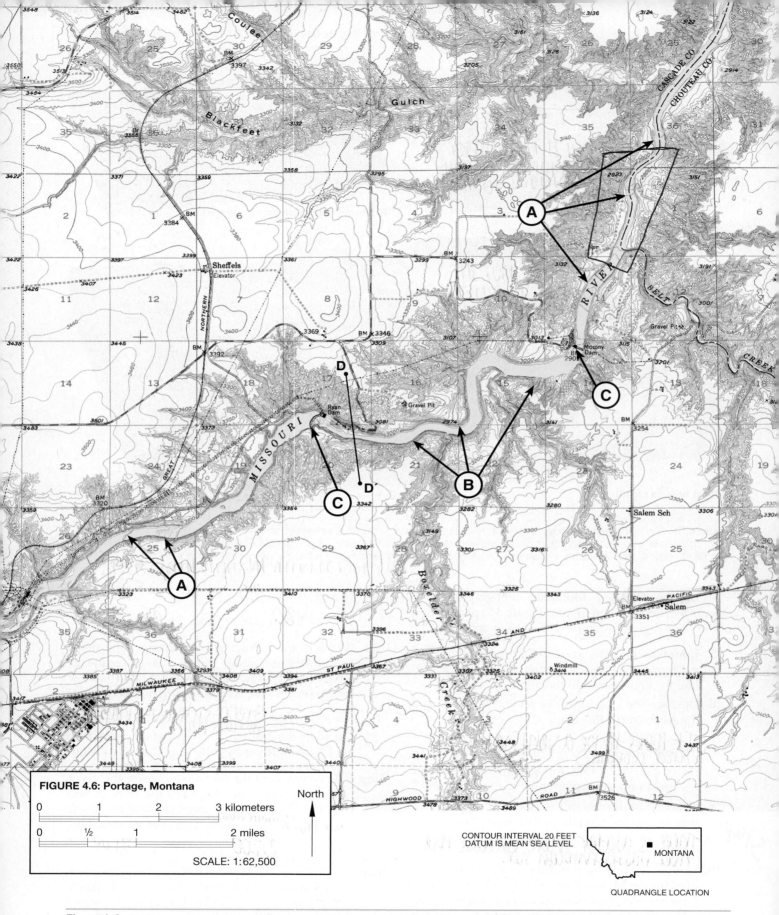

Figure 4.6 Portion of the Portage, Montana, topographic map which includes the Missouri River. (Map source: U.S. Geological Survey)

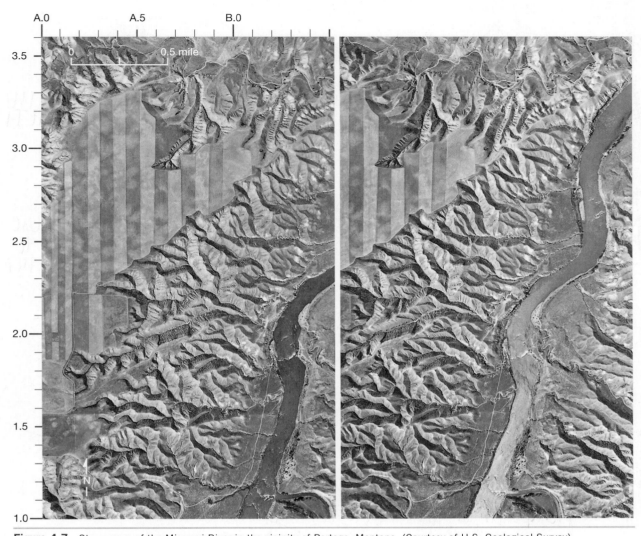

Figure 4.7   Stereogram of the Missouri River in the vicinity of Portage, Montana. (Courtesy of U.S. Geological Survey)

Step 3: Use a string and the bar scale to measure the approximate distance (to the nearest mile) between these two guide contours.

Distance: _0.5_ miles

Step 4: Calculate the gradient of the river using the following formula.

$$\text{Gradient} = \frac{\text{Difference in elevation (feet)}}{\text{Distance (miles)}}$$

Gradient: _40_ feet per mile

22. Approximately how many feet is the Missouri River from ultimate base level (sea level)?

Distance from ultimate base level: _____ feet

23. What are the features in the river labeled with the letter A? (*Hint:* See the inside cover.)

Levee

24. Based on your answer to question 23, are the Missouri River and its tributaries actively (eroding *or* depositing) in this region?

The Missouri River and its tributaries are _depositing_ material.

25. Over time, as tributaries erode and lengthen their courses in their headwaters, what will happen to the upland areas?

They would most likely
flatten out becoming more like
floodplains.

Refer to the Angelica, New York, topographic map (Figure 4.8) to complete the following.

26. Draw an arrow on the map, indicating the direction that the main river, the Genesee, is flowing. (*Hint:* Use the elevations of the two bench marks [BM] next to the river near the top and bottom of the map to determine your answer.)

27. Use the BM elevations from question 26 to calculate the approximate gradient of the Genesee River. (*Hint:* See question 21.)

Gradient: _____ feet per mile

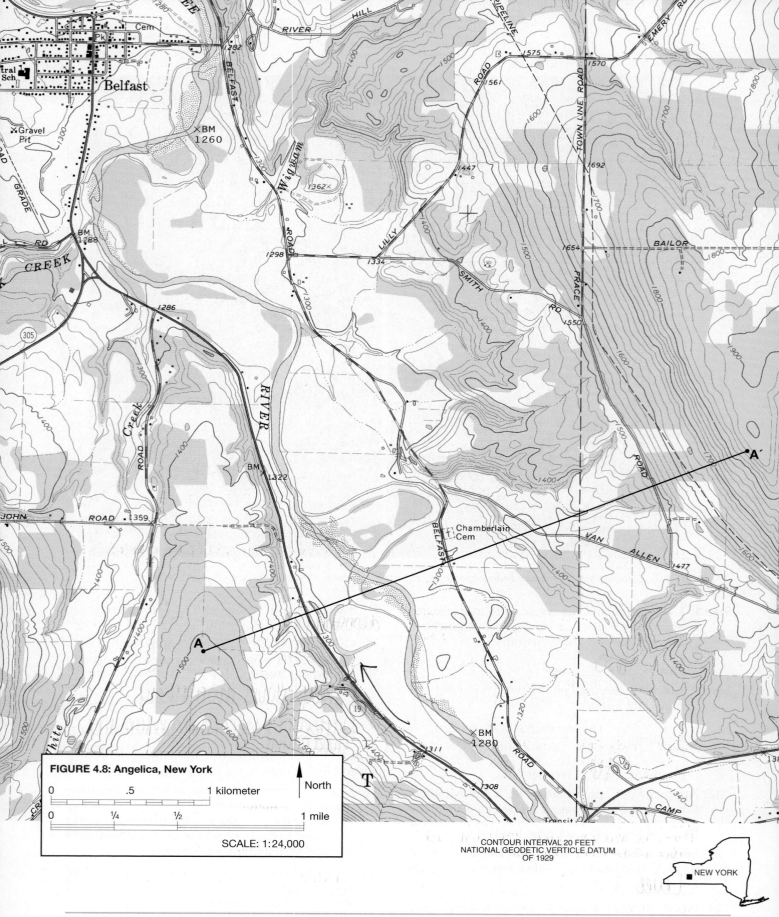

Figure 4.8 Portion of the Angelica, New York, topographic map, which includes the Genesee River. (Map source: U.S. Geological Survey)

Figure 4.9 Profile of the Genesee River Valley.

28. Use Figure 4.9 to draw a profile along line A–A'. Use only guide contour lines. (*Note:* Exercise 3 contains a detailed explanation for constructing topographic profiles.)

29. Approximately how many feet is the Genesee River from ultimate base level?

   Feet above base level (sea level): _____

30. The path of the Genesee River can best be described as which of the following: (a straight course *or* a meandering course?).

   The Genesee River follows a _____ course.

31. Which phrase most accurately describes most of the areas beyond the Genesee River Valley (very broad and flat *or* relatively hilly and dissected)?

   Most of this area is _____.

   Refer to the Campti, Louisiana, topographic map (Figure 4.10) and the stereogram of the same area (Figure 4.11) to complete the following. On the map, A indicates the width of the Red River floodplain and the dashed lines, B, mark the two sides of the meander belt of the river.

32. Approximately what percentage of the map area is flat and part of the Red River floodplain?

   Floodplain = _____ % of the map area

33. Find the guide contour that follows the Red River (look above the word *RIVER* in the upper portion of the map). The value of this contour line will tell you that the Red River is less than how many feet above sea level?

   The Red River is less than _____ feet above sea level.

34. Using Figure 4.5C as a reference, identify the type of feature found at each of the following letters on the map in Figure 4.10:

   Letter C (*Old River*): _____

   Letter D: _____

   Letter E: _____

   Letter F: _____

35. If no contour lines cross the Red River and the contour interval is 20 feet, what is the maximum gradient of this section of the Red River? (*Note:* Although the red lines that cross the river resemble contour lines, they are land boundaries established by the French, using the long-lot survey system.)

   Maximum gradient: _____ feet per mile

36. Identify and label examples of a point bar, a cutbank, and an oxbow lake on the stereogram in Figure 4.11.

37. Write a statement that compares the width of the meander belt of the Red River to the width of its floodplain.

   _____

   _____

38. What is the dominant activity of the Red River (downcutting *or* lateral erosion)?

   Dominant activity: _____

   Complete the following by comparing the Portage, Angelica, and Campti topographic maps.

39. On which of the three maps is the gradient of the main river steepest?

   The gradient is steepest on the _____ map.

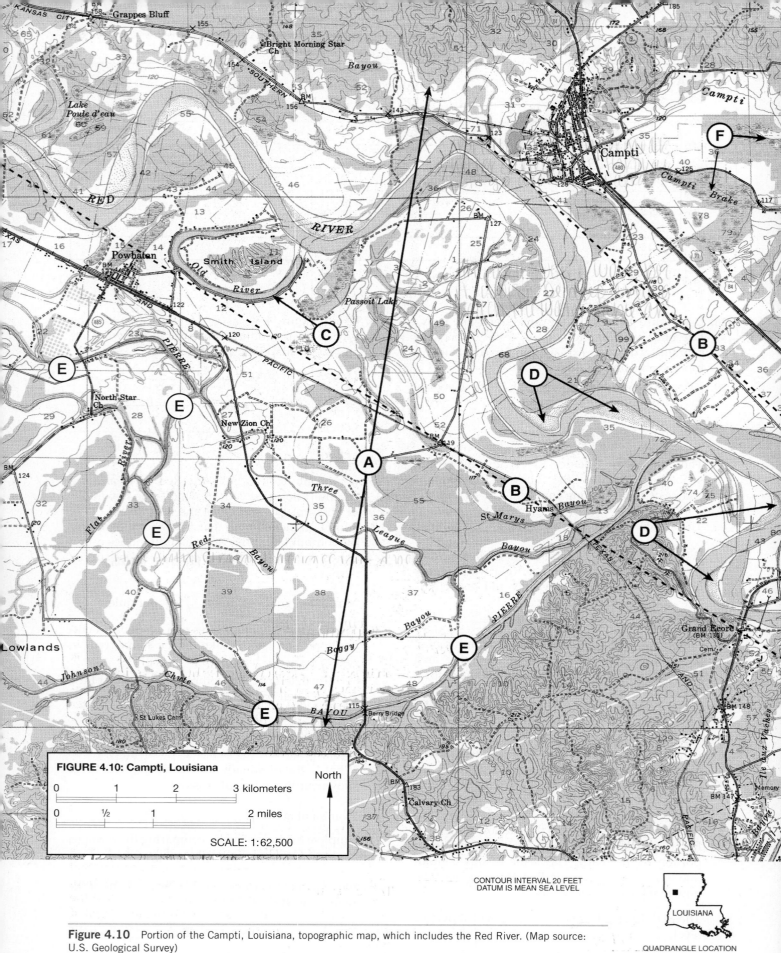

Figure 4.10 Portion of the Campti, Louisiana, topographic map, which includes the Red River. (Map source: U.S. Geological Survey)

FIGURE 4.10: Campti, Louisiana

0   1   2   3 kilometers

0   ½   1   2 miles

SCALE: 1:62,500

North

CONTOUR INTERVAL 20 FEET
DATUM IS MEAN SEA LEVEL

LOUISIANA

QUADRANGLE LOCATION

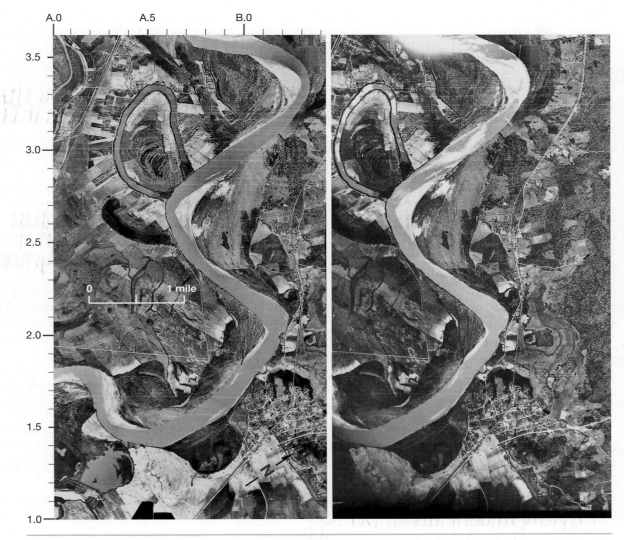

Figure 4.11 Stereogram of the Campti, Louisiana, area. (Courtesy of U.S. Geological Survey)

40. Choosing from Figure 4.6, 4.8, or ⟨4.10⟩ write the name of the map that is best described by each of the following statements.

4.10 **a.** Primarily floodplain: _Figure 4.8_

4.6 **b.** River valleys separated by broad, relatively flat upland areas: _Figure 4.10_

c. Greatest number of streams and tributaries: ___ _Figure 4.8_

d. Poorly drained lowland area with marshes and swamps: _Figure 4.10_

e. Downcutting is the dominant process: _Figure 4.10_

f. Nearest to base level: _Figure 4.10_

Groundwater

One of Earth's most valuable resources, groundwater is an important source of water for human use, irrigation, and industry. In many areas, overuse and contamination threaten the groundwater supply. Groundwater can also be a natural hazard, damaging building foundations and aiding the movement of materials during landslides and mudflows. Furthermore, many regions face problems where groundwater withdrawal results in land subsidence.

Water Beneath the Surface

Groundwater is water that has soaked into Earth's surface and completely fills the pore spaces in the soil and bedrock in the layer called the **zone of saturation**. The upper surface of this saturated zone is called the **water table**. Above the water table is the **unsaturated zone**, where the pore spaces are mainly filled with air (Figure 4.12).

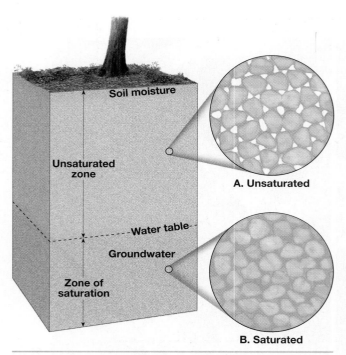

Figure 4.12 Idealized distribution of groundwater.

Use Figure 4.13, which illustrates the subsurface of a hypothetical area, to complete the following.

41. Label the zone of saturation, the unsaturated zone, and the water table.

42. Describe the shape of the water table in relationship to the shape of the land surface.

It is pretty straight across but does dip down with the land.

43. Whenever a substantial amount of water is withdrawn from a well, the water table forms a **cone of depression**. Label the cone of depression in Figure 4.13. What factors might cause a cone of depression to become larger or smaller in size?

The amount of water being taken up by the well.

44. Use a pencil to shade the area between the dashed lines labeled A. This zone represents an <u>impermeable lens of clay</u>. Describe what will happen to water that infiltrates to the depth of the clay lens at point A.

Because it is impermeable the water cannot pass through it. It will go around this area.

The blue dashed line in Figure 4.13 represents the level of the water table during the dry season.

45. How does the drop in the water table during the dry season affect the operation of the well?

The well won't work because it won't be deep enough to reach the water table anymore

Groundwater Movement

Use Figure 4.14, a hypothetical topographic map showing the location of several water wells, to complete the following. The numbers in parentheses indicate the depth of the water table below the surface in each well.

46. Calculate the elevation of the water table at each well location and write the approximate elevation next to each well. Next, use a colored pencil to draw smooth 10-foot contours that show the shape of the water table. (Start with the 1060-foot contour.) Use a different colored pencil to draw arrows on the map that indicate the direction of the slope of the water table.

a. Toward which direction does the water table slope?

_____

b. Referring to the site of the proposed water well, at approximately what depth below the surface should the proposed well intersect the water table?

_____

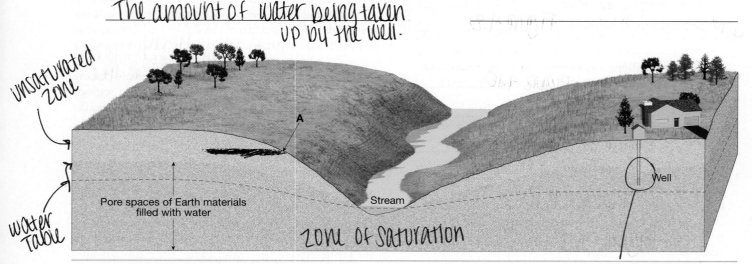

Figure 4.13 Subsurface of an area, showing saturated and unsaturated materials.

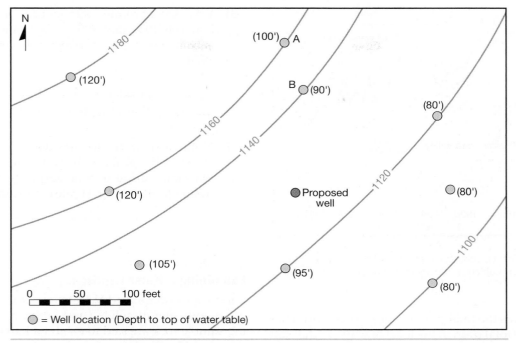

Figure 4.14 Hypothetical topographic map showing the location of several water walls.

47. Assume that a dye was put into well A on May 10, 2010, and detected in well B on May 25, 2011. What was the rate of groundwater movement between the two wells, in centimeters per day? (*Hint:* Convert feet to centimeters.)

    Velocity: _____ centimeters per day

Ground Subsidence

As demand for fresh water increases, surface subsidence caused by the withdrawal of groundwater from **aquifers** presents a serious problem in many areas. Several major urban areas, such as Las Vegas, Houston, Mexico City, and the Central Valley of California are experiencing subsidence caused by over-pumping wells (Figure 4.15). In parts of Mexico City, compaction of the subsurface material resulting from the reduction of fluid pressure as the water table is lowered has caused as much as 7 meters of subsidence.

A classic example of land subsidence due to groundwater withdrawal occurred in California's Santa Clara Valley, which borders the southern part of San Francisco Bay. Figure 4.16 is a graph that shows the relationship between ground subsidence in the valley and the level of water in a well in the same area. Use it to answer the following questions.

48. What is the general relationship between ground subsidence and the level of water in the well?

_____

Figure 4.15 The marks on this utility pole indicate the level of the surrounding land in preceding years. Between 1925 and 1977, this part of the San Joaquin Valley, California, subsided almost 9 meters because of the withdrawal of groundwater and the resulting compaction of sediments. (Photo courtesy of U.S. Geological Survey)

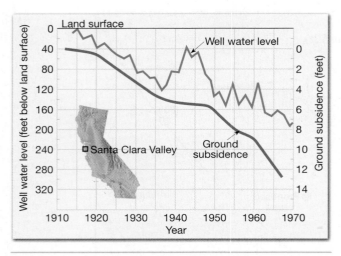

Figure 4.16 Ground subsidence and water level in a well in the Santa Clara Valley, California. (Courtesy of U.S. Geological Survey)

49. What was the total ground subsidence, and what was the total drop in the level of water in the well during the period shown on the graph?

   Total ground subsidence: _____ feet

   Total drop in well level: _____ feet

50. During the period shown on the graph, on an average, about how much land subsidence occurred with each 20-foot decrease in the water level in the well (1 foot, 5 feet, *or* 10 feet)?

   Subsidence: about _____ foot/feet

51. Was the ground subsidence that occurred between 1930 and 1950 (less *or* more) than the subsidence that occurred between 1950 and 1970?

   Ground subsidence from 1930 to 1950 was _____ than that from 1950 to 1970.

52. Notice that minimal subsidence occurred from 1935 to 1950. Refer to the well water level during the same period of time and suggest a possible reason for the reduced rate of subsidence.

   _____

   _____

Examining a Karst Landscape

Landscapes dominated by features that are the result of groundwater dissolving and removing the underlying soluble rock are said to exhibit a **karst topography** (Figure 4.17). On the surface, karst topography can be identified by *disappearing streams*, *solution valleys*, and depressions called *sinkholes* (Figure 4.17). Beneath the surface, removal of soluble rock may result in *caves* and *caverns*. A well-known karst region in the United States is in the vicinity of Kentucky's Mammoth Cave National Park (Figure 4.18). The bedrock in this area is mainly composed of soluble limestone.

Use the Park City topographic map (Figure 4.18) and the stereogram of the same area (Figure 4.19) to answer the following.

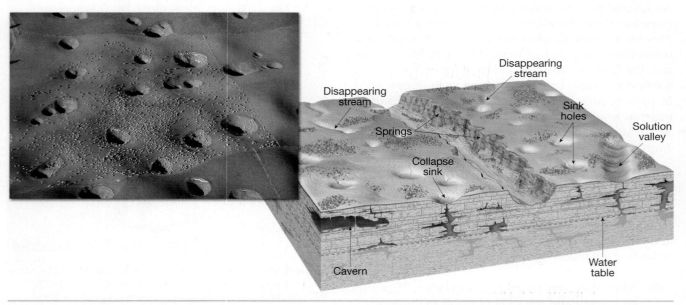

Figure 4.17 Features associated with well-developed karst topography. Inset photo shows sinkholes on New Zealand's South Island. (Photo by David Wall/Alamy)

Figure 4.18 Portion of the Park City, Kentucky, topographic map. (Courtesy of U.S. Geological Survey)

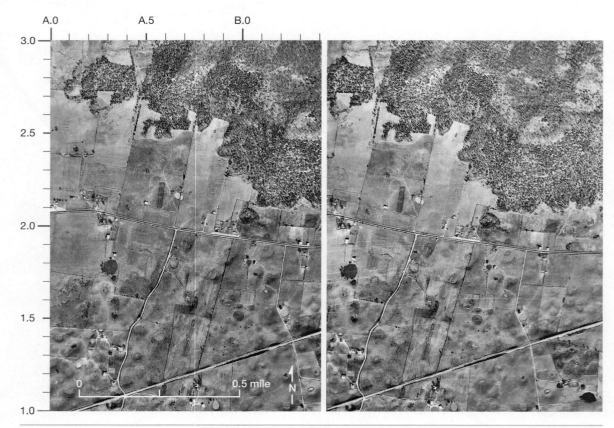

Figure 4.19 Stereogram of the Park City, Kentucky, area. (Courtesy of U.S. Geological Survey)

53. Locate three sinkholes (depressions) on the map and mark each with an X. (*Hint:* Look for closed contour lines with hachures.)

54. Notice that several sinkholes have water in them. What does this indicate about the depth of the water table in this area?

    The water table is near the surface.

55. Describe what is happening to Gardner Creek in the area indicated with the letter A on the map.

    The stream is lengthening. ?

    ↓
    underground stream

Companion Website

The companion website provides numerous opportunities to explore and reinforce the topics of this lab exercise. To access this useful tool, follow these steps:

1. Go to www.mygeoscienceplace.com.
2. Click on "Books Available" at the top of the page.
3. Click on the cover of *Applications and Investigations in Earth Science, 7e.*
4. Select the chapter you want to access. Options are listed in the left column ("Introduction," "Web-based Activities," "Related Websites," and "Field Trips").

Shaping Earth's Surface: Running Water and Groundwater

Name _____   **Course/Section** _____

Date _____   **Due Date** _____

1. Write a statement that describes the movement of water through the hydrologic cycle, citing several of the processes that are involved.

   _____
   _____
   _____

2. Assume that you need to determine the rate at which infiltration occurs in an area. What are the variables you must consider before you can formulate an answer?

   _____
   _____

3. Write a brief paragraph summarizing the results of your permeability experiment.

   _____
   _____
   _____

4. Why do soil-covered hillsides with sparse vegetation often experience severe soil erosion?

   _____
   _____
   _____

5. The graph in Figure 4.20 shows lag time between rainfall and peak flow (flooding) for an urban area and a rural area. Which graph, (A *or* B) represents a rural area? Explain your choice.

   Graph _____ represents a rural area.

   _____

6. On Figure 4.21, identify and label as many features of the river and valley as possible. Briefly describe the elevation of the stream as it relates to sea level.

   _____
   _____
   _____

7. Name two features you would expect to find on the floodplain of a widely meandering river near its mouth.

FEATURE	DESCRIPTION
_____	_____
_____	_____

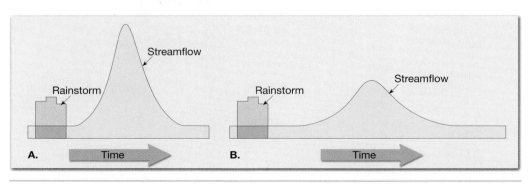

Figure 4.20   Art to be used with question 4.

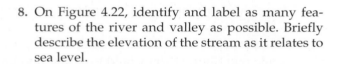

Figure 4.21 River and valley features. (Photo by Lane V. Erickson/Shutterstock)

8. On Figure 4.22, identify and label as many features of the river and valley as possible. Briefly describe the elevation of the stream as it relates to sea level.

_____

_____

_____

Figure 4.22 River and valley features. (Photo by Michael Collier)

9. Assume that you have decided to drill a water well. List two factors that should be considered prior to drilling.

_____

_____

10. Name two features you would expect to find in a region with karst topography.

FEATURE	DESCRIPTION
_____	_____
_____	_____

Shaping Earth's Surface: Arid and Glacial Landscapes

Objectives

Completion of this exercise will prepare you to:

A. Locate the desert and steppe regions of North America.

B. Describe the evolution of mountainous desert landforms found in the Basin and Range region of the western United States.

C. Describe the types of glacial deposits.

D. Identify and explain the formation of the features commonly found in areas once covered by continental ice sheets.

E. Describe the evolution of glaciated mountainous areas.

F. Identify and explain the features created by alpine glaciation.

Materials

calculator	ruler
hand lens	stereoscope
string	

Deserts and Desert Landscapes

Arid (**desert**) and semiarid (**steppe**) climates cover about 30 percent of Earth's land area (Figure 5.1). Despite the arid conditions, running water is the dominant agent of erosion in deserts. Wind erosion, although more significant in dry areas than elsewhere, is only of secondary importance.

Precipitation in dry climates is not only meager but sporadic. When rain occurs, it is often in the form of short, heavy downpours. Consequently, **flash floods** are a relatively common occurrence. Even when desert streams are full, they die out quickly because the water flowing in them readily soaks into the ground.

1. Indicate whether you think the following statements about the world's dry lands are *true* (T) or *false* (F).

   a. _____ Deserts are always hot.

   b. _____ Deserts are typically covered with sand dunes.

   c. _____ Deserts are rare, encompassing only a small percentage of Earth's land surface.

   d. _____ Deserts are practically lifeless.

   e. _____ Deserts are located away from the ocean.

Many people are surprised to learn that these statements are all common misconceptions (*false*). Sand dunes are widespread in only a few locations; plants and animals thrive where water is available; and the driest desert on Earth is located in coastal Chile.

Evolution of a Mountainous Desert Landscape

A classic region for studying the effects of running water in dry areas is the **Basin and Range** region, which includes portions of California, Nevada, Utah, Oregon, Arizona, and New Mexico (Figure 5.2, page 77). In this region, the erosion of mountain ranges and subsequent deposition of sediment in adjoining basins have produced a landscape characterized by several unique landforms.

The Basin and Range is characterized by **fault-block mountains** that formed when tensional force elongated and fractured the crust into numerous blocks (Figure 5.2A). The area's infrequent and intermittent precipitation typically results in streams that carry their eroded material from the mountains into interior basins. **Alluvial fans** and **bajadas** (features that result from the coalescence of alluvial fans) often form as streams deposit sediment on the less steep

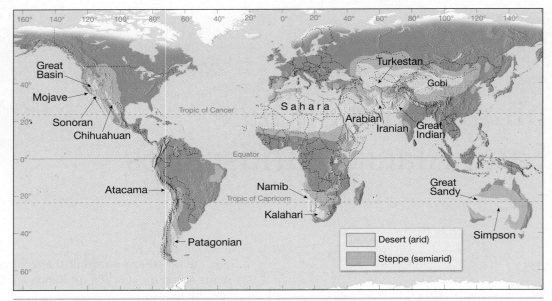

Figure 5.1 Arid and semiarid climates cover about 30 percent of Earth's land surface. No other climate group covers such a large area. Low-latitude dry climates are the result of the global distribution of air pressure and winds. Middle-latitude deserts and steppes exist principally because they are sheltered in the deep interiors of large landmasses.

slopes at the base of the mountains (Figure 5.2B). Occasionally, a shallow **playa lake** may develop near the center of a basin.

As the front of the mountain is worn down by erosion, a broad, sloping bedrock surface called a **pediment**, covered by a thin layer of sediment, often forms at its base (Figure 5.2C). Continuing erosion in the mountains and deposition in the basins may eventually fill the basin, leaving a nearly flat surface dotted with isolated peaks called **inselbergs**.

Figure 5.3 (page 78) is a portion of the Antelope Peak, Arizona, topographic map. It illustrates many of the features of a mountainous desert landscape. Use the map and accompanying stereogram (Figure 5.4, page 79) to complete the following. You will find the diagrams in Figure 5.2 helpful.

2. On the map, draw a line around the area that is illustrated in the stereogram.

3. Is the vegetation in the area (dense *or* sparse)? Are there (few *or* many) dry streams?

    Vegetation is _____, and there are _____ dry streams.

4. Determine the total relief of the map area (that is, the difference in elevation between highest and lowest points).

    Total relief: _____ feet

5. Are the streams that dominate the area (continuously flowing *or* intermittent)?

    Streams are _____

6. Which of the two lines, (A *or* B) follows the steepest slope?

    The steepest slope is line _____

7. Draw arrows on the map to indicate the directions that intermittent streams flow as they leave the mountains.

8. Identify the features indicated with the following letters and briefly describe how they formed. (*Hint:* Refer to Figure 5.2.)

    C: _____

    _____

    D: _____

    _____

9. The area labeled A on the map is a bedrock surface covered by a thin layer of sediment.

    The feature is called a(n) _____.

10. Assume that erosion continues in the area without interruption. How might the area look millions of years from now?

    _____

    _____

    _____

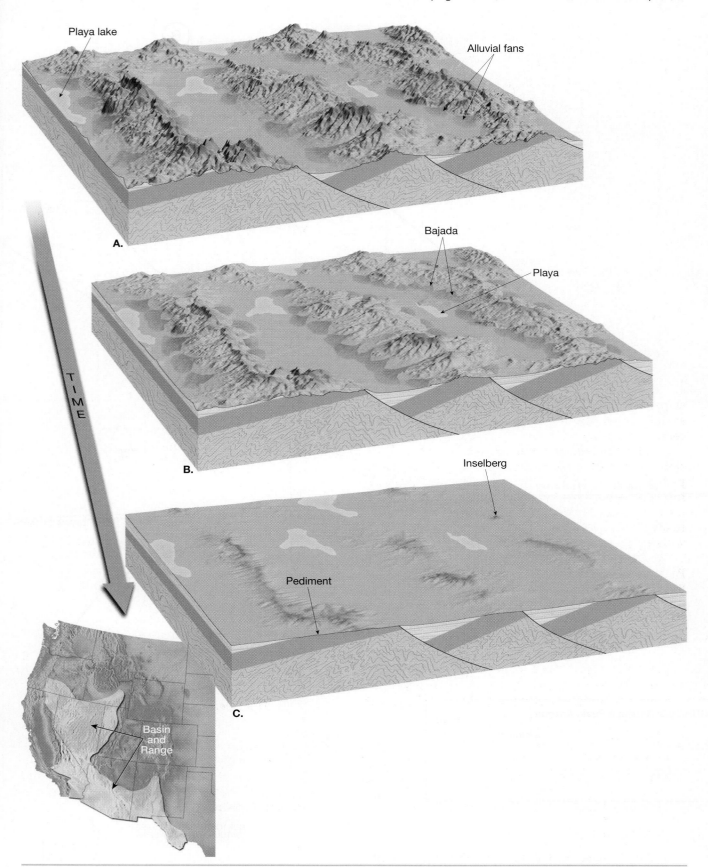

Figure 5.2 Stages of landscape evolution in a block-faulted, mountainous desert such as the Basin and Range region of the West. **A** Early stage; **B** Middle stage; **C** Late stage.

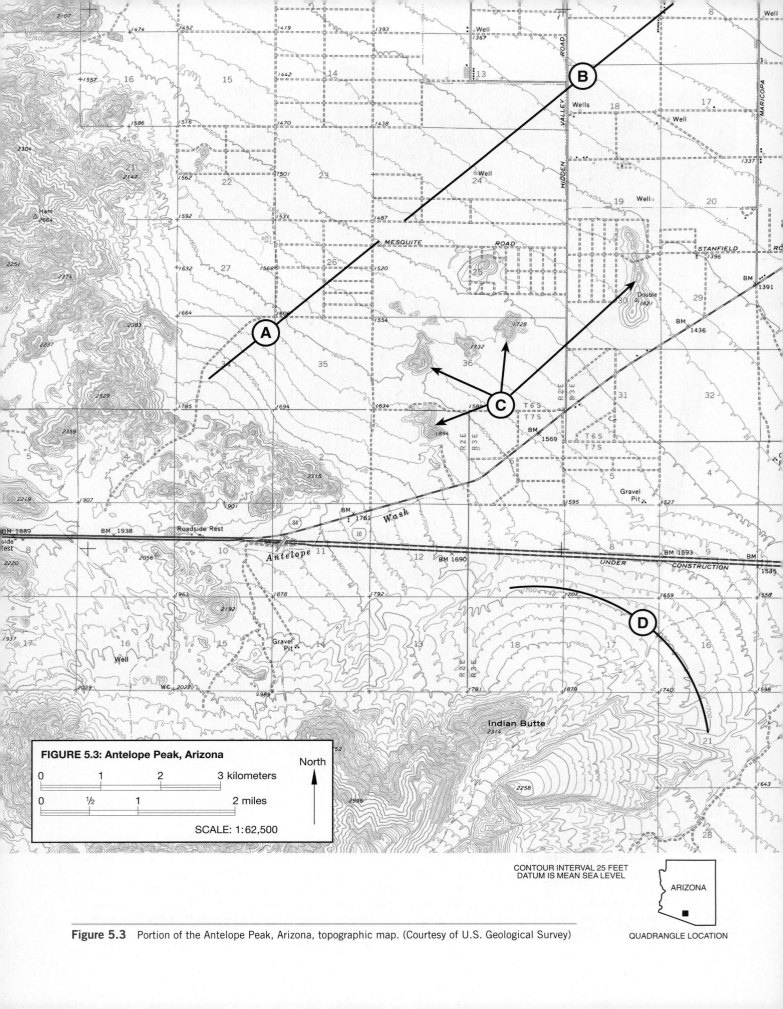

Figure 5.3 Portion of the Antelope Peak, Arizona, topographic map. (Courtesy of U.S. Geological Survey)

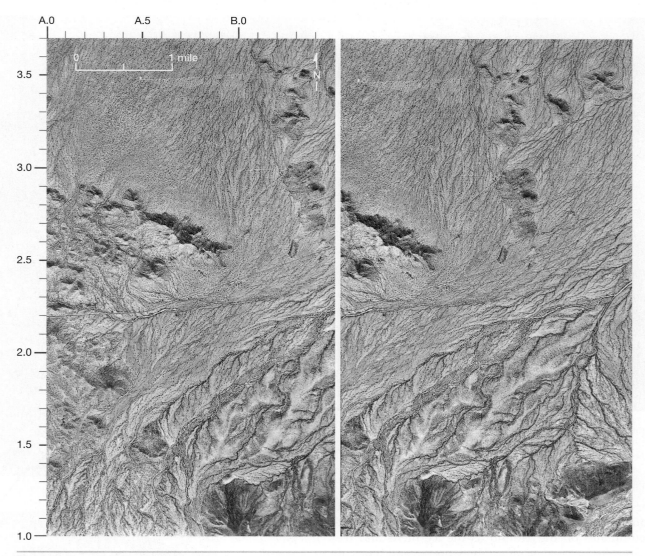

Figure 5.4 Stereogram of the Antelope Peak, Arizona, area. (Courtesy of U.S. Geological Survey)

Glacial Landscapes

Glaciers cover nearly 10 percent of Earth's land area. These moving masses of ice create many unique landforms and are part of an important link in the rock cycle in which the products of weathering are transported and deposited as sediment.

Literally thousands of glaciers exist today. They occur in regions where, over long periods of time, the yearly snowfall has exceeded the quantity of ice lost by melting. **Alpine glaciers**, or **valley glaciers**, develop from snow and ice that collects in mountainous areas. At high latitudes, enormous **ice sheets** cover much of Greenland and Antarctica.

Glacial erosion and deposition leave unmistakable imprints on Earth's surface. In regions once covered by continental ice sheets, glacially scoured surfaces and subdued terrain dominated by glacial deposits are clearly visible (Figure 5.5). In contrast, erosion by alpine glaciers accentuates the irregular mountainous topography, often producing spectacular scenery characterized by sharp, angular features.

Glacial Deposits and Depositional Features

The general term **drift** applies to all sediments of glacial origin, regardless of how, where, or in what form they were deposited. There are two types of glacial drift: (1) **till**, which is characteristically unsorted sediment deposited directly by a glacier (Figure 5.6), and (2) **stratified drift**, which is material that has been sorted and deposited by glacial meltwater.

Figure 5.5 Glacial striations in bedrock, Yellowknife, Northwest Territories, Canada. (Courtesy of Dr. Richard Waller, Keele University, United Kingdom)

Close up of cobble

Figure 5.6 Glacial till is an unsorted mixture of many different sediment sizes. A close examination often reveals cobbles that have been scratched as they were dragged along by the glacier. (Photos by E. J. Tarbuck)

The most widespread depositional features of glaciers are **moraines**, which are ridges of till that form along the edges of glaciers, and unsorted material that accumulates on the ground as the ice melts and recedes called *ground moraine*. There are several types of moraines, some common only to alpine glaciers.

Use Figure 5.7, which illustrates a hypothetical *retreating glacier*, to complete the following.

11. As a glacier retreats it sometimes stalls and deposits a recessional moraine. Label the recessional moraine in Figure 5.7.

12. On Figure 5.7, label an area covered by ground moraine.

Features of Continental Ice Sheets

During the **Pleistocene epoch**, ice sheets and alpine glaciers covered most of Canada, portions of Alaska, and much of the northern United States, as well as extensive portions of northern Europe and Asia. The impact that these ice sheets had on the landscape was significant.

Landforms produced by ice sheets, especially those that covered portions of the United States, are largely depositional. Some of the most extensive glacial deposits are present in the north-central United States. In many places, moraines, outwash plains, kettles, and other depositional features are prominent landscape features.

13. Briefly describe each of the following glacial deposits shown in Figure 5.7.

   Drumlin: _____

   _____

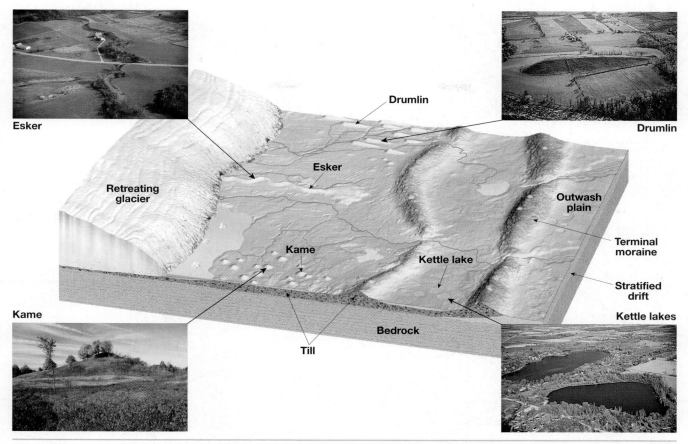

Figure 5.7 Characteristic depositional features of glaciers. (Drumlin photo courtesy of Ward's Natural Science Establishment; kame, esker, and kettle photos by Richard P. Jacobs/JLM Visuals)

Esker: _____

_____

Kame: _____

_____

Kettle: _____

_____

Outwash plain: _____

_____

Figure 5.8 is a portion of the Whitewater, Wisconsin, topographic map, which illustrates many of the depositional features associated with ice sheets. Use the map and the accompanying stereogram of the area (Figure 5.9, page 83) to complete the following.

14. After examining the map and stereogram, draw a line on the map to outline the area illustrated on the stereogram.

15. Is the general topography of the land in Sections 7 and 8 in the northwest portion of the region (higher *or* lower) in elevation than the land around the letter A located near the center of the map? Is it (more *or* less hilly)?

The topography is _____ in elevation and _____ hilly.

16. What evidence on the map indicates that portions of the area are poorly drained? On what part of the map are these features located?

_____

_____

17. Use Figure 5.10 (page 83) to draw a topographic profile along the A–A′ line on Figure 5.8.

18. Is the area that coincides with Kettle Moraine State Forest (higher *or* lower) in elevation than the land to the northwest and southeast?

The area is _____ in elevation.

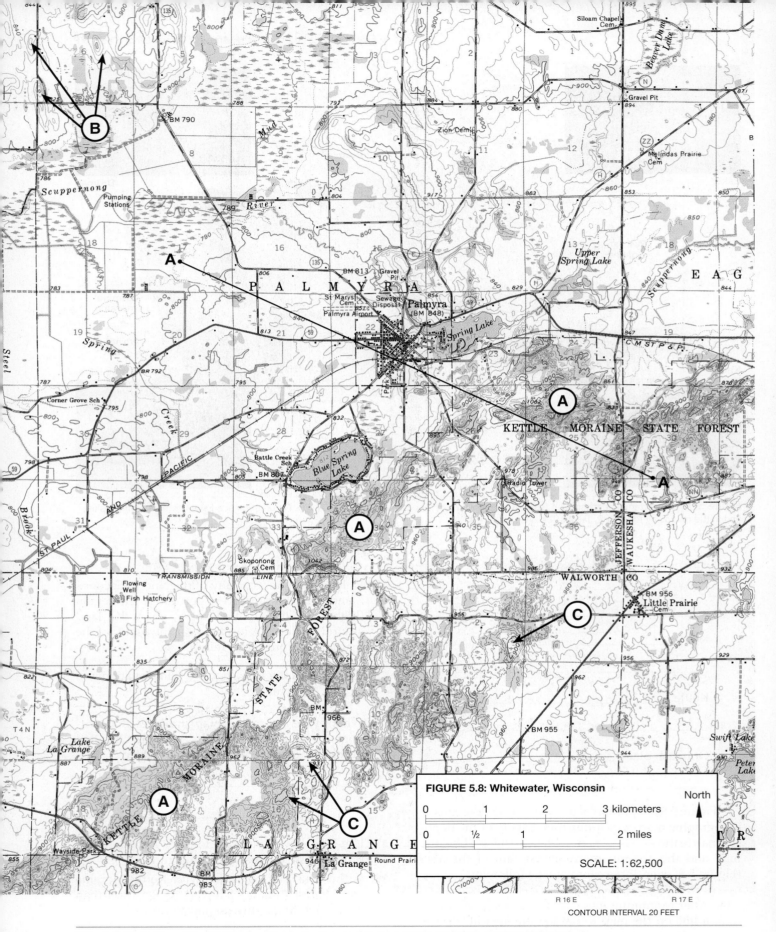

Figure 5.8 Portion of the Whitewater, Wisconsin, topographic map. (Courtesy of U.S. Geological Survey)

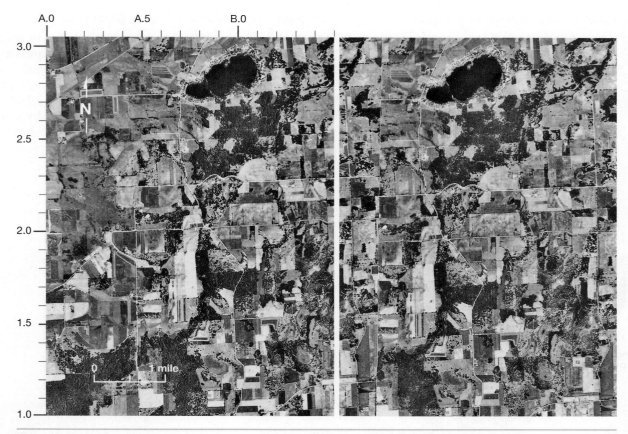

Figure 5.9   Stereogram of the Whitewater, Wisconsin, area. (Courtesy of U.S. Geological Survey)

Figure 5.10   Profile of the Whitewater map along line A–A'.

19. The feature labeled A on the map is a *long* ridge composed of till. Is this ridge called (an esker, a moraine, *or* a drumlin)?

    The long ridge is a(n) _____.

20. The streamlined, asymmetrical hills composed of till, labeled B, are what type of feature?

    The streamlined hills are _____

21. Examine the shape of the features labeled B on the map and in Figure 5.8. How can the features

be used to determine the direction of ice flow in a glaciated area?

_____

22. Using the feature labeled B in Figure 5.8 as a guide, draw an arrow on the map to indicate the direction of ice flow that occurred in this region.

23. What is the likely location of the outwash plain on the map? Identify and label the area "outwash plain." (*Hint:* Refer to Figure 5.7.)

_____

24. Label the area covered by ground moraine.

25. What term is applied to the numerous almost circular depressions designated with the letter C?

    Letter C points to a _____

26. What material is more likely being mined in the gravel pits north and northeast of the town of Palmyra (till *or* stratified drift)?

    Material being mined is _____

Features of Alpine or Valley Glaciation

Alpine glaciers often exaggerate the already mountainous topography of a region by eroding and deepening the valleys that they occupy. Figure 5.11 illustrates the changes that a formerly unglaciated mountainous area experiences as a result of alpine glaciation. Many of the landforms produced by glacial erosion, such as **arêtes**, **cirques**, **horns**, and **hanging valleys**, are identified in Figure 5.11.

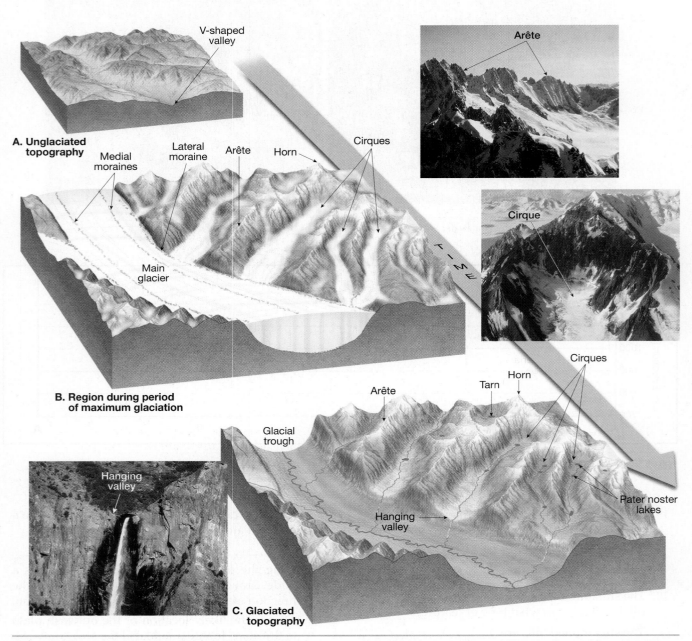

Figure 5.11 Erosional landforms created by alpine glaciers. The unglaciated landscape in Part **A** is modified by valley glaciers in Part **B**. After the ice recedes, in Part **C**, the terrain looks very different than before glaciation. (Arête photo from James E. Patterson Collection; cirque photo by Marli Miller; hanging valley photo by E. J. Tarbuck)

Figure 5.12 Aerial photograph of alpine glaciers and glacial features, Mont Blanc, France. (Courtesy of U.S. Geological Survey)

Refer to Figure 5.11 to complete the following.

27. How has glaciation changed the shape and depth of the main valley?

_____

_____

28. How has glacial erosion changed the gradients or slopes of tributary streams?

_____

_____

_____

29. Briefly describe the changes that occur in mountainous areas due to alpine glaciation.

_____

_____

_____

Use Figure 5.12, an oblique aerial photograph of an area around Mont Blanc, France to complete the following. (*Hint*: Refer to Figure 5.11.)

30. Draw arrows on the photograph to indicate the direction that the main glacier is flowing.

31. Give the name of the glacial feature described by each of the following statements and the letter on the photograph that labels an example of that feature. Use Figure 5.11C as a guide.

a. Sinuous, sharp-edged ridge:

Name: _____

Letter of example: _____

b. Hollowed-out, bowl-shaped depression that forms in the area of snow accumulation and ice formation:

Name: _____

Letter of example: _____

c. Moraines that form along the side of a valley:

Name: _____

Letter of example: _____

d. Moraines that form when two valley glaciers coalesce to form a single ice stream:

Name: _____

Letter of example: _____

Refer to Figure 5.13, a portion of the Holy Cross, Colorado, topographic map. This is a mountainous area that has been shaped by alpine glaciers.

32. Use Figure 5.14 to draw a topographic profile along the A-A' line from Sugar Loaf Mountain to Bear Lake. (Use only guide contours and mark the position of the Lake Fork stream on the profile.)

33. Describe the shape of Lake Fork Valley, based on your profile.

_____

34. Lake Fork Valley is a glacial _____.

35. Identify the glacial feature indicated by each of the following letters. Use Figure 5.11C as a reference.

B: _____

C: _____

36. Which letter(s) (D, E, F, *or* G) indicates *tarn*(s)—lakes that forms in a cirque?

Letter(s): _____

37. The feature marked H on the map is composed of glacial till. What type of glacial feature is it? How did it form?

_____

_____

38. Explain how Turquoise Lake formed.

_____

_____

Companion Website

The companion website provides numerous opportunities to explore and reinforce the topics of this lab exercise. To access this useful tool, follow these steps:

1. Go to www.mygeoscienceplace.com.

2. Click on "Books Available" at the top of the page.

3. Click on the cover of *Applications and Investigations in Earth Science, 7e.*

4. Select the chapter you want to access. Options are listed in the left column ("Introduction," "Web-based Activities," "Related Websites," and "Field Trips").

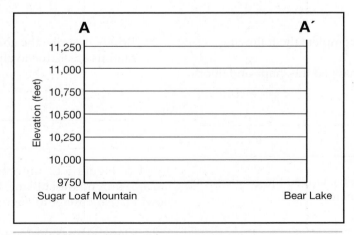

Figure 5.14 Topographic profile of the valley of Lake Fork on the Holy Cross, Colorado, map.

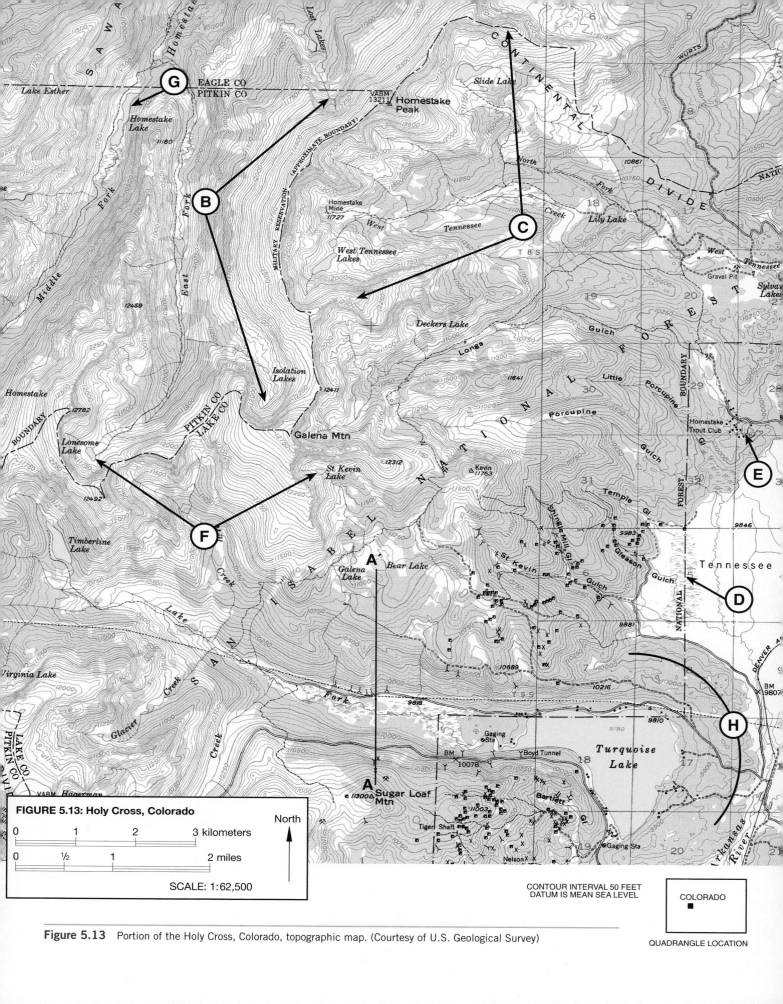

FIGURE 5.13: Holy Cross, Colorado

North

| 0 | | 1 | | 2 | | 3 kilometers |

| 0 | | ½ | | 1 | | 2 miles |

SCALE: 1:62,500

CONTOUR INTERVAL 50 FEET
DATUM IS MEAN SEA LEVEL

COLORADO

QUADRANGLE LOCATION

Figure 5.13 Portion of the Holy Cross, Colorado, topographic map. (Courtesy of U.S. Geological Survey)

Notes and calculations.

Shaping Earth's Surface: Arid and Glacial Landscapes

Name _____   **Course/Section** _____

Date _____   **Due Date** _____

1. What area of the United States is characterized by fault-block mountains with streams that drain into adjoining basins?

   _____

2. Name the feature shown in the lower half of Figure 5.15.

   _____

Figure 5.15 Photo to accompany question 2. (Photo by Michael Collier)

3. What type of feature is located at letters C and D on the Antelope Peak, Arizona, topographic map (Figure 5.3)?

   C: _____

   D: _____

4. Toward what direction does the pediment slope on the Antelope Peak topographic map (Figure 5.3)?

   _____

5. If you were working in the field, explain how you would determine whether a glacial feature is a moraine or an esker.

   _____

   _____

   _____

6. In the following space, sketch a simplified map-view (that is, area viewed from above) of the Whitewater, Wisconsin, topographic map (Figure 5.8). Include and label the outwash

plain, end moraine, drumlins, and a few kettles and kettle lakes.

7. Assume that you are hiking in the mountains. You suspect that the area was glaciated in the past. Describe some of the features you would look for to confirm your suspicion.

_____

_____

8. What reason did you give for the formation of Turquoise Lake on the Holy Cross, Colorado, topographic map (Figure 5.13)?

_____

9. On Figure 5.16, identify and label the alpine glacial features.

Figure 5.16 Athabaska Glacier in Canada's Jasper National Park. (Photo by Adina Tovy/Robert Harding)

Figure 6.12 Photo to accompany question 22. (Photo by Dennis Tasa)

Figure 6.13 Photo to accompany question 23.

18. What is the geologic range of plants that belong to the group *Ginkgo*?

    From _____ Period to _____ Period.

19. What is the geologic range of the group *Phacops*, a type of trilobite?

    From _____ Period to _____ Period.

20. What is the geologic range of *Lepidodendron*?

    From _____ Period to _____ Period.

21. Imagine that you have discovered an outcrop of sedimentary rock that contains fossil shark teeth and fossils of *Archimedes*. In which time periods might this rock have formed?

    From _____ Period to _____ Period.

22. What is the geologic range of the fossil shown in Figure 6.12?

    From _____ Period to _____ Period.

23. What is the geologic range of the fossil shown in Figure 6.13?

    From _____ Period to _____ Period.

24. Imagine that you have discovered a rock outcrop that contains the fossils identified in questions 22 and 23. What is the geologic range of this rock?

    From _____ Period to _____ Period.

25. Figure 6.14 illustrates two different sequences of sedimentary rock strata located some distance apart.

    a. Determine the geologic range/age of each rock layer by its fossil content and write your answer on the figure.

    b. Draw lines connecting the rock layers of the same age in outcrop 1 to those in outcrop 2.

    c. What term is used for the process of matching one rock unit with another of the same age?

    _____

    d. Based on the ages of the rock layers in outcrop 1, identify an unconformity. Remember that an unconformity is a break in the rock record. Draw a wavy line on the figure to denote the unconformity.

Radiometric Dating

Radioactivity has provided a reliable means for calculating the **numerical (absolute) age** of rocks, a procedure called **radiometric dating**. Radioactive atoms, such as those of uranium-238, emit particles from their nuclei that we call *radiation*. Ultimately, the process of decay produces an atom that is stable and no longer radioactive. For example, the radioactive decay of uranium-238 produces the stable element lead-206.

Determining Radiometric Ages

An unstable (radioactive) isotope that decays is referred to as the **parent**, while the stable isotope that results is called the **daughter product**. The rate of radioactive disintegration is expressed as the **half-life**—the amount of time it takes for half of the parent material to turn into its daughter product. Therefore, after one half-life, 50 percent of the parent isotope will have decayed. After the second half-life, half of the remaining radioactive atoms, or 25 percent ($\frac{1}{4}$) of the original amount will remain, and 75 percent will have decayed to form the daughter product. *With each successive half-life, the remaining parent isotope will be reduced by half.*

Among the many radioactive isotopes that exist in nature, six have proved particularly useful in providing radiometric ages. Rubidum-87, thorium-232, and the

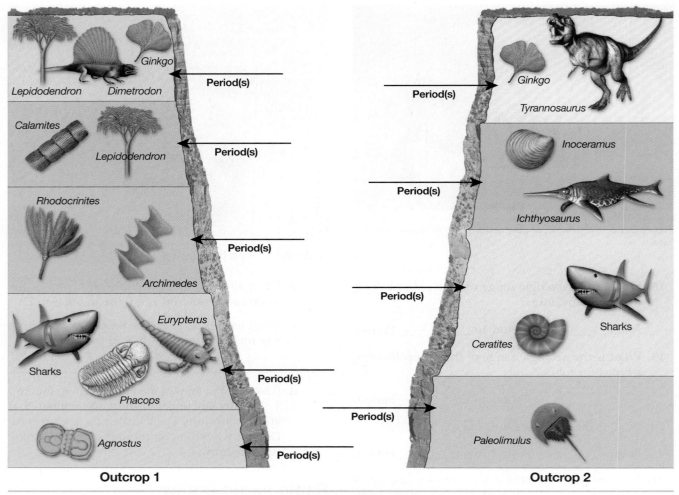

Outcrop 1

Outcrop 2

Figure 6.14   Sequences of sedimentary rock layers to accompany question 25.

two isotopes of uranium have very long half-lives and can be used only for dating rocks that are millions of years old. Although the half-life of potassium-40 is 1.3 billion years, analytical techniques make the detection of tiny amounts of its stable daughter product (argon-40) possible in rocks that are younger than 100,000 years. To date very recent events, the radioactive isotope of carbon (carbon-14) can sometimes be used.

Half-Life Experiment

This experiment is designed to help you understand the concept of half-life. You will need the following materials to complete this activity:

32 nickels
32 pennies
Red and blue colored pencils

Step 1: Place the 32 nickels heads-side up on the lab table. These represent atoms of the radioactive parent isotope.

Step 2: Place the nickels into a container, shake well, and dump them onto the lab table.

Step 3: Count and remove all of the nickels that are *tails-side up*. The nickels you remove represent parent atoms that have decayed. Replace each tails-side-up nickel with a penny. The pennies represent daughter atoms.

Step 4: On the chart, record the number of remaining nickels (heads-side up) and the pennies that represent the daughter atoms (Figure 6.15). The combined total should equal 32 coins.

Step 5: Place these coins in the container and repeat Steps 3 and 4 until only one or two nickels remain.

After you have completed the exercise (Trial 1), use a blue pencil to plot the number of nickels that remained at each interval and connect the points with a solid line on the graph (Figure 6.16). This line represents the change in parent atoms over time. Next, use a red pencil to plot the number of pennies at each interval on the same graph and connect the points with a solid line. This line represents the change in the number of daughter atoms over time.

Repeat this experiment (Trial 2), using the table in Figure 6.17 and the graph in Figure 6.18.

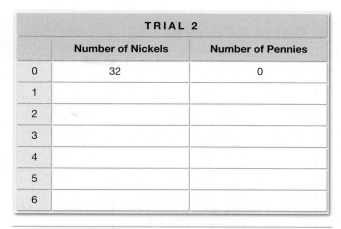

TRIAL 1		
	Number of Nickels	Number of Pennies
0	32	0
1		
2		
3		
4		
5		
6		

Figure 6.15 Chart to record data from Trial 1.

TRIAL 2		
	Number of Nickels	Number of Pennies
0	32	0
1		
2		
3		
4		
5		
6		

Figure 6.17 Chart to record data from Trial 2.

26. Compare the graphs you completed with the idealized half-life graph shown in Figure 6.19. What differences do you observe?

_____

27. If you had conducted this experiment 10 times and averaged your results, how do you think the results would have changed?

_____

Use Figure 6.19 to complete the following.

28. What percentage of the original parent isotope remains after each of the following half-lives have elapsed?

**PERCENTAGE OF
PARENT ISOTOPE REMAINING**

One half-life: _____

Two half-lives: _____

Three half-lives: _____

Four half-lives: _____

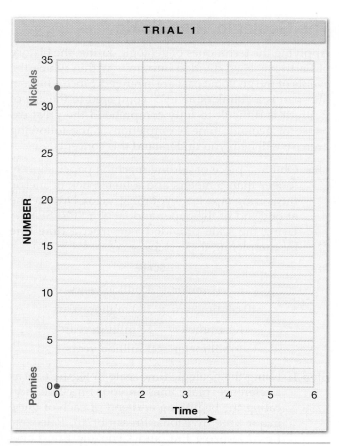

Figure 6.16 Graph to plot data from Trial 1.

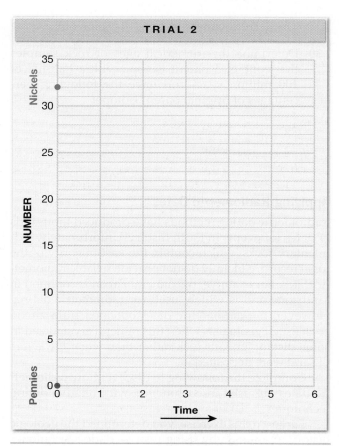

Figure 6.18 Graph to record data from Trial 2.

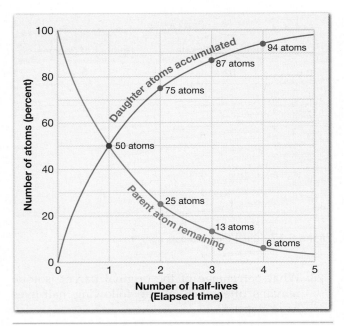

Figure 6.19 Idealized radioactive decay curves.

29. Geologists know that potassium-40 decays to argon-40 with a half-life of 1.3 billion years. Analysis of a hypothetical sample of granite reveals that 75 percent of the potassium-40 atoms have decayed to form argon-40. What is the age of sample of granite?

    The granite sample is _____ years old.

30. Determine the numerical ages of rock samples that contain a parent isotope with a half-life of 100 million years and have the following percentages of original parent isotope:

    50%: Age = _____

    25%: Age = _____

    6%: Age = _____

Applying Radiometric Dates

Radiometric dating methods have produced thousands of dates for events in Earth history. Unfortunately, sedimentary rocks, which contain the vast majority of the fossil record, generally do not provide reliable radiometric dates. For example, the mineral zircon is common in sandstone and is widely used for radiometric dating. However, the radiometric dates obtained from zircon crystals tell geologists when this mineral crystallized in some distant igneous rock rather than when the layers of sand were deposited.

Complete the following to discover how radiometric dates are used in conjunction with relative dating techniques.

Earlier in this exercise, you determined the geologic history of a region by using relative dating techniques (see Figure 6.9). Use Figure 6.9 to answer the following.

31. An analysis of a mineral sample from dike K indicates that 25 percent of the parent isotope is present in the sample.

    a. How many half-lives have elapsed since dike K formed?

    _____

    b. If the half-life of the parent isotope is 50 million years, what is the numerical age of dike K? Write your answer below and next to dike K on Figure 6.9.

    Dike K is _____ million years old.

32. An analysis of a mineral sample of intrusion J indicates its age to be 400 million years. Write the numerical age of rock J on Figure 6.9.

33. Are rock layers H and I (younger *or* older) than 100 million years? Explain.

    Layers H and I are _____ than 100 million years.

    _____

34. Determine the possible age range of rock layer E.

    Layer E is between _____ and _____ million years old.

35. Determine the age of rock layer A.

    Rock layer A is greater than _____ million years.

The Geologic Time Scale

Applying the techniques of geologic dating, the history of Earth has been divided into units within which the events of the geologic past are arranged. Since a human life span is a "blink of an eye" compared to the age of Earth, it is difficult to fully comprehend the great expanse of geologic time. Completion of the following exercise will help you understand this concept.

36. Obtain a piece of adding machine paper slightly longer than 5 meters and a meterstick or metric measuring tape from your instructor. Draw a line at one end of the paper and label it "PRESENT." Using the scale provided, construct a **time line** by completing these steps.

    SCALE

    1 meter = 1 billion years

    10 centimeters = 100 million years

    1 centimeter = 10 million years

    1 millimeter = 1 million years

Step 1: Using the geologic time scale in Figure 6.20 as a reference, place lines on your time line indicating the beginning and end of each of these major time divisions: Precambrian, Paleozoic Era, Mesozoic Era, and Cenozoic Era. Label each division and indicate its numerical age.

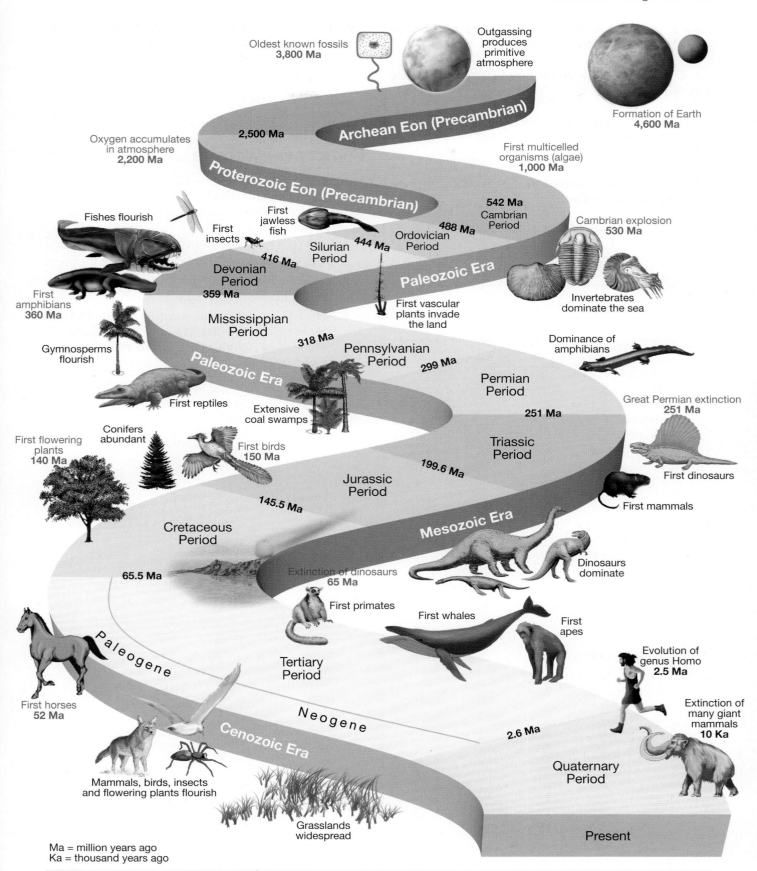

Figure 6.20 The geologic time scale. (Data from the Geological Society of America)

Step 2: Plot and label the events shown in red in Figure 6.20 on your time line.

Use your time line and Figure 6.20 to complete the following.

37. What fraction or percentage of geologic time is represented by Precambrian time?

    Approximately _____ of geologic time.

38. List the major animal groups in order of their appearance on Earth.

    _____

    _____

    _____

    _____

    _____

    _____

39. How many times longer is all geologic time than the time represented by recorded history (about 5000 years)? (Round geologic time to 5 billion years to simplify.)

    Geologic time is _____ times longer than recorded history.

Companion Website

The companion website provides numerous opportunities to explore and reinforce the topics of this lab exercise. To access this useful tool, follow these steps:

1. Go to www.mygeoscienceplace.com.
2. Click on "Books Available" at the top of the page.
3. Click on the cover of *Applications and Investigations in Earth Science, 7e.*
4. Select the chapter you want to access. Options are listed in the left column ("Introduction," "Web-based Activities," "Related Websites," and "Field Trips").

Geologic Time

Name _____

Course/Section _____

Date _____

Due Date _____

Use Figure 6.21 to complete questions 1–8.

1. Place the lettered features in proper sequence, from oldest to youngest, in the space provided on the figure.

2. What type of unconformity separates layer G from layer F?

   _____

3. Which principle of relative dating did you apply to determine that rock layer H is older than layer I?

   _____

4. Which principle of relative dating did you apply to determine that fault M is older than rock layer F?

   _____

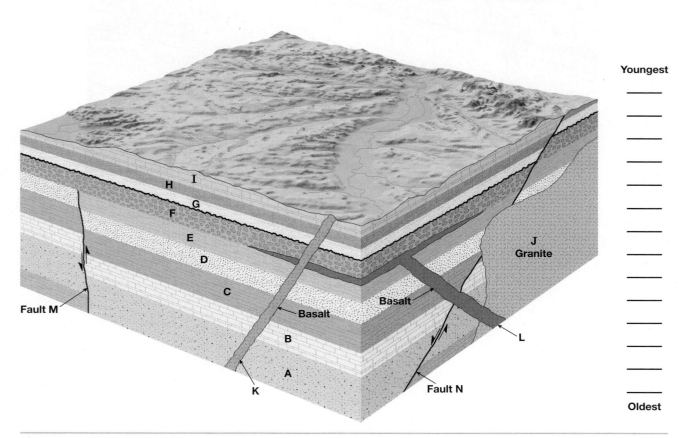

Figure 6.21 Geologic block diagram of a hypothetical region.

5. Explain how you know that fault N is older than igneous intrusion J.

   _____

6. If rock layer F is 150 million years old and layer E is 160 million years old, what is the approximate age of fault M?

   Approximate age of fault M: _____ million years

7. The analysis of samples from layers G and F indicates the following proportions of parent material to daughter product. If the half-life of the parent is known to be 75 million years, what are the ages of the two layers?

	PARENT	DAUGHTER	AGE
Layer G	50%	50%	_____
Layer F	25%	75%	_____

8. What time interval is represented by the disconformity at the base of rock layer G?

   Time interval: _____ to _____ million years

9. What fraction of time is represented by the Precambrian?

   Precambrian: _____ percent

10. Examine the photograph in Figure 6.22. Describe, as accurately as possible, the relative geologic history of the area, using the principles of relative dating covered in this exercise.

    _____

    _____

    _____

    _____

    _____

Figure 6.22 Photo of sedimentary beds to be used with question 10. (Photo by E. J. Tarbuck)

Geologic Maps and Structures

Objectives

Completion of this exercise will prepare you to:

A. Explain the orientation of folded rocks and faults, using the terms *strike* and *dip*.

B. Draw and interpret a simple geologic block diagram.

C. Recognize common structural features, including anticlines, synclines, domes, and basins.

D. Recognize the common types of faults.

E. Interpret a simplified geologic map and use it to construct a geologic cross section.

Materials

| ruler | hand lens |
| colored pencils | protractor |

Introduction

Structural geology is the study of the architecture and evolution of Earth's crust (Figure 7.1). The geologic features that result from the interactions of Earth's plates, called **rock** or **tectonic structures**, include folds, faults, and joints. **Geologic maps** show the ages, distribution, and orientation of these basic rock structures. These maps allow geologists to interpret the nature of the rocks below the surface and determine the forces that caused their deformation.

Stress is the term used to describe the force that acts on a rock unit to change its shape or volume. The forces that act to shorten a rock body are known as **compressional stresses**, whereas those that elongate or pull apart a rock unit are called **tensional stresses**. **Shear stresses** cause rock to **shear**—which involves the movement of one part of a rock body past another (Figure 7.2).

Figure 7.1 Deformed sedimentary strata near Palmdale, California. (Photo by E. J. Tarbuck)

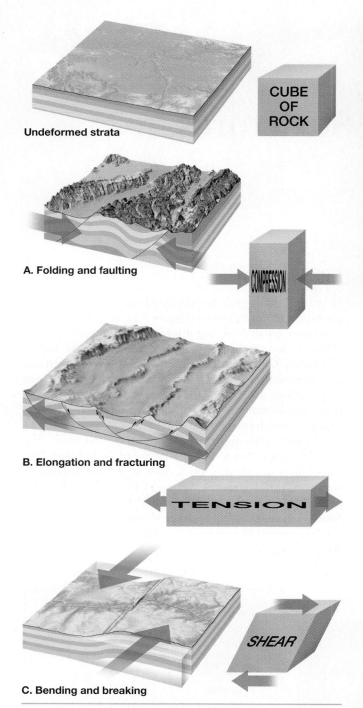

Figure 7.2 Simplified diagram showing the deformation of rock layers. **A.** Compressional stresses tend to shorten a rock body. **B.** Tensional stresses act to elongate, or pull apart, a rock unit. **C.** Shear stresses act to bend or break a rock unit.

If the stress applied to a rock unit exceeds its strength, the rock may undergo ductile deformation or fracture. **Ductile deformation** occurs at high temperatures and pressures deep within Earth. The result is often wavelike undulations called **folds** (see Figure 7.1).

Figure 7.3 Joints are fractures along which no appreciable displacement has occurred. (Photo by E. J. Tarbuck)

At lower temperatures and pressures found near the surface, most rocks behave like a brittle solid and fracture or break when the strength of the rock is surpassed. This type of deformation is called **brittle failure**. If the rocks on either side of the fracture move, the geologic feature is called a **fault** (see Figure 7.1). A **joint** is a fracture or break in a rock along which there has been no displacement (Figure 7.3).

Strike and Dip

Geologists use measurements called *strike* (trend) and *dip* (inclination) to help define the orientation of a rock layer or fault (Figure 7.4). By knowing the strike and dip of rocks at the surface, geologists can predict the nature and structure of rock units that are hidden beneath the surface.

Strike is the compass direction of the line produced by the intersection of an inclined rock layer or fault with a horizontal plane at the surface (Figure 7.4). The strike, or compass bearing, is generally expressed as an angle relative to north. For example, "north 10° east" (N10°E) means the line of strike is 10 degrees to the east of north. The strike of the rock units illustrated in Figure 7.4 is approximately north 55° east (N55°E).

Dip is the angle of inclination of the surface of the rock unit or fault from the horizontal plane. Dip includes both an angle of inclination and a direction toward which the rock is inclined. In Figure 7.4 the dip angle of the rock layer is 30°. A good way to visualize dip is to imagine that a water line will always run down the rock surface parallel to the dip. The direction of dip will always be at a 90° angle to the strike. (To illustrate, hold

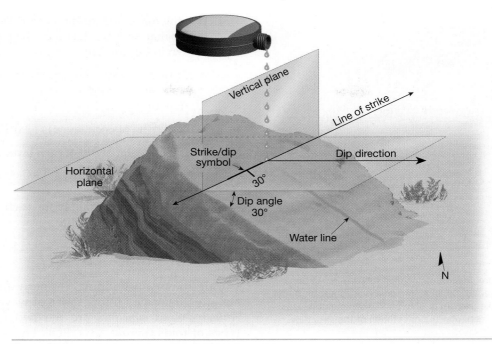

Figure 7.4 Illustration of the strike and dip of a rock layer.

your closed textbook at an angle to the tabletop. The upper edge of your text represents the strike. Regardless of the way you point the text, the direction of dip of the book is always at 90°, or a right angle, to the strike.)

Typically, the strike and dip of rock units and faults are shown on geologic maps with standard symbols that provide descriptive information about them—for example, $\overline{}_{20°}$. The long line shows the strike direction, the short line indicates the direction of the dip, and the dip angle is noted. In Figure 7.4, the strike-dip symbol indicates that the rocks are dipping toward the southeast at a 30° angle from the horizontal plane (30°SE).

1. Figure 7.5 illustrates geologic map views (views from directly overhead) of two hypothetical areas, showing a strike-dip symbol for a rock layer in each area. Complete the information requested below each map and draw a single large arrow on each map to indicate the direction of dip of the rock layer.

Block Diagrams

Block diagrams allow geologists to illustrate both the map view and **cross section** (the view from the side or beneath the surface) of an area (Figure 7.6).

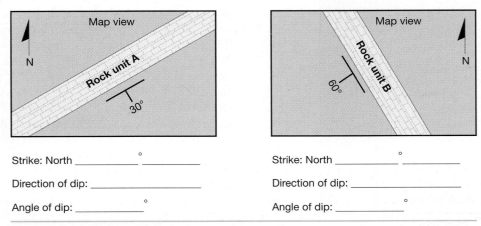

Strike: North _____°_____

Direction of dip: _____

Angle of dip: _____°

Strike: North _____°_____

Direction of dip: _____

Angle of dip: _____°

Figure 7.5 Geologic map views of two hypothetical areas.

A. Map view

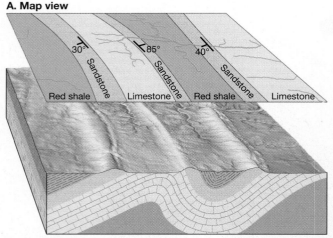

B. Block diagram

Figure 7.6 Diagram illustrating inclined rock layers, as they would appear on a geologic map **A.** and on a block diagram **B.** By establishing the strike and dip of outcropping sedimentary beds on a map, geologists can infer the orientation of the structure below ground.

Use the block diagram in Figure 7.6 as a guide to answer questions 2, 3, and 4.

2. Complete the cross sections on the front and side of the block diagram in Figure 7.7.

3. Complete each of the block diagrams in Figures 7.8 and 7.9.

4. Use the following guidelines to sketch both a geologic map view and a block diagram of the area shown in Figure 7.10.

   a. Draw four sedimentary layers of equal thickness on the surface. (*Hint:* Sketch the map view first.)

   b. The strike of each layer is N90°E.

   c. The direction of dip of each layer is to the south.

   d. The angle of dip of each layer is 60°.

5. Show strike-dip symbols for each rock layer on both parts of Figure 7.10.

Types of Folds

During mountain building, formerly flat-lying rocks are often deformed into a series of waves called *folds.* *Anticlines* and *synclines* are the two most common types of folds (Figure 7.11). Rock layers that fold upward, forming an arch, are called **anticlines** (Figure 7.12). Often associated with anticlines are downfolds, or troughs, called **synclines** (Figure 7.13).*

The **axial plane** is an imaginary plane drawn through the long axis of a fold that divides it as equally as possible into two halves, called *limbs* (Figure 7.14A). In a *symmetrical fold,* the limbs are mirror images of

*By strict definition, an anticline is a structure in which the oldest strata are found in the center. This most typically occurs when strata are upfolded. Furthermore, a syncline is strictly defined as a structure in which the youngest strata are found in the center. This occurs most commonly when strata are downfolded.

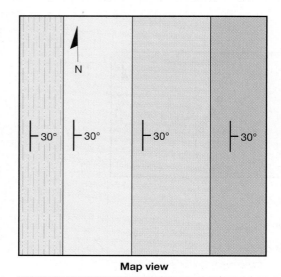

Map view

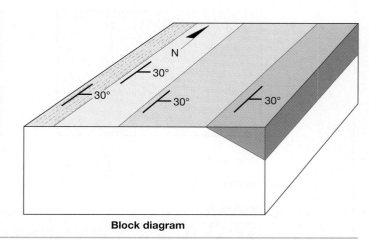

Block diagram

Figure 7.7 Map view and block diagram of a hypothetical area.

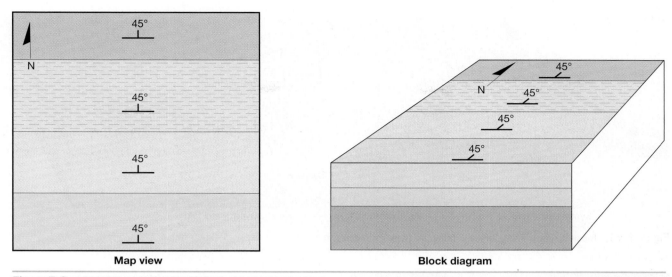

Figure 7.8   Map view and block diagram of a hypothetical area.

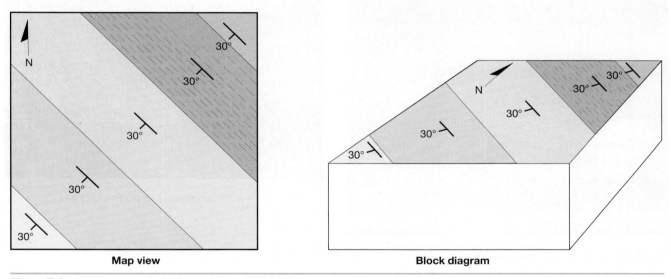

Figure 7.9   Map view and block diagram of a hypothetical area.

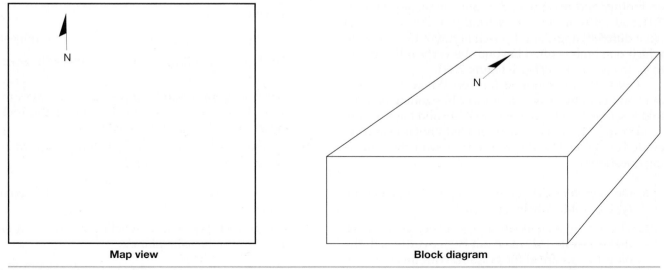

Figure 7.10   Map view and block diagram of a hypothetical area.

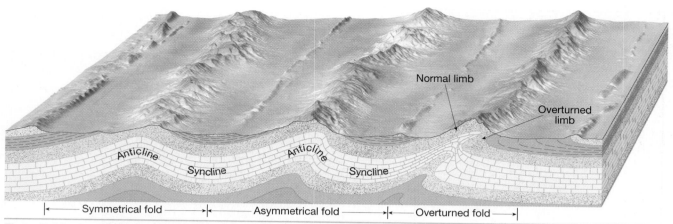

Figure 7.11 Block diagram illustrating the principal types of folded geologic structures.

Figure 7.12 An asymmetrical anticline. (Photo by E. J. Tarbuck)

Figure 7.13 A nearly symmetrical syncline. (Photo by E. J.

each other and diverge at the same angle (see Figures 7.11 and 7.13). In an *asymmetrical fold*, the limbs each have different angles of dip (see Figure 7.12). A fold in which one limb is tilted beyond the vertical is referred to as an *overturned fold* (see Figure 7.11).

Folds do not continue forever. Where folds "die out" and end, the axis is no longer horizontal, and the fold is said to be *plunging* (Figure 7.14B and Figure 7.15).

Figure 7.16 illustrates an eroded anticline and an eroded syncline. Use this figure to answer the following questions.

6. On each block diagram in Figure 7.16, label the type of fold, anticline or syncline.

7. The *law of superposition* (see Exercise 6) states that, in strata that have not been overturned, the oldest rocks are at the bottom. With this in mind, list by number the rock layers in each block diagram from oldest to youngest.

Anticline:

Oldest _____ Youngest

Syncline:

Oldest _____ Youngest

8. The plane extending through each block diagram is called _____.

9. Draw appropriate strike-dip symbols on the surface of each block diagram for one of the rock layers on each side of the axial plane.

10. Do the rock layers in an anticline dip (toward *or* away from) the axial plane?

The rock layers dip _____ the axial plane.

11. Do the rock layers in a syncline dip (toward *or* away from) the axial plane?

The rock layers dip _____ the axial plane.

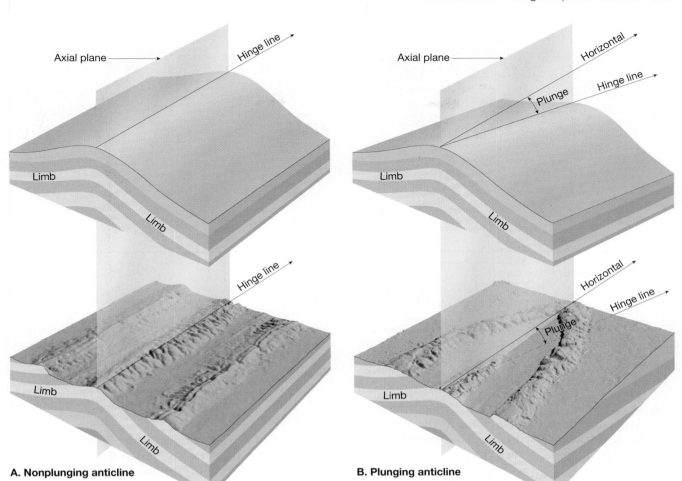

Figure 7.14 Features of simple folds. **A.** Nonplunging (horizontal) anticline **B.** Plunging anticline.

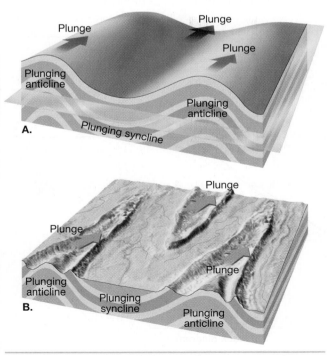

Figure 7.15 Plunging folds. **A.** Idealized view. **B.** View after extensive erosion.

12. Is the anticline shown in Figure 7.16 (symmetrical *or* asymmetrical)? Is the syncline (symmetrical *or* asymmetrical)?

    The anticline is _____.

    The syncline is _____.

13. Are these folds (plunging *or* nonplunging) folds?

    These folds are _____.

14. Label the oldest exposed layers on the surface of both block diagrams, on both sides of the axial plane, with X.

15. Complete the following statements that describe what happens to the ages of the surface rocks as you move away from the axial plane of each of the following structures.

    a. On an *eroded anticline*, do the rocks exposed at the surface get (older *or* younger) moving away from the axial plane?

       Rocks exposed at the surface get _____.

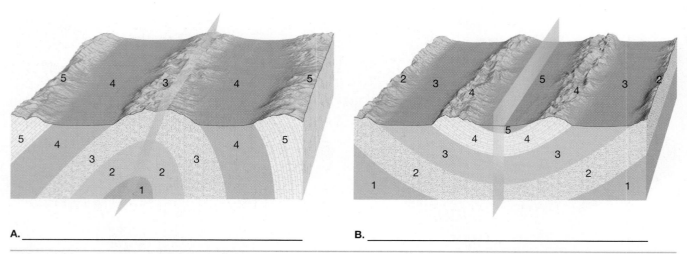

A. _____

B. _____

Figure 7.16 Idealized eroded folds.

b. On an *eroded syncline*, do the rocks exposed at the surface get (older *or* younger) moving away from the axial plane?

Rocks exposed at the surface get _____.

16. On Figure 7.17, complete the block diagram, using the information provided on the map view. Rocks illustrated with the same pattern/colors are part of the same rock layer.

17. Write the names of the two types of geologic structures illustrated at the appropriate place on the block diagram in Figure 7.17.

Anticlines and synclines are somewhat linear features caused by compressional forces. Two other types of folds, **domes** and **basins**, are roughly circu-

lar features that result from vertical displacement (Figure 7.18). Upwarping of sedimentary rocks produces a dome, whereas a basin is a downwarped structure.

18. Draw several appropriate strike-dip symbols on Figure 7.18.

19. On Figure 7.19, complete the geologic block diagram by drawing an eroded basin. Draw four rock layers, with appropriate strike-dip symbols, and label the layers from oldest (1) to youngest (4).

20. In a basin, are the oldest surface rocks found (near the center *or* at the flanks)?

The oldest rocks are found _____.

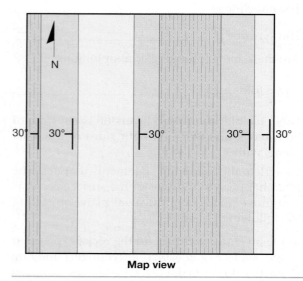

Map view

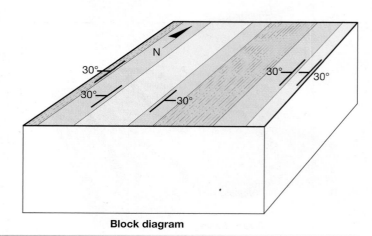

Block diagram

Figure 7.17 Map view and block diagram of a hypothetical area.

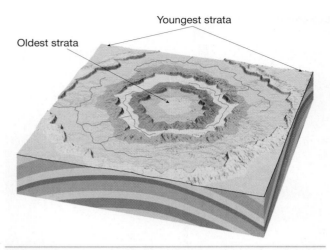

Figure 7.18 A typical eroded dome has an outcrop pattern that is roughly circular, with the rock layers dipping away from its center.

Types of Faults

Faults are fractures or breaks in rocks along which movement has occurred. As with folds, the terms *strike* and *dip* are used to describe the orientation of faults. Faults in which the relative movement of rock units is primarily vertical (along the dip of the fault plane) are called **dip-slip faults** (Figure 7.20). When the dominant motion of rock units is horizontal (in the direction of the strike), the faults are called **strike-slip faults**.

Dip-Slip Faults

Geologists describe dip-slip faults by noting the relative motion of the blocks on top of and below the fault. If the **hanging wall block** (above the fault, where miners who are extracting minerals deposited along a fault zone would hang lanterns) moves down relative to the **footwall block** (the bottom of the fault, where miners stand), the fault is classified as a **normal fault** (Figure 7.21A). Normal faults result from tensional forces.

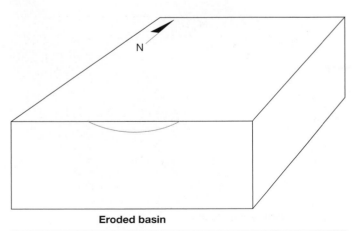

Eroded basin

Figure 7.19 Block diagram of an eroded basin.

Figure 7.20 A typical dip-slip fault in which the relative movement of the rock units is primarily vertical. (Photo by E. J. Tarbuck)

Compressional forces may produce a **reverse fault**, where the hanging wall moves up relative to the foot-wall (Figure 7.21B).

21. On each large block diagram in Figure 7.21, write the word *older* or *younger* to indicate the relative ages of the surface rocks on both sides of the fault trace.

22. Complete Figure 7.22 by illustrating an eroded reverse fault on diagram B. Place arrows on both sides of the fault to illustrate relative motion.

Strike-Slip Faults

Strike-slip faults result from horizontal displacement of rock units along the strike or trend of a fault (Figure 7.23). Some strike-slip faults, such as the famous San Andreas fault system in California, are near-vertical faults that exhibit tens or hundreds of kilometers of displacement.

A strike-slip fault is described as a **right-lateral fault** or a **left-lateral fault**, depending on the relative motion of the blocks. When facing the fault, if the crustal block on the opposite side is displaced to the *right*, it is a *right-lateral fault*. Likewise, if the opposite side is displaced to the *left*, it is a *left-lateral fault*.

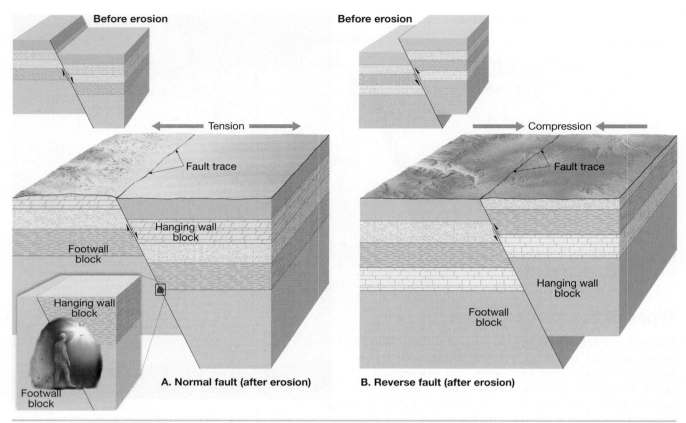

Figure 7.21 Block diagrams illustrating two common types of dip-slip faults. **A.** Normal fault. **B.** Reverse fault.

23. Is the fault illustrated in Figure 7.23 a (right- *or* left)-lateral fault?

The fault is a _____-lateral fault.

24. Complete the right-hand block diagram in Figure 7.24 by drawing a left-lateral fault. Draw arrows on both sides of the fault to illustrate the relative motion of each block.

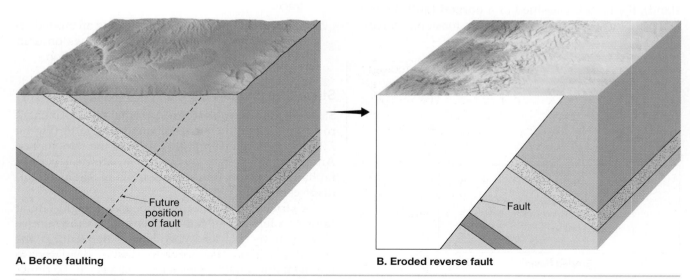

A. Before faulting

B. Eroded reverse fault

Figure 7.22 Block diagram illustrating a reverse fault.

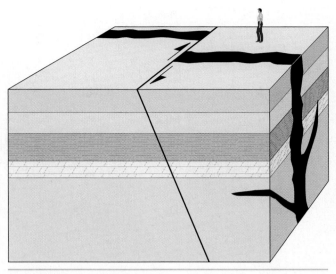

Figure 7.23 Block diagram illustrating a strike-slip fault.

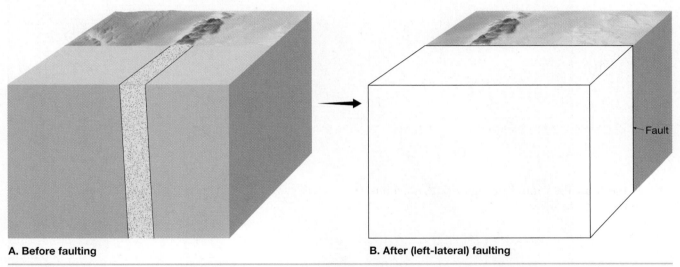

A. Before faulting

B. After (left-lateral) faulting

Figure 7.24 Block diagram illustrating a left-lateral strike-slip fault.

Examining a Geologic Map

Examine Figure 7.25, which is a simplified geologic map of Devils Fence, Montana, and Figure 7.26, an explanation of the Devils Fence map. Then answer questions 25–33. (*Note:* You may find the geologic time scale in Figure 6.20, in Exercise 6, a helpful reference.)

25. What are the names and approximate ages, in years, of the youngest and oldest *sedimentary* rock units shown in the map explanation in Figure 7.26 (page 117)?

Youngest sedimentary rock unit: _____

Age: _____ million years

Oldest sedimentary rock unit: _____

Age: _____ million years

26. On the map, write the word *oldest* where the oldest sedimentary rock unit is exposed at the surface and the word *youngest* where the youngest sedimentary rocks occur.

27. Are the intrusive igneous rocks near the center of the structure (younger *or* older) than the adjacent sedimentary rocks? (*Hint:* See Figure 7.26.)

The igneous rocks are _____ than the sedimentary rocks.

28. Examine the strike and dip of the rock units on the Devils Fence geologic map. Draw several large arrows on the map, pointing in the direction of dip of several of the rock units.

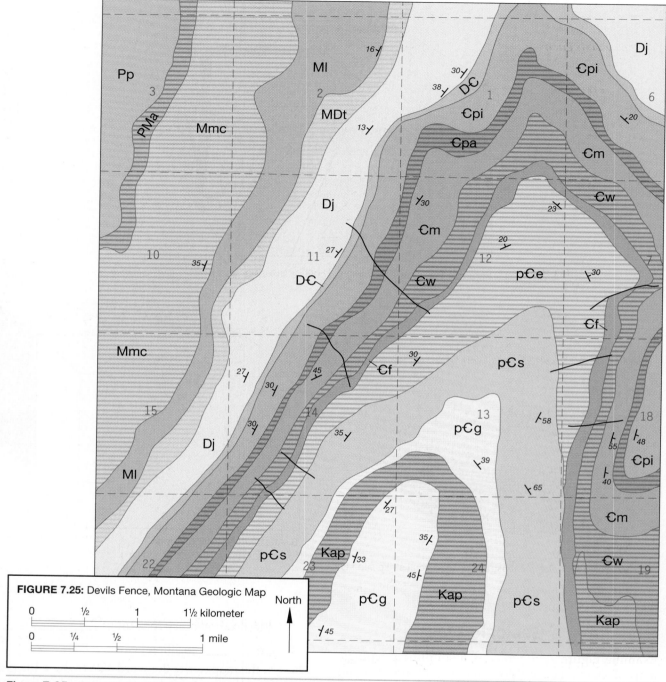

FIGURE 7.25: Devils Fence, Montana Geologic Map

North

0   ½   1   1½ kilometer

0   ¼   ½   1 mile

Figure 7.25 Simplified geologic map—Devils Fence, Montana.

a. Do the rock layers located near the center of the map in Section 14 dip toward the (northwest *or* southeast)?

The rocks dip toward the _____.

b. The same rocks in Section 14 are also found in Section 18. Do the rocks in Section 18 dip toward the (east *or* west)?

The rocks dip toward the _____.

29. What is the approximate angle of dip of the rock units in Section 18?

Angle of dip: _____

30. Draw a dashed line representing the hinge line of the large geologic structure that occupies most of the map. Label the line *hinge line*.

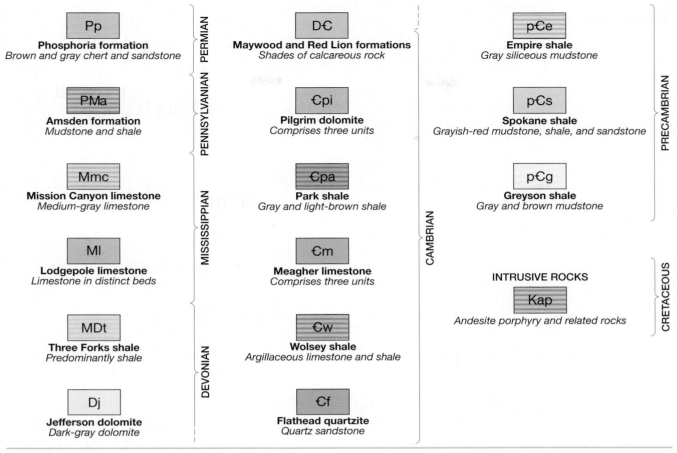

Figure 7.26 Explanation for the geologic map of Devils Fence, Montana, in Figure 7.25.

31. Are the rocks getting (older *or* younger) away from the hinge line?

The rocks are getting _____.

32. Is this geologic structure a plunging (anticline *or* syncline)?

The structure is a plunging _____.

33. List two lines of evidence that support your answer to question 32.

_____

_____

Companion Website

The companion website provides numerous opportunities to explore and reinforce the topics of this lab exercise. To access this useful tool, follow these steps:

1. Go to www.mygeoscienceplace.com.
2. Click on "Books Available" at the top of the page.
3. Click on the cover of *Applications and Investigations in Earth Science 7e.*
4. Select the chapter you want to access. Options are listed in the left column (*Introduction, Web-based Activities, Related Websites, and Field Trips*).

Notes and calculations.

Geologic Maps and Structures

Name _____

Course/Section _____

Date _____

Due Date _____

1. Refer to Figure 7.27 to describe the strike and dip of the limestone layer.

   Strike: North _____°_____

   Direction of dip: _____

   Angle of dip: _____°

2. Use Figure 7.27 to draw a map view of an eroded syncline. Show at least four different sedimentary beds and label the dip and strike of each bed with the proper symbol.

3. Name the geologic structures shown in Figure 7.28. What was the dominant type of stress that acted on this rock body?

   Structure: _____

   Stress: _____

Figure 7.28 Photograph to accompany question 3. (Photo by E. J. Tarbuck)

4. What type of geologic structure is illustrated in the Devils Fence geologic map (Figure 7.25)? List two lines of evidence in support of your choice.

   _____

   _____

5. What type of stress, (compression *or* tension), was responsible for producing the geologic structure in Figure 7.25?

   Type of stress: _____

6. Where are the youngest rocks located in an eroded syncline?

   _____

   _____

7. Complete the block diagram in Figure 7.29. Show at least four sedimentary rock layers and add arrows to show the relative movement on both sides of the fault.

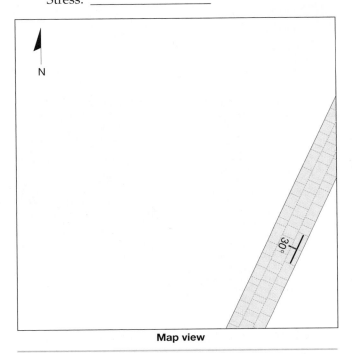

Map view

Figure 7.27 Diagram to accompany questions 1 and 2.

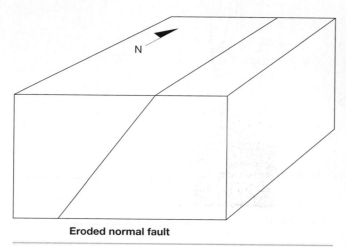

Eroded normal fault

Figure 7.29 Block diagram of an eroded normal fault.

8. What type of strike-slip fault (right-lateral fault *or* left-lateral fault) is shown in Figure 7.30?

   The fault is a _____ strike-slip fault.

9. Label the hanging wall and footwall on each of the two faults illustrated in Figure 7.31. On each photo, draw arrows showing the relative movement on both sides of the fault. Name the type of fault illustrated by each photo and describe the type of stress that produced it.

   A. Name of fault type: _____

   _____ stress

   B. Name of fault type: _____

   _____ stress

10. A large geologic structure, centered in Michigan, is shown on the simplified geologic map in Figure 7.32. What are geologic structures such as this called?

    Geologic structure is called a(n) _____.

A.

B.

Figure 7.31 Photographs of two faults to accompany question 9. (Photo by Marli Miller)

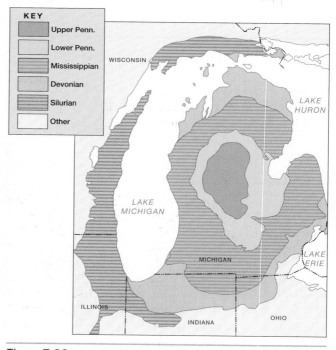

KEY
- Upper Penn.
- Lower Penn.
- Mississippian
- Devonian
- Silurian
- Other

WISCONSIN

LAKE HURON

LAKE MICHIGAN

MICHIGAN

LAKE ERIE

ILLINOIS

INDIANA

OHIO

Figure 7.32 Simplified geologic map to accompany question 10.

Figure 7.30 Photo to accompany question 8.

Earthquakes and Earth's Interior

Objectives

Completion of this exercise will prepare you to:

A. Examine an earthquake seismogram and identify the P waves, S waves, and surface waves.

B. Use a seismogram and travel–time graph to measure the distance to the epicenter of an earthquake and determine the time it occurred.

C. List the name, depth, and state of matter of each of Earth's layers.

D. Describe the temperature gradient in Earth's interior.

E. Explain why the asthenosphere likely consists of very weak (less rigid) material.

F. Explain how the distribution of earthquakes provides evidence that large, rigid slabs of the lithosphere descend into the mantle.

Materials

calculator   colored pencils
drawing compass   ruler

Earthquakes

Earthquakes are vibrations caused by the sudden and rapid movement of a large volume of rock along a fault. The energy generated by earthquakes radiates outward in the form of **seismic waves** from the source of the quake, called the **focus** (Figure 8.1). **Seismographs** located throughout the world record the ground motions produced by passing seismic waves (Figure 8.2). Seismic recordings, called **seismograms**, are used to determine the time and location of an earthquake, as well as to study Earth's internal structure (Figure 8.3).

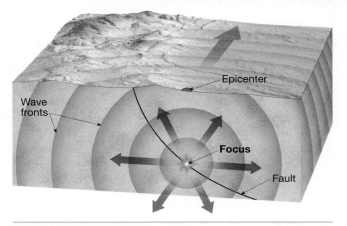

Figure 8.1 Earthquake focus and epicenter. The focus is the zone within Earth where the initial displacement occurs. The epicenter is the surface location directly above the focus.

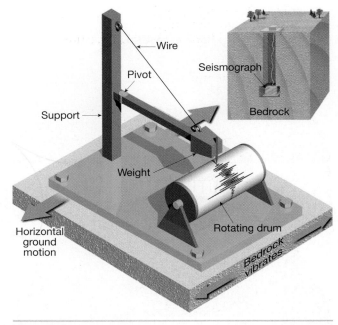

Figure 8.2 Principle of the seismograph. The inertia of the suspended mass tends to keep it motionless while the recording drum, which is anchored to bedrock, vibrates in response to seismic waves.

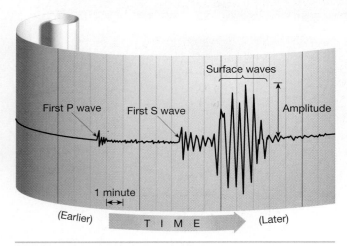

Figure 8.3 Typical earthquake seismogram.

Examining Seismograms

The three basic types of seismic waves generated by an earthquake are **P waves**, **S waves**, and **surface waves**. P and S waves travel through Earth's interior, and surface waves are transmitted along the surface (Figure 8.4). Of the three wave types, P waves have the greatest velocity and, therefore, reach the seismograph station first. Surface waves arrive at the seismograph station last.

Use Figure 8.3, which shows a typical seismogram, to complete the following.

1. How many minutes elapsed between the arrival of the first P wave and the arrival of the first S wave?

   Minutes elapsed: __5 Minutes__

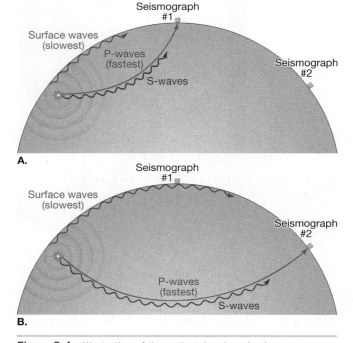

Figure 8.4 Illustration of the paths taken by seismic waves.

2. How many minutes elapsed between the arrival of the first P wave and the first surface wave?

   Minutes elapsed: __~7 Minutes__

3. How much time elapsed between the arrival of the first S wave and the first surface wave?

   Minutes: __2 Minutes__

4. Is the maximum amplitude (wave height) of the surface waves (less than *or* greater than) the maximum amplitude of the P waves?

   The maximum amplitude of surface waves is __greater__ than that of P waves.

Locating an Earthquake

The focus of an earthquake is the place within Earth where an earthquake originates. Earthquake foci have been recorded to depths as great as 600 km. When locating an earthquake on a map, seismologists plot the **epicenter**, the point on Earth's surface directly above the focus (see Figure 8.1).

The difference in the velocities of P and S waves provides a method for locating the epicenter of an earthquake. Both P and S waves leave the earthquake focus at the same instant. Since P waves have greater velocity, the further away the focus is from the recording instrument, the greater will be the difference in the arrival times of the first P wave compared to the first S wave.

To determine the distance between a recording station and an earthquake epicenter, follow these steps:

Step 1: Determine the P–S interval (in minutes) from the seismogram.

Step 2: Find the place on the travel–time graph, Figure 8.5, where the vertical separation between the P and S curves is equal to the P–S interval determined in Step 1.

Step 3: From this position, draw a vertical line that extends to the top or bottom of the graph and read the distance to the epicenter.

To locate an earthquake epicenter, seismograms from three different stations are needed. First, determine the distance of each station from the epicenter, using the procedure described in the preceding paragraph. Then draw a circle around each station with a radius equal to its distance from the epicenter. The point where all three circles intersect is the earthquake epicenter (Figure 8.6).

5. Examine Figure 8.5. Does the difference in the arrival times of the first P wave and the first S wave on a seismogram (increase *or* decrease), the farther a station is from the epicenter?

   The difference in arrival times __increase__ .

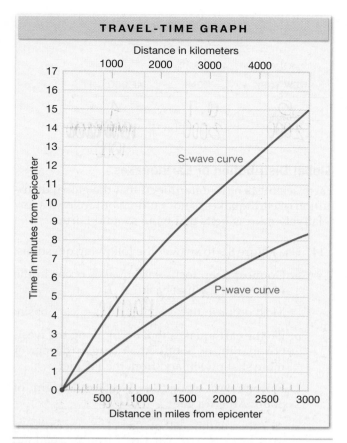

Figure 8.5 Travel–time graph used to determine the distance between an earthquake epicenter and a seismic station.

6. Use Figure 8.5 to determine the difference in arrival times, in minutes, between the first P wave and first S wave for stations that are the following distances from an epicenter:

1000 miles:	3.1	minutes
2400 km:	5.7	minutes
3000 miles:	11.5	minutes

On the seismogram in Figure 8.3, you determined the difference in the arrival times between the first P waves and the first S waves to be 5 minutes. Use this figure to complete the following.

7. Refer to the travel–time graph (Figure 8.5). What is the distance between the epicenter and the station for the earthquake recorded in Figure 8.3?

Distance: _____2000_____ miles

8. For the earthquake recorded in Figure 8.3, about how long did it take the first P wave to reach the station, (3, 6, *or* 14) minutes? (*Hint:* Use the travel–time graph in Figure 8.5 to answer this question.)

The first P wave arrived _____6_____ minutes after the earthquake began.

9. If the first P wave was recorded at 10:39 P.M. local time at the station in Figure 8.3, what was the local time when the earthquake occurred?

_____10:33 PM_____ P.M. local time

Figure 8.7 shows seismograms for the same earthquake recorded at three locations—New York, Nome,

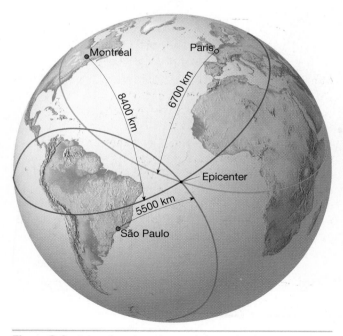

Figure 8.6 An earthquake epicenter is located using the distances obtained from three or more seismic stations.

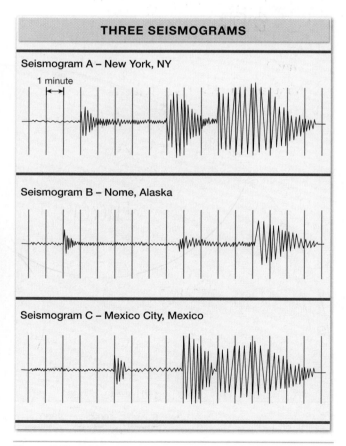

Figure 8.7 Seismograms of the same earthquake, recorded in three different cities.

Table 8.1   Epicenter Data Table

	NEW YORK	NOME	MEXICO CITY
Elapsed time between first P waves and first S waves	5	4.7	4
Distance from epicenter, in miles	2000	3000	1300 1200 1500

Alaska, and Mexico City. Use this information to complete the following.

10. Use Figure 8.7 and the travel–time graph, Figure 8.5, to determine the distance of each station from the epicenter. Write your answers in Table 8.1.

11. Use Figure 8.8 and a compass to draw a circle around each of the three stations with a radius, in miles, equal to its distance from the epicenter. (*Note:* Use the distance scale provided on the map.)

12. What are the approximate latitude and longitude of the epicenter of the earthquake that was recorded by the three stations?

    _____40_____ latitude and _____109_____ longitude

13. The first P wave was recorded in New York at 9:01 EST. At what time (EST) did the earthquake actually occur?

    _____8:56_____ EST

Global Distribution of Earthquakes

Earth scientists have determined that the global distribution of earthquakes is not random but follows a few relatively narrow belts that wind around Earth.

14. Use Figure 8.9 to answer the following questions.

    a. Do most deep focus earthquakes occur in the (Atlantic *or* Pacific) basin?

       Most occur in the _____Pacific_____ basin.

    b. Do the earthquakes that occur along the western margin of South America get (deeper *or* more shallow) inland from the Pacific?

       Earthquakes along the Western Margin of South America get _____deeper_____.

    c. Are most of the earthquakes along the western margin of North America (deep *or* shallow) focus?

       Most earthquakes are _____shallow_____ focus.

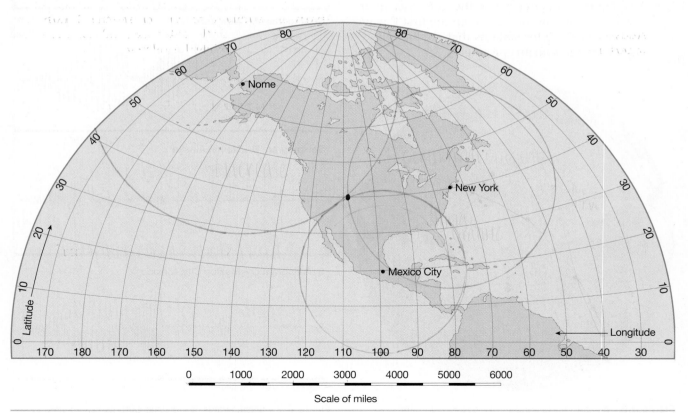

Figure 8.8   Map for locating an earthquake epicenter.

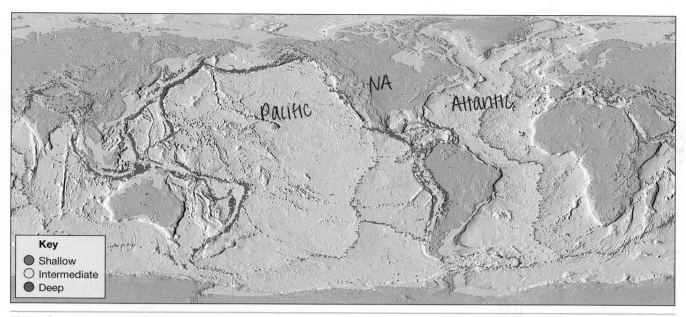

Figure 8.9 World distribution of earthquakes. (Data from NOAA).

15. Describe the shape and depth of the belt of earthquakes belt that extends down the middle of the Atlantic Ocean basin.

They are all shallow and they are in a curved line that goes through the middle of the Atlantic ocean.

The Earth Beyond Our View

The study of earthquakes has contributed greatly to our understanding of Earth's internal structure. Varia-

tions in the travel times of P and S waves provide scientists with evidence of changes in the properties of rocks at depth. Seismic waves travel faster through rock that is *stiff* (rigid). Further, because S waves cannot travel through fluids, it is inferred that a fluid layer exists near Earth's center (Figure 8.10).

The study of seismic waves has led researchers to conclude that Earth's interior is divided into several layers in addition to the **crust**—the **lithosphere**, **asthenosphere**, **mantle**, **outer core**, and **inner core**. Use Figure 8.10 to complete the following.

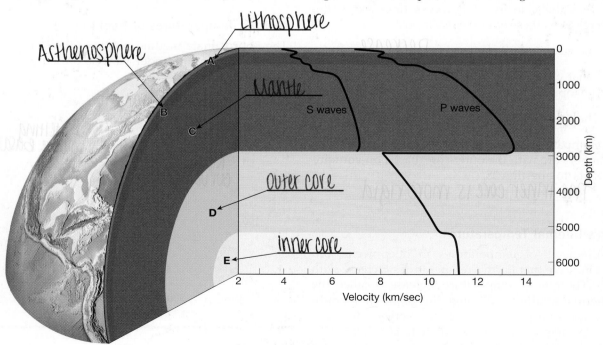

Figure 8.10 Illustration showing how P and S wave velocities vary with depth.

16. Place the names of the major zones of Earth's interior, labeled A–E on the illustration.

17. What are the approximate velocities of P and S waves at the bottom of layer A?

    P wave velocity: _____4.8_____ km/sec

    S wave velocity: _____8.2_____ km/sec

18. Does the velocity of P waves and S waves (increase *or* decrease) with increased depth in layer A?

    Velocity _____Increase_____.

19. Does the velocity of P waves and S waves (increase *or* decrease) immediately below layer A? Decrease

20. Does the change in velocity of seismic waves as they enter layer B indicate that the top portion of layer B is (more *or* less) rigid than layer A?

    Layer B is _____less_____ rigid.

21. What is layer C called?

    Layer C is the _____Mantle_____.

22. What fact concerning S waves indicates that layer C is not liquid? Only

    S waves can't travel through solids

23. What happens to S waves when they reach layer D, and what does this indicate about this layer?

    The S waves no longer travel indicating the outer core is liquid.

24. Does the velocity of P waves (increase *or* decrease) as these waves enter layer D?

    The velocity of P waves _____Decrease_____.

25. What change in velocity do P waves exhibit at the top of layer E?

    The velocity of P waves _____Increase_____.

26. Compare the rigidity of the material in Earth's inner core with the outer core, based on your answer to question 25.

    The inner core is more rigid

Earth's Internal Temperature

Measurements of temperatures in deep wells and mines indicate that Earth's temperatures increase with depth. The rate of temperature increase is called the **geothermal gradient**. Although the geothermal gradient varies from place to place, an average rate for a particular region can be calculated. Use Table 8.2, which shows an idealized temperature gradient for the upper Earth, to complete the following.

Table 8.2 Idealized Internal Temperatures of Earth, Compiled from Several Sources

DEPTH (KM)	TEMPERATURE (CELSIUS)
0	20°
25	600°
50	1000°
75	1250°
100	1400°
125	1525°
150	1600°

27. Plot the temperature values from Table 8.2 on the graph in Figure 8.11. Then draw a line to connect the points. Label the line *geothermal gradient*.

28. Referring to the graph, does the Earth's internal temperature increase at a (constant *or* changing) rate with increasing depth?

    Temperature increases at a _____changing_____ rate.

29. Is the rate of temperature increase from the surface to 100 km (greater *or* less) than the rate of increase below 100 km?

    The rate of temperature increase is _____greater_____.

30. Is the temperature at the base of the lithosphere, about 100 kilometers below the surface, approximately (600°C, 1400°C *or* 1800°C)?

    The temperature is approximately _____1400°C_____.

Melting Temperatures of Rocks

The approximate melting points of the igneous rocks granite and basalt, under various pressures (depths),

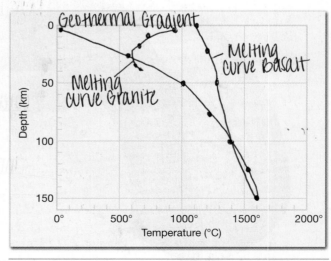

Figure 8.11 Graph for plotting temperature and melting point curves.

Table 8.3 Idealized Melting Temperatures of Granite (with water) and Basalt at Various Depths within Earth

GRANITE (WITH WATER)		BASALT	
DEPTH (KM)	MELTING TEMP. (CENTIGRADE)	DEPTH (KM)	MELTING TEMP. (CENTIGRADE)
0	950°	0	1100°
10	700°	25	1160°
20	660°	50	1250°
30	625°	100	1400°
40	630°	150	1600°

have been determined in the laboratory and are shown in Table 8.3. Granite that contains water and basalt have been selected because they are common materials in Earth's crust.

31. Plot the melting temperatures from Table 8.3 on the graph you completed in Figure 8.11. Draw a different colored line for each set of points and label them *melting curve for wet granite* and *melting curve for basalt*.

    Refer to the graphs you completed in Figure 8.11 to complete the following.

32. At approximately what depth would wet granite reach its melting temperature and generate magma?

    _____ 25 - 30 _____ km within Earth

33. Evidence suggests that the oceanic crust and the underlying rocks to a depth of about 100 km have a basaltic composition. Does the melting curve for basalt indicate that the lithosphere above approximately 100 km (has *or* has not) reached the melting temperature for basalt? Therefore, at those depths, should basalt be (solid *or* molten)?

    The melting temperature **has not** been reached.

    Basalt should be _____ **solid** _____.

34. Figure 8.11 indicates that basalt reaches its melting temperature at a depth of approximately _100 - 150_ km.

35. Referring to Figure 8.10, what is the name of the layer that begins at a depth of about 100 km and extends to approximately 450 km?

    _____ Asthenosphere _____

36. Does the graph you constructed (support *or* refute) the concept of a weak asthenosphere that is capable of "flowing"?

    The graph _____ **support** _____ this concept.

Earthquakes—Evidence for Plate Tectonics

The study of earthquakes has contributed significantly to the understanding of plate tectonics. The plate tectonic theory states that large, rigid slabs of the lithosphere are descending into the mantle. It is along these zones that deep focus earthquakes are generated.

Figure 8.12 illustrates the distribution of earthquake foci during a one-year period in the vicinity of the Tonga Islands. Use the figure to complete the following.

37. At approximately what depth do the deepest earthquakes occur?

    _____ 610 _____ kilometers

38. Are the earthquake foci in the area distributed (in a random manner *or* nearly along a line)?

    Foci are distributed **nearly along a line**.

39. On the illustration, outline the zone of earthquakes.

40. Draw a line on Figure 8.12 at a depth of 100 kilometers to indicate the top of the *asthenosphere*—the zone of partly melted and weak Earth material. Label the line *top of asthenosphere*.

41. The elastic rebound theory predicts that earthquakes can be generated only in the lithosphere—the layer of solid, rigid material. However, based on studies in the Tonga Islands, earthquakes occur within the weak, ductile asthenosphere. How do Earth scientists explain this contradiction?

    The earthquakes in the lithosphere are all relatively shallow, so they might occur more often.

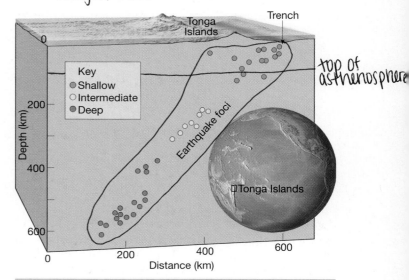

Figure 8.12 Distribution of earthquake foci in the vicinity of the Tonga Islands.

Companion Website

The companion website provides numerous opportunities to explore and reinforce the topics of this lab exercise. To access this useful tool, follow these steps:

1. Go to www.mygeoscienceplace.com
2. Click on "Books Available" at the top of the page.

3. Click on the cover of *Applications and Investigations in Earth Science 7e.*
4. Select the chapter you want to access. Options are listed in the left column (*Introduction, Web-based Activities, Related Websites, and Field Trips*).

Earthquakes and Earth's Interior

Name _____   **Course/Section** _____

Date _____   **Due Date** _____

1. Use Figure 8.13 to sketch a typical seismogram where the first P wave arrives 3 minutes ahead of the first S wave and 8 minutes ahead of the first surface wave. Label each type of wave.

2. How far from the earthquake epicenter is the seismic station that recorded the seismogram you constructed in question 1?
   _____1000_____ miles from the epicenter.

3. Describe the location where most deep-focus earthquakes occur.
   Deep in the asthenosphere

4. The velocity of S waves decreases as the waves enter the asthenosphere. What does this suggest about the asthenosphere compared to the overlying lithosphere?
   The asthenosphere is less rigid compared to the lithosphere

5. Explain why S waves do not travel through Earth's outer core.
   S waves do not travel through the outer core because S waves can only go through solids. The outer core is liquid.

6. List the depths of the following layers of Earth:
   The mantle depth is from _____450_____ to _____2900_____ km.
   The outer core depth is from _____2900_____ to _____5200_____ km.
   The inner core depth is from _____5200_____ to _____6500_____ km.

7. On the graph you constructed in Figure 8.11, at what depth did you determine that wet granite would melt?
   _____30_____ km

8. Why do Earth scientists think that rigid slabs of the lithosphere are descending into the mantle near the Tonga Islands?
   Because earthquakes are occurring in the asthenosphere near the Tonga islands.

9. What is the location directly above an earthquake focus called?
   Earthquake _Epicenter_

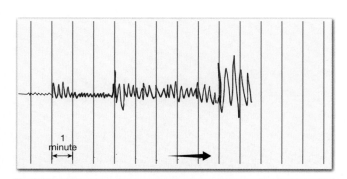

Figure 8.13   Illustration to accompany question 1.

10. Label and describe the properties of Earth's major layers on Figure 8.14.

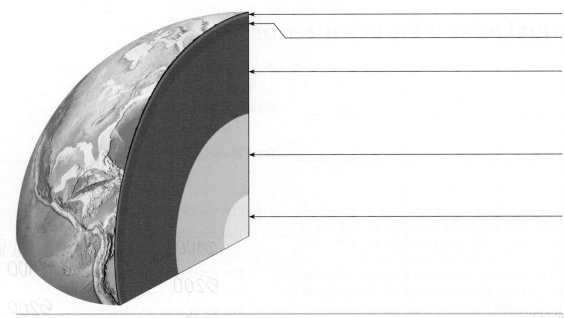

Figure 8.14 Illustration to accompany question 10.

PART 2
Oceanography

(The Photo Library Wales/Alamy)

Introduction to Oceanography

Objectives

Completion of this exercise will prepare you to:

  A. Locate and name the major water bodies on Earth.

  B. Discuss the distribution of land and water in each hemisphere.

  C. Locate and describe the general features of ocean basins.

  D. Explain the relationship between salinity and density of seawater.

  E. Describe how salinity varies with latitude and depth in the oceans.

  F. Describe how seawater temperature varies with latitude and depth in the oceans.

  G. Explain how temperature and salinity affect the density of seawater.

Materials

colored pencils	ruler	dye
graduated cylinder	world map	test tubes
salt solutions	rubber band	salt
beaker	ice	

Extent of the Oceans

The global ocean covers over seventy percent of Earth's surface and **oceanography** is an important focus of Earth science studies. This exercise investigates some of the physical characteristics of the oceans.

  1. Refer to a globe, a wall map of the world, or an atlas and locate each of the oceans and major water bodies in the following list. Label each on the world map in Figure 9.1.

  2. Which ocean covers the greatest area?

    _____

OCEANS	OTHER MAJOR WATER BODIES	
A. Pacific	1. Caribbean Sea	7. North Sea
B. Atlantic	2. Mediterranean Sea	8. Black Sea
C. Indian	3. Coral Sea	9. Baltic Sea
D. Arctic	4. Sea of Japan	10. Bering Sea
	5. Gulf of Mexico	11. Red Sea
	6. Persian Gulf	12. Caspian Sea

  3. Which ocean is almost entirely in the Southern Hemisphere?

    _____

Distribution of Land and Water

The total area of Earth is about 510 million square kilometers (197 million square miles). Of this, approximately 360 million square kilometers (140 million square miles) are covered by oceans and adjacent marginal seas.

  4. What percentage of Earth's surface is covered by oceans and marginal seas?

$$\frac{\text{Area of oceans and marginal seas}}{\text{Area of Earth}} \times 100\% =$$

    _____ % ocean

  5. What percentage of Earth's surface is land?

    _____ % land

  6. Referring to Figure 9.1, which hemisphere, Northern or Southern, could be called the "water" hemisphere and which the "land" hemisphere?

    Water hemisphere: _____

    Land hemisphere: _____

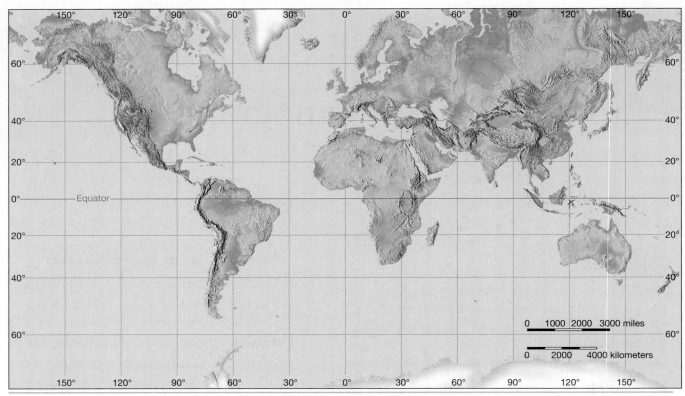

Figure 9.1 World map.

7. Use Figure 9.2 to estimate the percentage of Earth's surface that is ocean at the following latitudes.

NORTHERN HEMISPHERE	SOUTHERN HEMISPHERE
40°: _____ % ocean	_____ % ocean
60°: _____ % ocean	_____ % ocean

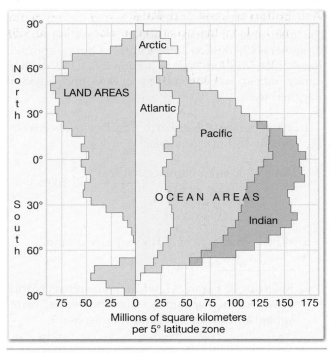

Figure 9.2 Distribution of land and water in each 5° latitude belt. (After M. Grant Gross, *Oceanography*; Prentice Hall, 1977)

Measuring Ocean Depths

Charting the shape or topography of the ocean floor is a fundamental task of oceanographers. In the 1920s, a technological breakthrough for determining ocean depths occurred with the invention of electronic depth-sounding equipment. The **echo sounder** (also referred to as *sonar*) measures the precise time that a sound wave, traveling at about 1500 meters per second, takes to reach the ocean floor and return to the instrument (Figure 9.3). Today, in addition to using sophisticated echo sounders such as *multibeam sonar*, oceanographers are also using satellites to map the ocean floor.

8. Use the formula in the caption for Figure 9.3 to calculate the depth of the ocean, in meters, for the following echo soundings.

5.2 seconds: _____ meters

6.0 seconds: _____ meters

2.8 seconds: _____ meters

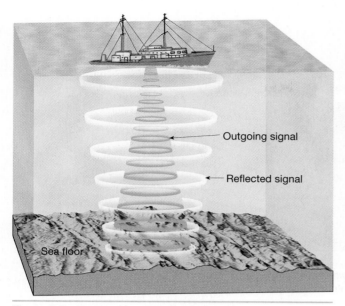

Figure 9.3 An echo sounder determines water depth by measuring the time interval required for an acoustic signal to travel from a ship to the seafloor and back. The speed of sound in water is 1500 m/sec. The ocean depth = 1/2(1500 m/sec × echo travel

Ships generally don't make single depth soundings. Rather, as a ship traverses from one location to another, it continually sends out sound impulses and records the echoes. In this way, oceanographers obtain many depth recordings from which a *profile* (cross-sectional view) of the ocean floor can be drawn.

The data in Table 9.1 were gathered by a ship equipped with an echo sounder as it traveled the North Atlantic Ocean eastward from Cape Cod, Massachusetts, to a point somewhat beyond the center of the Atlantic Ocean.

9. Use the data in Table 9.1 to construct a generalized profile of the ocean floor in the North Atlantic on Figure 9.4. Begin by plotting the dis-

Table 9.1 Echo Sounder Depths Eastward from Cape Cod, Massachusetts

POINT	DISTANCE (KM)	DEPTH (M)
1	0	0
2	180	150
3	270	2700
4	420	3300
5	600	4000
6	830	4800
7	1100	4750
8	1150	2500
9	1200	4800
10	1490	4750
11	1750	4800
12	1800	3100
13	1860	4850
14	2120	4800
15	2320	4000
16	2650	3000
17	2900	1500
18	2950	1000
19	2960	2700
20	3000	2700
21	3050	1000
22	3130	1900

tance of each point from Cape Cod, at the indicated depth. Complete the profile by connecting the points.

Ocean Basin Topography

Various features are located along the continental margins and on the ocean basin floor (Figure 9.5). **Continental shelves**, flooded extensions of the continents, are gently sloping submerged surfaces that extend from the shoreline toward the ocean basin. The seaward edge of the continental shelf is marked by the

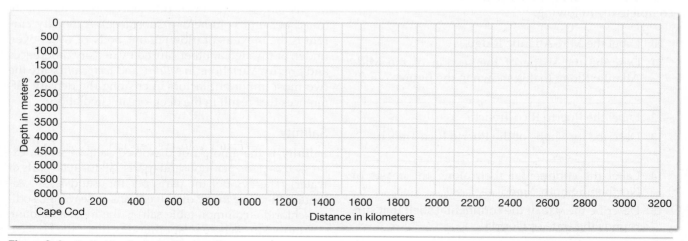

Figure 9.4 North Atlantic Ocean floor profile.

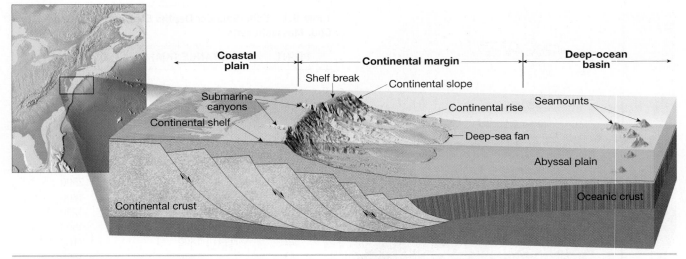

Figure 9.5 Generalized continental margin. Note that the slopes shown for the continental shelf and continental slope are greatly exaggerated. The continental shelf has an average slope of 0.1 degree, while the continental slope has an average of about 5 degrees.

continental slope, a relatively steep structure (compared to the shelf) that marks the boundary between continental crust and oceanic crust. Deep, steep-sided valleys known as **submarine canyons**, partially eroded by the periodic downslope movements of dense, sediment-laden water called **turbidity currents**, are often cut into the continental slope. The ocean basin floor, which constitutes almost 30 percent of Earth's surface, includes remarkably flat areas known as **abyssal plains**, tall volcanic peaks called **seamounts**, **oceanic plateaus** generated by mantle plumes, and **deep-ocean trenches**, which are deep linear depressions that are found primarily in the Pacific Ocean basin. Near the center of most ocean basins is a topographically elevated feature, characterized by extensive faulting and numerous volcanic structures, called an **oceanic ridge**, or **mid-ocean ridge**.

Refer to Figure 9.6, which shows the western portion of the North Atlantic basin, and a world map to complete the following.

10. Label the Mid-Atlantic Ridge.
11. Describe the shape of the section of the Mid-Atlantic Ridge that is shown in the figure.

_____

12. Label the Puerto Rico Trench.
13. Describe the shape of the Puerto Rico Trench.

_____

14. Label the continental shelf along the coasts of North and South America.
15. Describe the size of the continental shelf that surrounds the state of Florida.

_____

16. At what location along the east coast of the United States is the continental shelf narrowest?

_____

17. Label two abyssal plains.
18. Label a submarine canyon.
19. On the profile that you constructed in Figure 9.4, label the continental shelf, the continental slope, an abyssal plain, the Mid-Atlantic Ridge, and a seamount.

Characteristics of Ocean Water

Ocean circulation has two primary components: surface ocean currents and deep-ocean circulation. Both are examined in greater detail in Exercise 11. While surface currents such as the Gulf Stream are driven primarily by the prevailing global winds, the deep-ocean circulation is largely a result of differences in water *density* (mass per unit volume of a substance). A **density current** is the movement (flow) of one body of water over, under, or through another caused by density differences and gravity. Variations in *salinity* and *temperature* are the two most important factors in creating the density differences that result in the deep-ocean circulation.

Salinity

Salinity is the amount of dissolved salts in water, expressed as parts per thousand (parts salt for 1000 parts of water). The symbol for parts per thousand is ‰. Although there are many dissolved salts in seawater, sodium chloride (common table salt) is the most abundant.

In regions where evaporation is high, the concentration of salt increases as the water evaporates. The

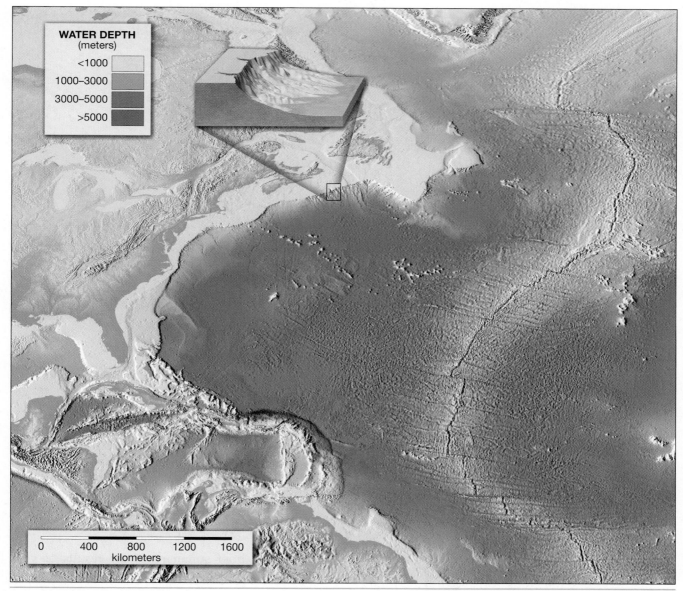

WATER DEPTH
(meters)

<1000

1000–3000

3000–5000

>5000

0   400   800   1200   1600
kilometers

Figure 9.6   Portion of the North Atlantic basin.

opposite occurs in areas of high precipitation and high runoff—where large rivers enter the ocean. The addition of fresh water dilutes the seawater and lowers its salinity. Since the factors that determine the concentration of salts in seawater are not constant from the equator to the poles, the salinity of seawater also varies with latitude (Figure 9.7).

Salinity–Density Experiment

To better understand how salinity affects the density of water, use the equipment supplied by your instructor (Figure 9.8) to conduct the following experiment.

Step 1: Fill the graduated cylinder with *cool* tap water up to the rubber band or other mark near the top of the cylinder.

Step 2: Fill a test tube about half full of solution A (saltwater) and pour it slowly into the cylinder. Record your observations.

Observations: _____

_____

_____

Step 3: Repeat Steps 1 and 2 twice and measure the time required for the front edge of the saltwater to travel from the rubber band to the bottom of the cylinder. Record the times for each trial in Table 9.2. *Make certain* that you drain the cylinder and refill it with fresh water and use the same amount of solution each time.

Figure 9.7 Processes affecting seawater salinity. Processes that *decrease* seawater salinity include precipitation, fresh water runoff, icebergs melting, and sea ice melting. Processes that *increase* seawater salinity include formation of sea ice and evaporation. (Upper left, Bernhard Edmaier/Photo Researchers, Inc.; upper right, Photoshot Holdings Ltd/Alamy; lower left, NASA; lower right, Shutterstock.com)

Step 4: Determine the travel time for solution B exactly as you did with solution A. Enter your times in Table 9.2.

Step 5: Fill a test tube about half full of solution B and add about $\frac{1}{2}$ teaspoon of salt. Shake the test tube vigorously. Determine the travel time of this solution and enter your results in Table 9.2.

Step 6: Clean all your glassware.

20. Write a brief summary of the results of your salinity–density experiment.

_____

_____

21. Which solution is the most dense, (A *or* B)?

Solution _____ is most dense.

Table 9.3 lists the approximate surface water salinity at various latitudes in the Atlantic and Pacific Oceans. Using these data, construct a salinity curve for

Figure 9.8 Lab setup for salinity–density experiment.

Table 9.2 Salinity–Density Experiment Data Table

SOLUTION	TIMED TRIAL #1	TIMED TRIAL #2	AVERAGE OF THE TWO TRIALS
A			
B			
Solution B plus salt		XXXX	XXXX

Table 9.3 Ocean Surface Water Salinity, in Parts per Thousand (‰), at Various Latitudes

LATITUDE	ATLANTIC OCEAN	PACIFIC OCEAN
60°N	33.0	31.0
50°N	33.7	32.5
40°N	34.8	33.2
30°N	36.7	34.2
20°N	36.8	34.2
10°N	36.0	34.4
0°	35.0	34.3
10°S	35.9	35.2
20°S	36.7	35.6
30°S	36.2	35.7
40°S	35.3	35.0
50°S	34.3	34.4
60°S	33.9	34.0

each ocean on the graph in Figure 9.9. *Use a different colored pencil for each ocean.* Then answer the following questions.

22. Which latitudes in the Atlantic Ocean have the highest surface salinities?

23. What are two factors that control the salinity of seawater?

_____ and _____

24. Why is the salinity of the surface waters of the equatorial and subtropical regions (about 30° latitude) in the Atlantic Ocean different?

25. Does the (Atlantic *or* Pacific) Ocean have a higher average surface salinity?

The _____ Ocean has a higher salinity.

Use Figure 9.10, which shows how ocean water salinity varies with depth at different latitudes, to complete the following.

26. In general, does salinity (increase *or* decrease) with depth in equatorial regions? Does salinity (increase *or* decrease) with depth in polar regions?

Salinity _____ with depth in equatorial regions.

Salinity _____ with depth in polar regions.

27. In the subtropics, why are the surface salinities much higher than the deepwater salinities?

The *halocline* (*halo* = "salt" and *cline* = "slope") is a layer of ocean water where there is a rapid change in salinity with depth.

28. Label the halocline on Figure 9.10.

29. Does the salinity of ocean water below the halocline (increase rapidly, remain fairly constant, *or* decrease rapidly)?

The salinity of ocean water _____ below the halocline.

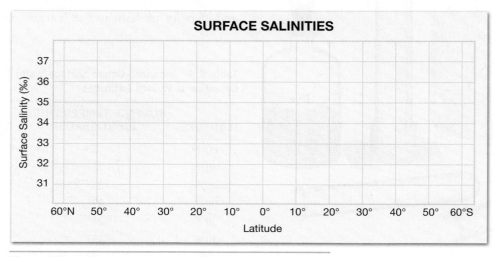

Figure 9.9 Graph for plotting surface salinities.

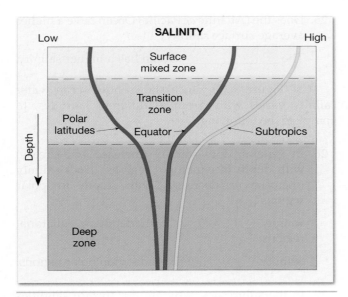

SALINITY

Low   High

Surface
mixed zone

Transition
zone

Polar
latitudes   Equator   Subtropics

Deep
zone

Depth

Figure 9.10 Changes in ocean water salinity with depth.

Water Temperatures

Temperature, like salinity, varies from the equator to the poles, changes with depth, and affects the density of seawater.

Temperature–Density Experiment

To illustrate the effects of temperature on the density of water, use the equipment provided by your instructor (Figure 9.11) to conduct the following experiment.

Figure 9.11 Lab setup for temperature–density experiment.

Step 1: Fill a graduated cylinder with *cold* tap water up to the rubber band or other mark near the top of the cylinder.

Step 2: Fill a test tube half full with *hot* tap water and add 2–3 drops of dye.

Step 3: Pour the contents of the test tube *slowly* into the cylinder. Record your observations.

Observations: _____

_____

_____

Step 4: Empty the graduated cylinder and refill it with *hot* water.

Step 5: Fill a test tube full of cold water and add 2–3 drops of dye. Then pour it into a beaker of ice and stir for several seconds. Refill the test tube three-fourths full with some of the liquid (but no ice) from your beaker. Pour this cold dye mixture *slowly* into the cylinder. Record your observations.

Observations: _____

_____

_____

Step 6: Clean the glassware and return it and the other materials to your instructor.

30. Write a brief summary of the temperature–density experiment.

_____

_____

_____

31. Given equal salinities, does (cold *or* warm) seawater have greater density?

_____ seawater has greater density.

Table 9.4 shows the average surface temperature and density of seawater at various latitudes. Using

Table 9.4   Idealized Ocean Surface Water Temperatures and Densities at Various Latitudes

LATITUDE	SURFACE TEMPERATURE (CENTIGRADE)	SURFACE DENSITY (g/cm³)
60°N	5°	1.0258
40°N	13°	1.0259
20°N	24°	1.0237
0°	27°	1.0238
20°S	24°	1.0241
40°S	15°	1.0261
60°S	2°	1.0272

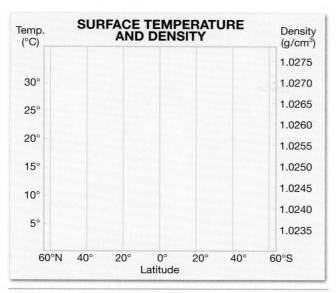

Figure 9.12 Graph for plotting surface temperatures and

these data, plot a colored line on the graph in Figure 9.12 for temperature and a different-colored line for density. Then complete the following.

32. Do the highest surface densities occur in (equatorial regions *or* at polar latitudes)?

    The highest surface densities occur _____ .

33. Describe the relationship between ocean surface temperatures and density.

    _____

34. Ocean surface temperatures are colder at 60°S latitude than at 60°N latitude. How does this difference affect the surface densities at these locations?

    _____

In question 22, you should have concluded that surface *salinities* are greatest in the subtropics at 30°N and 30°S latitudes.

35. Refer to the density curve in Figure 9.12 and the surface salinity graph in Figure 9.9. What evidence supports the fact that the density of seawater is influenced more by temperature than by salinity?

    _____

    _____

Companion Website

The companion website provides numerous opportunities to explore and reinforce the topics of this lab exercise. To access this useful tool, follow these steps:

1. Go to www.mygeoscienceplace.com
2. Click on "Books Available" at the top of the page.
3. Click on the cover of *Applications and Investigations in Earth Science 7e*.
4. Select the chapter you want to access. Options are listed in the left column (*Introduction, Web-based Activities, Related Websites, and Field Trips*).

Notes and calculations.

Introduction to Oceanography

Name _____   **Course/Section** _____

Date _____   **Due Date** _____

1. Write a brief statement comparing the distribution of land and water in the Northern Hemisphere to the distribution in the Southern Hemisphere.

   _____

   _____

2. On the ocean basin profile in Figure 9.13, label the following: continental shelf, continental slope, abyssal plain, seamounts, mid-ocean ridge, and deep-ocean trench.

3. Explain how an echo sounder is used to determine the topography of the ocean floor.

   _____

   _____

   _____

   _____

4. Complete the following statements with the correct response.

a. Does an increase in the salinity of seawater result in an (increase *or* decrease) in density?

   An increase in salinity results in a(n) _____ in density.

b. Does a decrease in the temperature of seawater cause its density to (increase *or* decrease)?

   A decrease in temperature causes a(n) _____ in density.

c. Are the highest surface salinities found in (polar, subtropical, *or* equatorial) regions?

   The highest surface salinities are found in _____ regions.

d. Does (temperature *or* salinity) have the greatest influence on the density of seawater?

   _____ has the greatest influence on the density of seawater.

e. Would (warm *or* cold) seawater with (high *or* low) salinity have the greatest density?

   _____ seawater with _____ salinity has the greatest density.

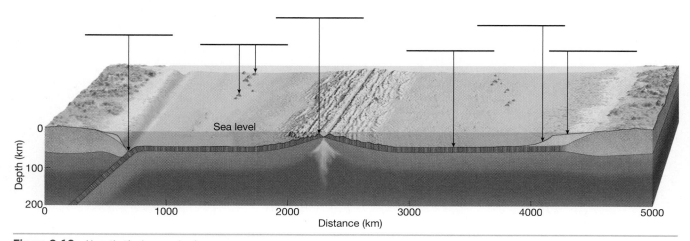

Figure 9.13 Hypothetical ocean basin.

5. Summarize the results of the salinity–density and temperature–density experiments.

   Salinity–density experiment: _____

   _____

   _____

   Temperature–density experiment: _____

   _____

   _____

6. Why is the surface salinity of the ocean higher in the subtropics than in equatorial regions?

   _____

   _____

   _____

7. Given your understanding of the relationship between ocean water temperature, salinity, and density, where would you expect surface water to sink and initiate a deepwater current flow in the Atlantic Ocean? Explain your choice.

   _____

   _____

   _____

8. Refer to the salinity–density experiment you conducted. Did solution (A *or* B) have the greatest density?

   Solution _____ had the greatest density

9. Describe the change in salinity with depth that occurs at the equator.

   Salinity: _____

   _____

   _____

10. Are the following statements true or false? Circle your response. If the statement is false, correct it to make it true.

    T F **a.** The Atlantic Ocean covers the greatest area of all the world's oceans.

    T F **b.** Continental shelves are part of the deep-ocean floor.

    T F **c.** Deep-ocean trenches are located mostly in the middle of ocean basins.

    T F **d.** High evaporation rates in the subtropics (about 30° latitude) cause the surface ocean water to have a lower-than-average salinity.

The Dynamic Ocean Floor

Objectives

Completion of this exercise will prepare you to:

A. List and explain several lines of evidence that support the theory of plate tectonics.

B. Locate and describe the mid-ocean ridge system and deep-ocean trenches.

C. Describe the relationship between earthquakes and plate boundaries.

D. Determine the rate of seafloor spreading along a mid-ocean ridge, using paleomagnetic evidence.

E. Use the rate of seafloor spreading to determine the age of an ocean basin.

F. Describe the three types of plate boundaries and the motion that occurs along them.

Materials

calculator   ruler
colored pencils   atlas, globe, or world wall map

Introduction

The idea that our present-day continents were once part of a single supercontinent was proposed in 1912 by Alfred Wegener. His hypothesis, called **continental drift**, proposed that about 200 million years ago, a supercontinent called **Pangaea** began to break apart into the present continents (Figure 10.1). Over time, each landmass slowly drifted across Earth's surface to its current position.

Following World War II, advanced technologies in oceanographic research provided new evidence that supported the continental drift hypothesis. Data collected from studies of rock magnetism, the distribution of earthquakes, and mapping of the seafloor led to the development of a more comprehensive theory, known as the **theory of plate tectonics**.

A. 200 Million Years Ago
(Late Triassic Period)

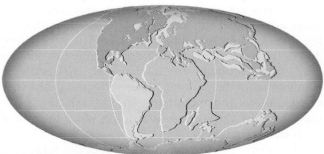

B. 90 Million Years Ago
(Cretaceous Period)

C. Present

Figure 10.1   Pangaea, as it is thought to have appeared 200 million years ago when it began to break apart into the present continents.

The theory of plate tectonics states that Earth's **lithosphere** is broken into several large, rigid slabs

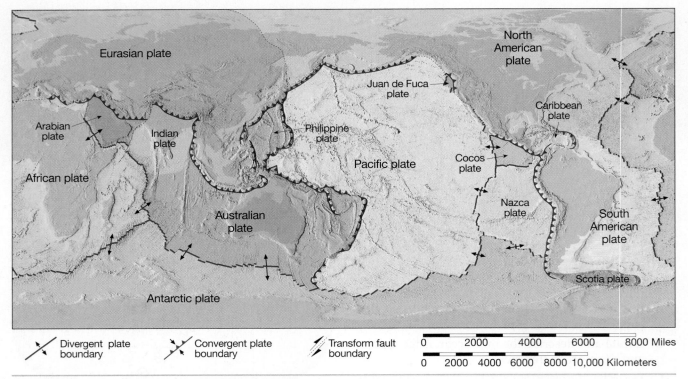

Divergent plate boundary

Convergent plate boundary

Transform fault boundary

| 0 | 2000 | 4000 | 6000 | 8000 Miles |

| 0 | 2000 | 4000 | 6000 | 8000 | 10,000 Kilometers |

Figure 10.2 The mosaic of rigid plates that constitutes Earth's outer shell.

called **lithospheric plates**, or simply **plates** (Figure 10.2). Plates are continually moving, and where they move apart along the **mid-ocean ridges**, upwelling of hot mantle rock creates new seafloor. The axis (a line down the center) of some ridge segments contains deep down-faulted structures called **rift valleys**. Along other plate margins, plates slide past each other or collide to form mountains. In addition, one plate may descend beneath an overriding plate to form a **deep-ocean trench**.

The discoveries of the global mid-ocean ridge system and deep-ocean trenches were of major importance in the development of plate tectonic theory.

1. On Figure 9.6 in Exercise 9 (page 137), outline the Mid-Atlantic Ridge and label the rift valley. (*Note:* The ridge is at least 2000 kilometers wide in most of the regions shown.)

2. Using an atlas or a world map for reference, draw the axis of the global mid-ocean ridge system on the map in Figure 10.3.

3. Label the following deep-ocean trenches on Figure 10.3. Draw a line to represent each trench.

1. Puerto Rico	**6.** Japan
2. Cayman	**7.** Mariana
3. Peru–Chile	**8.** Tonga
4. Aleutian	**9.** Kermadec
5. Kuril	**10.** Java (Sunda)

Plate Boundaries

Earth scientists recognize three distinct types of plate boundaries, each distinguished by the relative movement of the plates on the opposite sides of the boundary. **Divergent plate boundaries** are located along the crests of oceanic ridges and are *constructive plate margins* where new oceanic lithosphere is generated. Although new lithosphere is constantly being produced at the oceanic ridges, our planet is not growing larger; its total surface area remains the same. To balance the addition of newly created lithosphere, older portions of oceanic lithosphere descend into the mantle along **convergent plate boundaries**. Because lithosphere is "destroyed" at convergent boundaries, these boundaries are also called *destructive plate margins*. Although all convergent zones have the same basic characteristics, they are highly variable, each controlled by the type of crustal material and tectonic setting. Convergent boundaries can form between two oceanic plates, between one oceanic plate and one continental plate, or between two continental plates. Along **transform fault boundaries**, plates slide horizontally past one another, without the production or destruction of lithosphere. Transform faults were first identified where they join offset segments of an ocean ridge. Although most are located within the ocean basins, a few cut through continental crust. An example of such a boundary is California's earthquake-prone San Andreas fault.

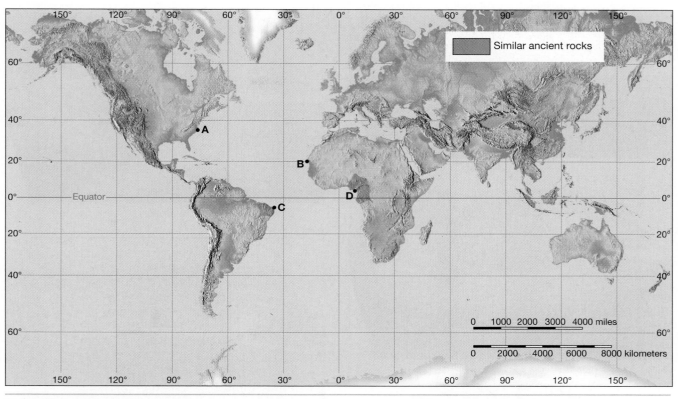

Figure 10.3 World map.

Use Figure 10.4, which illustrates the plate boundaries, to complete the following.

4. Does Figure 10.4A represent a (convergent *or* divergent) plate boundary?

   This is a _____ plate boundary.

   a. Are the plates along this boundary (moving apart *or* moving together)?

      The plates are _____.

   b. Does this plate boundary occur at (deep-ocean trenches *or* mid-ocean ridges)?

      The plate boundary occurs at _____.

   c. Does this plate boundary result in the (construction *or* destruction) of lithosphere?

      This plate boundary results in _____ of lithosphere.

5. Does Figure 10.4B represent a (convergent, divergent, *or* transform plate boundary)?

   The figure represents a _____ plate boundary.

   a. Are the plates along this type of boundary (moving apart *or* moving together)?

      The plates are _____.

b. Examine Figure 10.5, which shows the three types of convergent plate boundaries. Label each of the three diagrams with the appropriate term: oceanic–continental convergence, oceanic–oceanic convergence, or continental–continental convergence.

c. Find one example of each type of convergent boundary on the plate map in Figure 10.2. Label each boundary with the terms from question 5b.

6. Does Figure 10.4C represent a (convergent, divergent, *or* transform) plate boundary?

   The figure is a _____ plate boundary.

   a. Is lithospheric material (being created, being destroyed, *or* remaining unchanged) along this type of boundary?

      Lithospheric material is _____ along this type of boundary.

   b. What is the name given to faults that facilitate plate movement along this type of boundary?

      _____ faults

   c. Which type of fault, (dip-slip *or* strike) is illustrated in Figure 10.4?

      _____ fault

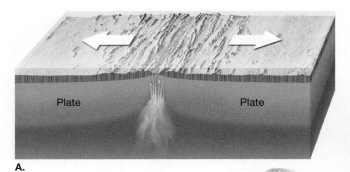

A.

B.

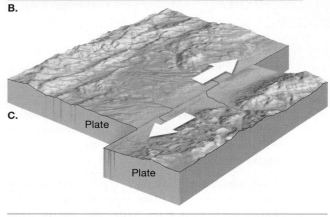

C.

Figure 10.4 Illustrations of the three types of plate boundaries.

Evidence for Plate Tectonics

Evidence: Fit of the Continents

Examine the east coast of South America and the west coast of Africa on the world map in Figure 10.3. Then complete the following.

7. Does the shape of the east coast of South America conform to the shape of the (east *or* west) side of the mid-ocean ridge in the Atlantic Ocean?

   It conforms to the _____ side.

8. Does the shape of the west coast of Africa conform to the shape of the (east *or* west) side of the mid-ocean ridge in the Atlantic Ocean?

   It conforms to the _____ side.

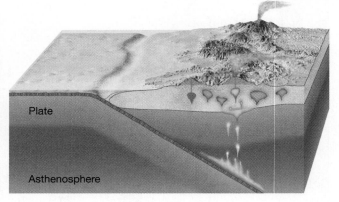

A. _____

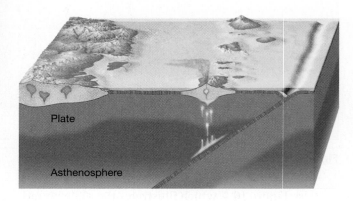

B. _____

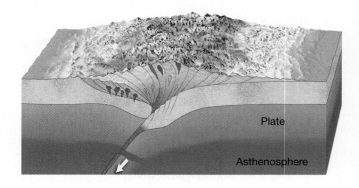

C. _____

Figure 10.5 Illustration of the three types of convergent plate boundaries.

9. If South America and Africa were moved to the mid-ocean ridge, would their shapes fit together along the ridge? (*Note:* The fit of the continents is more exact on a globe because flat maps have distortions.)

   The margins _____ fit together.

Figure 10.3 shows the location of an outcrop of ancient rocks in northeastern South America that are similar in rock type and age of rocks to western Africa.

10. Does your answer to question 9 support the idea that these continents were once joined?

_____

11. In Figure 10.1A, label the present-day continents that comprised Pangaea.

Evidence: Earthquakes

The distribution of earthquakes provides some of the most convincing evidence for plate tectonics.

Compare the plate map shown in Figure 10.2 to the earthquake distribution map shown in Figure 8.9 in Exercise 8 (page 125).

12. Are most deep-focus earthquakes associated with (mid-ocean ridges *or* deep-ocean trenches)?

Most deep-focus earthquakes are associated with

_____ .

13. Are the earthquakes that occur along mid-ocean ridges (shallow-focus *or* deep-focus) earthquakes?

These are _____ earthquakes.

14. What occurs along ocean trenches (but not along mid-ocean ridges) that accounts for the occurrence of deep-focus earthquakes?

_____

_____

Evidence: Paleomagnetism

The iron-rich mineral magnetite, common in small amounts in lava flows, becomes slightly magnetized and aligned with Earth's magnetic field when molten rock cools and crystallizes. These mineral grains are similar to tiny compass needles because they point toward the magnetic poles. The ancient magnetism, called **paleomagnetism**, present in rocks on the ocean floor can be used to determine the rate at which plates are forming and moving away from the ridge crest.

Scientists have also discovered that the polarity of Earth's magnetic field has periodically reversed, such that the North Magnetic Pole becomes the South Magnetic Pole and vice versa.

As plates move apart along the mid-ocean ridge, magma from the mantle rises to the surface and creates new ocean floor. As the magma cools, the minerals align themselves with the prevailing magnetic field. In the event that Earth's magnetic field reverses polarity, new material forming at the ridge will be magnetized in an opposite (or *reversed*) direction.

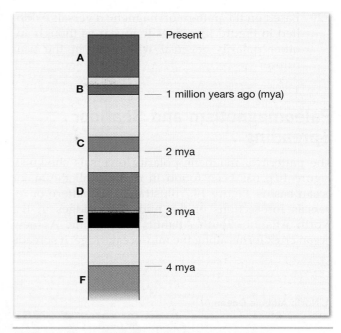

Figure 10.6 Chronology of magnetic polarity during the past 4 million years. Periods of normal polarity, when a compass needle would have been oriented as it is today, are shown in color. (Data from Allan Cox and G. Dalrymple)

Scientists have reconstructed Earth's magnetic polarity reversals over the past several million years, as shown in Figure 10.6. The periods of normal polarity (when a compass needle would have pointed north, as it does today) are shown in various colors and labeled A–F. Use Figure 10.6 to answer the following questions.

15. How many intervals (3, 5, *or* 7) of *reversed polarity* (shown in tan color) have occurred during the past 4 million years?

_____ intervals of reversed polarity

16. Approximately how many years ago did the current period of normal polarity (labeled A) begin?

_____ years ago

17. Did the Earth experience normal or reverse polarity 1.5 million years ago?

Earth experienced _____ polarity 1.5 million years ago.

18. Did the period of normal polarity, C, begin (1, 2, *or* 3) million years ago?

Normal polarity began _____ million years ago.

19. During the past 4 million years, has each interval of reverse polarity lasted (more *or* less) than 1 million years?

Reverse polarity lasted _____ than 1 million years.

20. Based on the pattern of magnetic reversals exhibited in Figure 10.6, does it appear as though another polarity reversal will occur in the near future?

Paleomagnetism and Seafloor Spreading

The pattern of magnetic polarity reversals shown in Figure 10.6 has been found in rocks of all the major ocean basins. Figure 10.7 illustrates this pattern of reversals for sections of the mid-ocean ridges in the North Atlantic, South Atlantic, and Pacific. As new ocean crust forms along the mid-ocean ridge, it spreads out equally on both sides of the ridge. Therefore, the pattern of polarity reversals that occurs on one side of the ridge axis can be matched with the polarity of the rocks on the other side of the ridge (Figure 10.7).

Use Figures 10.6 and 10.7 to complete the following.

21. On Figure 10.7, identify and mark the periods of normal polarity with the letters A–F (as in Figure 10.6). Begin at the ridge crest and label along both sides of each ridge. (*Note:* The left side of the South Atlantic has been completed as a guide.)

22. Using the South Atlantic as an example, label the *beginning* of the normal polarity period C that began *2 million years ago* on the left sides of the Pacific and North Atlantic diagrams.

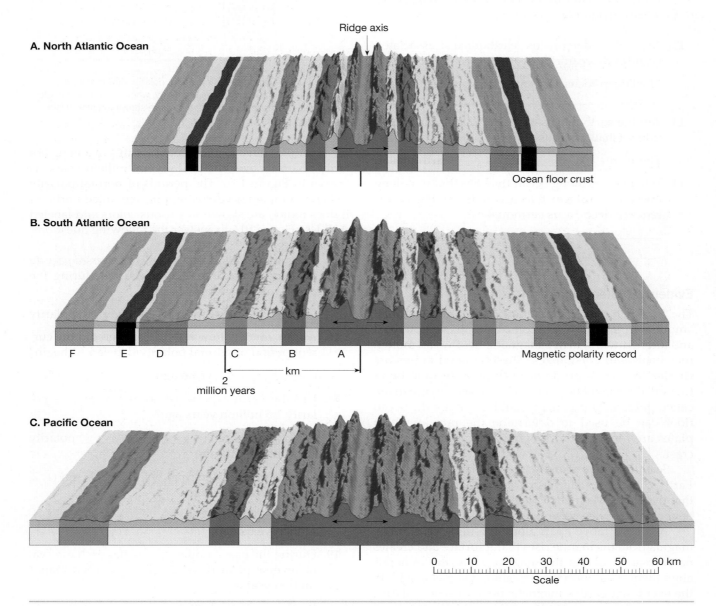

Figure 10.7 Generalized record of the magnetic polarity reversals across the mid-ocean ridge in the North Atlantic, South Atlantic, and Pacific Oceans. Periods of normal polarity are shown in color and correspond to those illustrated in Figure 10.6.

23. Using the distance scale at the bottom of Figure 10.7, measure the distance from the ridge axis to the beginning of the normal polarity period C for each ridge. Place the distances on Figure 10.7. How many kilometers has the left side of each of these ocean basins spread during the past 2 million years?

    South Atlantic basin: _____ km

    North Atlantic basin: _____ km

    Pacific basin: _____ km

    The distances you obtained in question 23 are for only one side of the ridge. Assuming that a ridge spreads equally on both sides, the actual distance each ocean basin has opened would be twice this amount.

24. How many kilometers has each ocean basin opened in the past 2 million years?

    South Atlantic basin: _____ km

    North Atlantic basin: _____ km

    Pacific basin: _____ km

    When the *distance* that an ocean basin has opened and the *time* it took for it to open are known, the *rate* of **seafloor spreading** can be calculated. To determine the rate of spreading in centimeters per year for each ocean basin, convert the distance the basin has opened from kilometers to centimeters and then divide this distance by the time—2 million years in this example.

25. Determine the rate of seafloor spreading for the Pacific and North Atlantic Ocean basins. As an example, the South Atlantic has been completed.

    a. South Atlantic: distance = 72 km
    $\times$ 100,000 cm/km = 7,200,000 cm

    $$\text{Rate of spreading} = \frac{7,200,000 \text{ cm}}{2,000,000 \text{ yr}} = 3.6 \text{ cm/yr}$$

    b. North Atlantic: distance = _____ km

    $\times$ 100,000 cm/km = _____ cm

    Rate of spreading = _____ cm/yr

    c. Pacific: distance = _____ km

    $\times$ 100,000 cm/km = _____ cm

    Rate of spreading = _____ cm/yr

Determining the Ages of Ocean Basins

The following procedure will allow you to calculate the age of a portion of the North Atlantic and South Atlantic basins.

26. Using Figure 10.3, measure the distance from Point A located off the Carolina coast to Point B off the African coast. Determine the distance in kilometers and then convert that distance into centimeters.

    Distance: _____ km

    Distance: _____ cm

27. Divide the distance in centimeters separating the continents obtained in question 26 by the rate of seafloor spreading for the North Atlantic basin that you calculated in question 25b. Your answer is the approximate age, in years, of the North Atlantic basin at that location.

    Distance/Rate of seafloor spreading =

    Age of the North Atlantic basin: _____ years

    Repeat the procedure above to determine the age of the South Atlantic basin from Point C along the eastern edge of Brazil to Point D off the coast of Africa.

28. How many years ago did South America and Africa begin to separate at that location?

    Age of the South Atlantic basin: _____ years

29. Based on your answers to questions 27 and 28, which part of the Atlantic basin appears to have opened first?

    The _____ part of the basin opened first.

Hot Spots and Plate Velocities

Researchers have determined that the volcanism on the island of Hawaii, known as a **hot spot**, arises from a plume of molten material moving upward from the mantle (Figure 10.8). It is believed that the position of this hot spot has remained constant over the past few million years.* In the past, as the Pacific plate moved over the hot spot, the successive volcanic islands of the Hawaiian chain were built. The island of Hawaii continues to grow as lava from this mantle plume flows to the sea.

Radiometric dating of the volcanoes in the Hawaiian chain has revealed that the islands increase in age with increasing distance from the island of Hawaii

*Recent evidence indicates that the mantle plume is gradually moving not stationary, as previously thought.

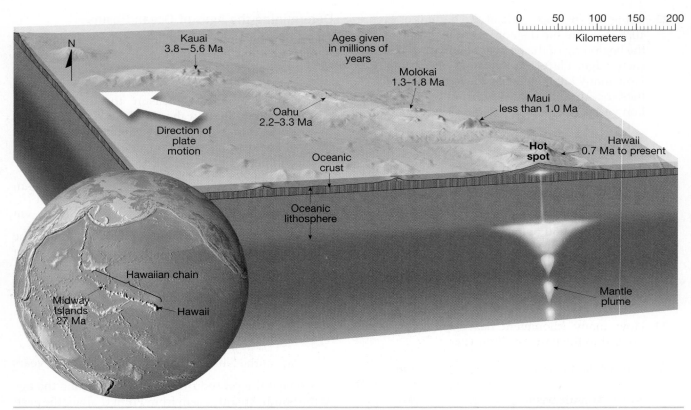

Figure 10.8 Movement of the Pacific plate over a hot spot and radiometric ages of the Hawaiian Islands, in millions of years.

(Figure 10.8). Based on the age of an island and its distance from the hot spot, the velocity of the plate can be calculated.

Use Figure 10.8 to answer the following questions.

30. What are the minimum and maximum ages of the island of Kauai?

    Minimum age: _____ million years

    Maximum age: _____ million years

31. What is the approximate distance (in kilometers) from the hot spot to the center of Kauai? Convert your answer to centimeters.

    _____ kilometers

    _____ centimeters

32. Using the data in questions 30 and 31, calculate the approximate maximum and minimum velocities of the Pacific plate as it moved over the Hawaiian hot spot in centimeters per year (cm/yr).

    Maximum velocity: _____ cm/yr

    Minimum velocity: _____ cm/yr

Companion Website

The companion website provides numerous opportunities to explore and reinforce the topics of this lab exercise. To access this useful tool, follow these steps:

1. Go to www.mygeoscienceplace.com
2. Click on "Books Available" at the top of the page.
3. Click on the cover of *Applications and Investigations in Earth Science 7e.*
4. Select the chapter you want to access. Options are listed in the left column (*Introduction, Web-based Activities, Related Websites,* and *Field Trips*).

The Dynamic Ocean Floor

Name _____   **Course/Section** _____

Date _____   **Due Date** _____

1. The distribution of earthquakes defines the boundaries of which major Earth feature?

   _____

2. Deep-focus earthquakes are associated with what prominent ocean-floor feature?

   _____

3. Shallow-focus earthquakes occur along what prominent ocean-floor features?

   _____

4. Describe how paleomagnetism is used to calculate the rate of seafloor spreading.

   _____

   _____

   _____

5. Based on your calculations for question 25, what are the rates of seafloor spreading for the following ocean basins?

   North Atlantic basin: _____ cm/yr

   Pacific basin: _____ cm/yr

6. Based on your calculations in question 32, what is the maximum velocity for the Pacific plate?

   _____ cm/yr

7. Based on your calculations in questions 27 and 28, what are the ages for the North and South Atlantic basins at the locations examined?

   North Atlantic basin: _____ million years old

   South Atlantic basin: _____ million years old

8. Complete the block diagrams in Figure 10.9 to illustrate the types of plate boundaries listed below each diagram. Include arrows to indicate relative plate motion.

9. Use Figure 10.10, which illustrates a generalized cross section of the plate boundary along the

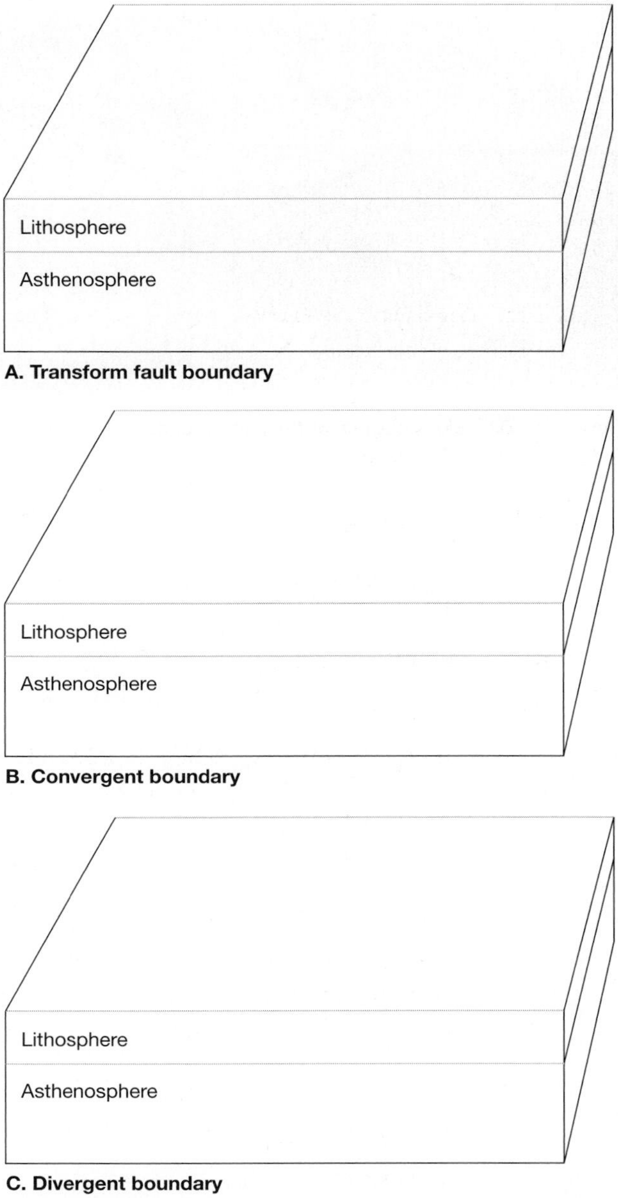

Lithosphere

Asthenosphere

A. Transform fault boundary

Lithosphere

Asthenosphere

B. Convergent boundary

Lithosphere

Asthenosphere

C. Divergent boundary

Figure 10.9   Block diagrams to accompany question 8.

western edge of South America, to answer the following questions.

a. Label each of the following features on the block diagram.

Asthenosphere	Continental crust
Deep-ocean trench	Oceanic crust
Continental lithosphere	Oceanic lithosphere

b. What type of plate boundary is illustrated in the figure? Be as specific as possible.

_____

_____

10. List and describe two lines of evidence from this exercise that support the theory of plate tectonics.

_____

_____

_____

_____

_____

_____

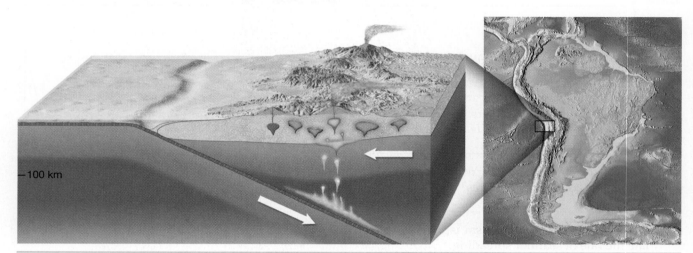

Figure 10.10 Block diagram of the plate boundary along the western edge of South America.

EXERCISE
11

Waves, Currents, and Tides

Objectives

Completion of this exercise will prepare you to:

A. Explain how waves and currents are generated.

B. Name the parts of a wave and describe the motion of water particles.

C. Calculate the velocity, wavelength, and wave period of a deep-water wave.

D. Explain what causes waves to break and form surf.

E. Identify the major surface ocean currents.

F. List the names and characteristics of the principal deep-water currents in the Atlantic.

G. Identify erosional and depositional features along shorelines.

H. Identify the types and causes of tides.

Materials

colored pencils   hand lens
calculator   atlas or world wall map

Waves

Most ocean waves derive their energy and motion from wind. When a breeze is less than 3 kilometers per hour, small wavelets appear. At greater wind speeds, more stable waves form and advance with the wind.

Wave Characteristics

Typically, waves are set in motion when wind causes water particles to move in circular orbits (Figure 11.1). If you watched a ball floating on the surface of the ocean, you would notice that the waveform moves forward, while the ball, and hence the water, does not. As water particles reach the highest point in their circular orbits, a **wave crest** forms, while particles at their lowest positions produce **wave troughs**. **Wave height** is the vertical distance between the crest and trough of a wave.

Beneath the surface, the circular orbits of water particles become progressively smaller with depth. At a depth equal to about *half* the **wavelength** (the horizontal distance separating two successive wave crests), the circular motion of water particles becomes negligible.

1. Refer to Figure 11.1 and select the letter that identifies each of the following.

	LETTER		LETTER
wave crest	_____	wavelength	_____
wave trough	_____	depth of negligible water particle motion	_____
wave height	_____		

2. Below what depth would a submarine have to submerge so that it would not be swayed by surface waves with a wavelength of 24 meters?

   Below _____ meters

Wave Properties

For **deep-water waves** traveling at depths greater than half the wavelength, the wave **velocity (V)** is related to the **wave period (T)** (the time interval between the arrival of successive wave crests) as well as the **wavelength (L)**. The mathematical equation that expresses this relationship is:

$$\text{Velocity} = \frac{\text{Wavelength}}{\text{Wave period}}$$

or

$$V = \frac{L}{T}$$

As a wave approaches shore, the ocean bottom begins to interfere with the orbital motion of the water particles, and the wave begins to "feel bottom" (Figure 11.1). Interference between the bottom and water

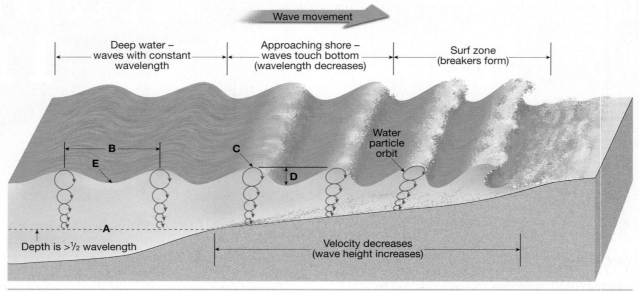

Figure 11.1   Features of deep- and shallow-water waves.

particle motion causes changes in the waveform. At a depth equal to about one-twentieth of the deep-water wavelength $\left(\frac{1}{20},\text{ or } 0.05L\right)$, the top of the wave begins to fall forward, and the wave breaks. This action creates the **surf zone**, where waves break and their erosive force is unleashed. In the surf zone, a significant amount of water is transported toward the shoreline.

Refer to Figure 11.1 to answer the following questions.

3. Is the shape of the orbits of water particles in deep-water waves (circular *or* elliptical)?

The shape of the orbits is _____.

4. Near the shore in shallow water, do the orbital shapes become (circular *or* elliptical)?

The shapes become _____.

5. In shallow water, are water particles in the wave crest (ahead of *or* behind) those at the bottom of the wave?

The wave particles are _____ those at the bottom of the wave.

6. As waves approach the shore, do their heights (increase *or* decrease)? Do wavelengths become (longer *or* shorter)?

Their heights _____, and their wavelengths become _____.

7. In the surf zone, is the water in the crest of a wave (falling forward *or* standing still)?

Water in the surf zone is _____.

8. What is the velocity of deep-water waves that have a wavelength of 40 meters and a wave period of 6.3 seconds?

$$\text{Velocity} = \frac{\text{Wavelength } (L)}{\text{Wave period } (T)} = \frac{40 \text{ m}}{6.3 \text{ sec}}$$

$$= \underline{\hspace{2cm}} \text{ m/sec}$$

9. At what water depth will a deep-water wave with a wavelength of 40 meters begin to "feel bottom"?

_____ meters

10. At what water depth will the wave described in question 8 begin to break?

_____ meters

11. What two factors determine the distance from the shoreline at which waves begin to break?

_____

_____

12. Imagine that you are standing on a beach but cannot swim. Your friend encourages you to walk into the surf zone created by incoming deep-water waves that have a wavelength of 30 meters. Would it be safe to walk out to where the waves are breaking? Explain how you arrived at your answer.

_____

_____

13. Along some shorelines, incoming waves cause the water to simply rise and fall rather than form a surf zone. What does this tell you about these shorelines?

_____

Tsunamis, ocean waves produced by submarine earthquakes, travel at very high velocities in deep water.

14. What is the velocity of a tsunami with a wavelength of 125 miles and a period of 15 minutes?

    Velocity = _____ miles per hour

Wave Refraction

Waves that approach a shoreline are **refracted** (bent) because the part of the wave that touches bottom first is slowed down, while the sections of the wave in deeper water continue to advance at the same velocity. The result of refraction is that waves tend to gradually align themselves parallel to the shoreline.

Figure 11.2 is a map view of a **headland** along a coastline, with water depths shown by blue contour lines. Assume that waves with a wavelength of 60 feet are approaching the shoreline from the bottom left of the figure.

15. At approximately what water depth (10, 20, 30, or 40 feet) will the approaching waves begin to touch bottom and slow down?

    Water depth: _____ feet

16. Using the wave shown in Figure 11.2 as a starting point, sketch a series of lines to illustrate the wave refraction that will occur as the waves approach the shore by following these steps:

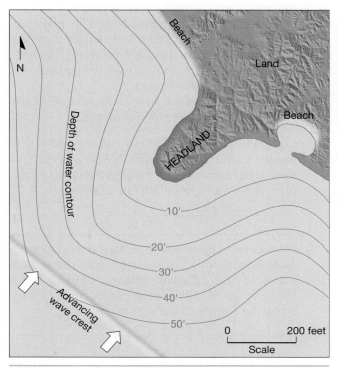

Figure 11.2 Coastline with water depth contours and an approaching wave.

Step 1: Mark the position on the 30-foot contour line where the wave front will first feel bottom.

Step 2: Knowing that the section of the wave that touches bottom will slow down first, sketch the shape of the wave front when it reaches the 20-foot contour line.

Step 3: Using the same methodology as in Step 2, sketch the wave front as it approaches the 10-foot contour line.

17. At approximately what water depth (3, 6, 12, or 24 feet) will the waves begin to break?

    Water depth: _____ feet

18. Use a dashed line to indicate where the waves will begin to break. Write *surf zone* along the line.

19. Will wave erosion be most severe (on the headland *or* in the bays)?

    Erosion will be most severe _____.

20. What effect will the concentrated energy from wave erosion eventually have on the shape of the coastline?

Ocean Currents

Surface Circulation

Ocean currents develop where the prevailing winds are persistent and cause the surface layer of the ocean to move as a large mass. Once set in motion, surface currents are influenced by the **Coriolis effect**, which deflects the path of the moving water to the right in the Northern Hemisphere and to the left in the Southern Hemisphere. *Warm currents* carry equatorial water toward the poles, while *cold currents* move water from higher latitudes toward the equator.

21. On the world map in Figure 11.3, draw arrows representing each of the major ocean currents listed below. Use an atlas or a wall map that depicts surface currents as a reference. Show warm currents with red arrows and cold currents with blue arrows. Label each current with the appropriate name.

MAJOR SURFACE OCEAN CURRENTS

1. Equatorial	7. Kuroshio (Japan)
2. Gulf Stream	8. West Wind Drift
3. California	9. Labrador
4. Canaries	10. North Atlantic Drift
5. Brazil	11. North Pacific Drift
6. Benguela	12. Peruvian

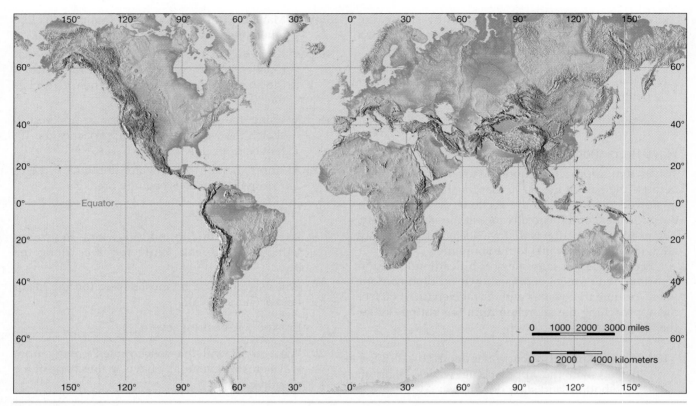

Figure 11.3 World map.

22. Which surface ocean current travels completely around the globe, west to east, without interruption?

    Name of current: _____

23. Which surface ocean current flows along the Atlantic coast of the United States? Is the current a (warm *or* cold) current?

    Name of current: _____

    This current is a _____ current.

24. What is the name of the surface ocean current located along the Pacific coast of the United States? Is the current a (warm *or* cold) current?

    Name of current: _____

    This current is a _____ current.

25. Is the general circulation of the surface currents in the North Atlantic Ocean (clockwise *or* counterclockwise)?

    The general circulation is _____.

26. In the South Atlantic, is the general circulation (clockwise *or* counterclockwise)?

    The general circulation is _____.

Deep Currents

Deep currents result when water of greater density sinks beneath water of lower density. The density of

water is determined by its salinity and temperature (covered in Exercise 9, "Introduction to Oceanography"). Furthermore, temperature has a greater influence on density than does salinity. In general, cold seawater with high salinity is the densest.

Most deep-ocean circulation begins in high latitudes where the water is cold, and its salinity increases as sea ice forms. When this surface water becomes dense enough, it sinks and moves throughout the ocean basins in sluggish currents.

Use Figure 11.4, which is a cross section of the Atlantic Ocean, to complete the following.

27. Compare Antarctic Bottom Water (ABW) to North Atlantic Deep Water (NADW). Which of these deep currents has the greatest density? Explain how you arrived at your answer.

    _____

    _____

28. What process causes the Mediterranean Intermediate Water (MIW) to become more dense than water in the adjacent Atlantic Ocean? (*Hint*: Water in the Mediterranean Sea is greatly affected by the climate of the region.)

    _____

    _____

Oceanographers estimate that when cold, dense water sinks, it will not reappear at the surface for an average of 500 to 2000 years. Figure 11.5 illustrates a

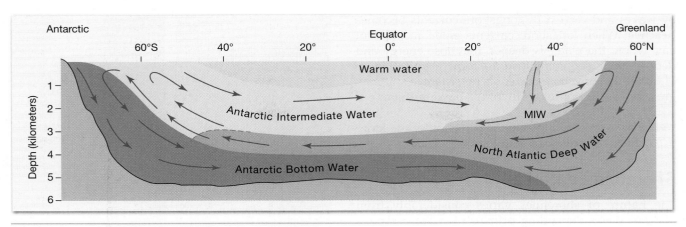

Figure 11.4 Cross section of the deep circulation in the Atlantic Ocean.

simplified model of deep-ocean circulation. Like a conveyor belt, the water travels from the Atlantic Ocean, through the Indian and Pacific oceans, and returns to the Atlantic. Use Figures 11.4 and 11.5 to answer the following questions.

29. What is the name of the deep-water current that begins in the North Atlantic Ocean south of Greenland?

30. Assume that water that sinks in the North Atlantic near Greenland takes 1000 years to resurface in the Indian Ocean, a distance of about 18,000 kilometers. What would be the approximate velocity of this deep current in km/yr?

Velocity: _____ km/yr

Longshore Currents

Waves are the dominant agents of erosion and deposition along shorelines. The sediment derived from wave erosion combined with that carried to the oceans by

streams is redistributed by waves and wave-generated currents. Waves that approach the shore at an angle produce currents within the surf zone that flow parallel to the shore. Because the water is turbulent, these **longshore currents** easily transport fine, suspended sand and roll larger sand and gravel particles along the bottom (Figure 11.6A). Much of the sediment reworked

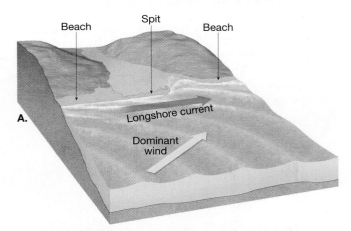

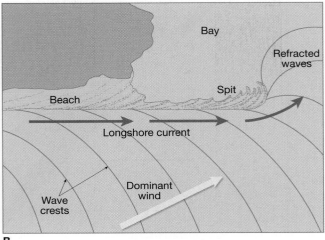

Figure 11.6 Longshore currents. **A.** Waves that approach the shore at an angle produce longshore currents that parallel the shore. **B.** Formation of a spit by sediment transported by longshore currents.

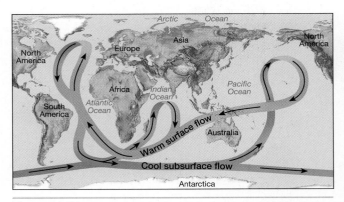

Figure 11.5 Idealized "conveyor belt" model of ocean circulation, which is initiated in the North Atlantic Ocean, where water radiates its heat to the atmosphere, cools, and sinks. This deep current moves southward and joins another deep current that encircles Antarctica. From here, this water spreads into the Indian and Pacific oceans, where it slowly rises and completes its journey by traveling along the surface into the North Atlantic Ocean.

by waves and carried by longshore currents becomes **beaches**. When longshore currents enter an adjacent bay, where the water is deeper, they lose energy and deposit their sediment load to form a **spit** (Figure 11.6B). If the longshore current is persistent, a spit may grow across the entire bay to form a **baymouth bar**.

31. Use arrows to indicate the direction of the longshore currents shown in Figure 11.7A,B.

Shoreline Features

The nature of shorelines varies considerably from place to place, depending primarily on the composition of coastal bedrock, the degree of exposure to ocean waves, and the position of the shorelines in relation to sea level. Geologists classify coasts into two broad categories—*submergent* and *emergent*.

Submergent coasts result from a rising sea level or subsiding land. Most U.S. Atlantic and Gulf coasts are submergent coasts due to a recent increase in sea level caused by melting of Ice Age glaciers (Figure 11.8). Submergent coasts tend to be irregular because the rising sea floods the lower reaches of river valleys,

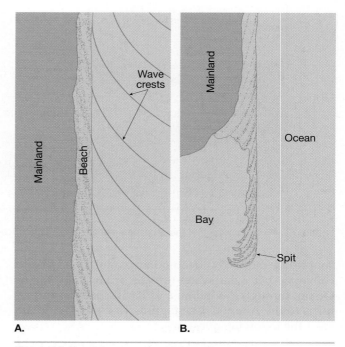

Figure 11.7 Illustrations to accompany question 31.

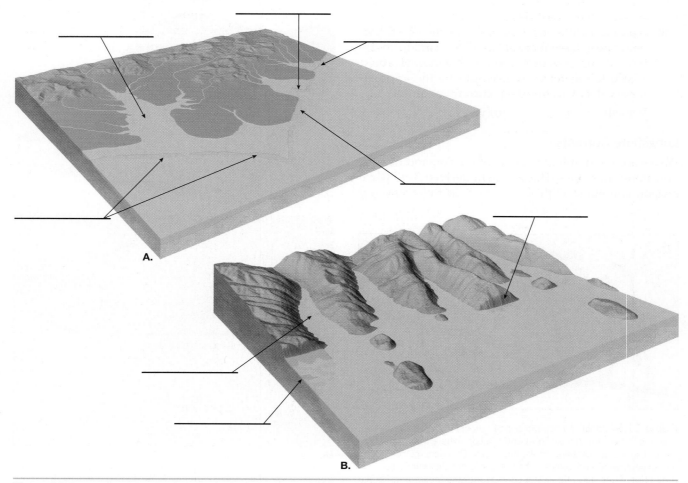

Figure 11.8 Illustrations showing general features of two different types of submergent coastlines.

producing bays called **estuaries**. In addition to drowned river mouths, the Atlantic and Gulf coasts, which are composed of easily erodible sedimentary rock, are characterized by wide beaches and **barrier islands** that parallel the shorelines (Figure 11.8A). By contrast, the coast of Maine, which is composed of highly resistant igneous and metamorphic rocks, is very rugged. During the Ice Age, glaciers scoured the landscape, and the weight of the ice sheet caused the underlying land to subside. When the ice retreated about 14,000 years ago, the rising ocean flooded the coastal lowlands, giving Maine its distinctive rocky coastline (Figure 11.8B). The coasts of Norway, British Columbia, and Alaska, which are dissected by fiords (long, steep-sided inlets carved by glaciers), represent the most dramatic type of submergent shoreline.

Emergent coasts result from rising land or falling sea level and are characterized by *wave-cut cliffs* and *wave-cut terraces*. Despite the general rise in sea level over the past several thousand years, tectonic forces have sufficiently uplifted much of the Pacific coast of the United States, producing what is mainly an emergent coastline. Relatively narrow beaches backed by steep cliffs and mountainous topography characterize much of the Pacific coast (Figure 11.9).

Despite the differences between submergent and emergent coasts, the effects of wave erosion and sediment deposition by currents cause them to share many features. Shoreline features associated with submergent and emergent coasts are described below:

- **Headland** A headland is land, possibly containing cliffs, that projects into an ocean.
- **Wave-cut platform** A wave-cut platform is a flat bedrock surface along a shore cut by erosion.
- **Delta** A delta is an accumulation of sediment that was deposited where a stream enters a lake or an ocean.
- **Tied island** A tied island is an island that is connected to the mainland or another island.
- **Estuary** An estuary is a river valley or mouth that is flooded by a rise in sea level or subsidence of the land.
- **Marine terrace** A marine terrace is a wave-cut platform that has been uplifted above sea level by tectonic forces.
- **Wave-cut cliff** A wave-cut cliff is a cliff cut by the surf into the coastal rocks.
- **Spit** A spit is an elongated ridge of sand that projects from the land into the mouth of an adjacent bay.
- **Sea stack** A sea stack is an isolated rock formation that resembles a tiny island left from the collapse and erosion of surrounding rocks.
- **Tombolo** A tombolo is a ridge of sand that connects a sea stack or small island to the mainland.
- **Beach** A beach is a gently sloping deposit of sand or gravel along the shoreline.
- **Barrier island** A barrier island is a low ridge of land that parallels the coast.

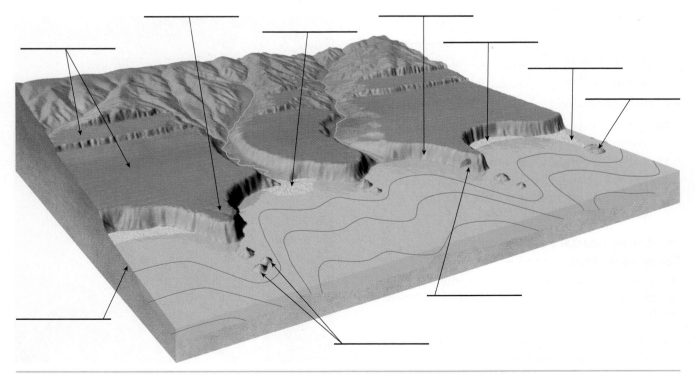

Figure 11.9 Illustration of an emergent coastline, common in southern California.

A. _____ B. _____ C. _____

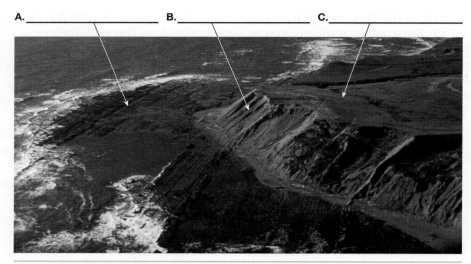

Figure 11.10 Image to accompany question 34. (Photo by John S. Shelton)

- **Sea arch** A sea arch is a headland that has been cut by erosion into an arch.
- **Baymouth bar** A baymouth bar is a bar that extends across the mouth of a bay.

32. Use the descriptions of shoreline features provided above to label the structures in Figures 11.8 and 11.9. The same feature may appear more than once.

33. Next to each of the features listed below, indicate whether it is the result of *erosional* or *depostional* processes.

   Sea stack: _____

   Wave-cut cliff: _____

   Spit: _____

   Barrier island: _____

   Baymouth bar: _____

   Marine terrace: _____

34. Label a marine terrace, a wave-cut cliff, and a wave-cut platform in Figure 11.10.

35. Label a baymouth bar and a spit in Figure 11.11.

36. Label a sea arch, a tombolo, a spit, a tied island, a wave-cut cliff, and a sea stack on the lines below each image in Figure 11.12.

Identifying Coastal Features on a Topographic Map

Refer to Figure 11.13 (page 164), a portion of the Point Reyes, California, topographic map, to complete the following.

37. What type of shoreline feature is Drakes Estero (located near the center of the map)?

   Drakes Estero is a(n) _____.

38. Point Reyes, located near the bottom of the map, is a typical headland undergoing severe wave erosion. What type of feature is Chimney Rock, located off the shore of Point Reyes?

   Chimney Rock is a(n) _____.

39. Several depositional features near Drakes Estero are related to the movement of sediment by longshore currents. What type of depositional feature is labeled A on the map?

   _____

A. _____ B. _____

Figure 11.11 Image to accompany question 35.

A. _____   B. _____   C. _____

D. _____   E. _____

_____

Figure 11.12 Image to accompany question 36. (Part E by Chris Cheadle/Photolibrary.com)

40. Follow Limantour Spit from its western tip at the mouth of Drakes Estero eastward to where it joins the mainland. Use an arrow to mark the direction of the longshore current that produced Limantour Spit.

41. What geologic feature is Estero de Limantour?

_____

42. Limantour Spit may eventually extend across the entire mouth of Estero de Limantour. If the west end of Limantour Spit reaches the mainland, will it be called a (tombolo *or* baymouth bar)?

It will be called a _____.

43. Locate the bench mark (BM) with an elevation of 552 feet on Point Reyes. If you walked from that bench mark southward to the shore of the Pacific, what landform would you cross?

_____

The land shown on the Point Reyes quadrangle is located directly to the southwest of the San Andreas Fault, in a very tectonically active region. As a result, some of this region has recently been uplifted and exhibits characteristics of an emergent coastline. On the other hand, because of the general rise in sea level over the past several thousand years, other areas exhibit features associated with submergent coastlines.

44. What feature indicates that the headland called Point Reyes has recently been uplifted?

_____

45. List at least two features on the map that are characteristic of submergent coasts—similar to the Atlantic coast.

_____

_____

Tides

Tides are the cyclical rise and fall of sea level caused by the gravitational attraction of the Moon and, to a lesser extent, the Sun. Gravitational pull creates a bulge in the ocean on the side of Earth nearest the Moon and on the opposite side of Earth from the Moon. The monthly tidal cycle lasts 29.5 days—the length of time it takes the Moon to complete its orbit around Earth. During its cycle, however, the position of Earth continually changes relative to the Sun and Moon. During the new moon and full moon phases, the Moon and Sun are aligned, and the tide-generating forces combine. The **tidal range** is greatest during this period, resulting in *higher* high tides and relatively *low* low tides, called **spring tides** (Figure 11.14A, page 165). Conversely, when the Moon is in the first- or third-quarter phase, the Sun's tidal force is working at right angles to the Moon's tidal force. This results in *lower* high tides and *higher* low tides, called **neap tides** (Figure 11.14B). Ideally, each tidal cycle should consist of two spring tides and two neap tides, each about one week in duration. However, many factors, including the shape of the coastline, the configuration of the

FIGURE 11.13: Point Reyes, California

0   1   2   3 kilometers

0   ½   1   2 miles

CONTOUR INTERVAL 80 FEET   SCALE: 1:62,500

North

Figure 11.13   Portion of the Point Reyes, California, topographic map. (Courtesy of U.S. Geological Survey)

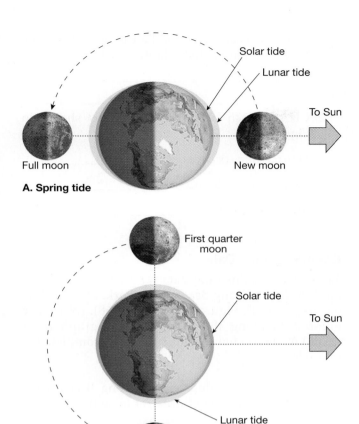

A. Spring tide

B. Neap tide

Figure 11.15 Tidal patterns for the month of September at various locations. (Data courtesy of U.S. Navy Hydrographic Office)

Figure 11.14 Earth–Moon–Sun positions and tides. **A.** When the Moon is in the full or new phase, the tidal bulges created by the Sun and Moon are aligned, causing a large tidal range called a *spring tide*. **B.** When the Moon is in the first- or third-quarter position, the tidal bulges produced by the Moon are at right angles to the bulges created by the Sun. Tidal ranges are smaller, and a *neap tide* is experienced.

ocean basin, and water depth, can significantly influence the tidal pattern observed at any particular location.

Use Figure 11.15, which illustrates the tidal patterns for three locations, to complete the following.

46. Use lines to divide each tidal pattern into approximately one-week segments that represent the two weeks of spring tides and two weeks of neap tides.

47. Write the word *spring* or *neap* above the period of time representing the appropriate type of tide.

48. Which coastal area experiences that largest tidal range (difference in height between the high tide and low tide)?

_____

49. Which coastal area experiences the smallest tidal range?

_____

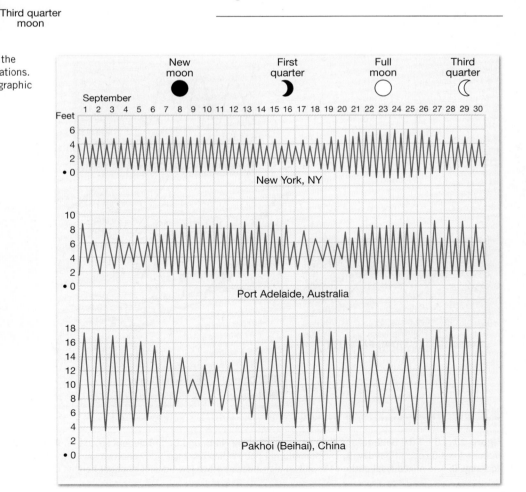

50. Notice that the phases of the Moon are shown at the top of Figure 11.15. Which coastal location experiences spring and neap tides that align most closely to those predicted by the phases of the Moon?

_____

Although all coastlines should experience two high tides and two low tides each tidal day, the complexities stated earlier result in three basic tidal patterns:

- **Diurnal (daily) tides** are characterized by a single high tide and a single low tide each tidal day (Figure 11.16A).

- **Semidiurnal (twice daily) tides** exhibit two high tides and two low tides each tidal day (Figure 11.16B).

- **Mixed tides** are similar to semidiurnal tides except that they are characterized by large inequalities in high-water heights, low-water heights, or both (Figure 11.16C). There are typically two high tides and two low tides each day, with both high and low tides of various heights.

51. Referring to Figure 11.16, write a general statement comparing the type of tide that occurs

along the Pacific coast of the United States to the type found along the Atlantic coast.

_____

_____

52. Classify each of the tidal patterns shown in Figure 11.17.

Diurnal tides occur at _____.

Semidiurnal tides occur at _____.

Mixed tides occur at _____.

Examining Tidal Data

Table 11.1 contains tidal data for a 20-day period in January 2003 at Long Beach, New York. Plot the data on the graph in Figure 11.18 (page 168). (*Note:* The number *12* along the bottom of the graph is 12 noon.) After you have plotted the data, complete the following.

53. What type of tide occurs at Long Beach, New York (diurnal, semidiurnal, *or* mixed)?

_____

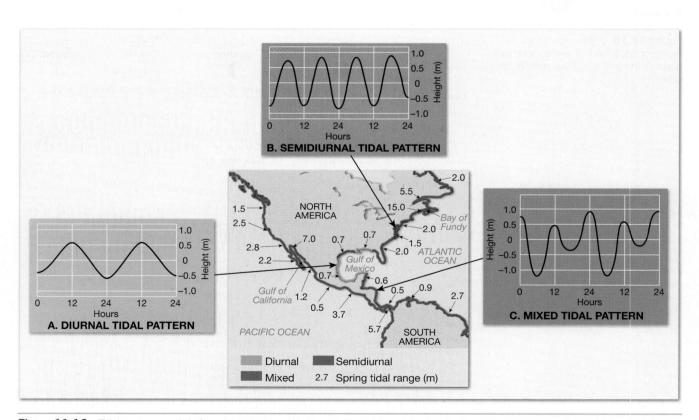

Figure 11.16   Tidal patterns and their occurrence along North and Central American coasts. **A**. The diurnal tidal pattern has one high tide and one low tide each tidal day. **B**. The semidiurnal pattern has two high tides and two low tides of approximately equal heights during each tidal day. **C**. The mixed tidal pattern shows two high tides and two low tides of unequal heights during each tidal day.

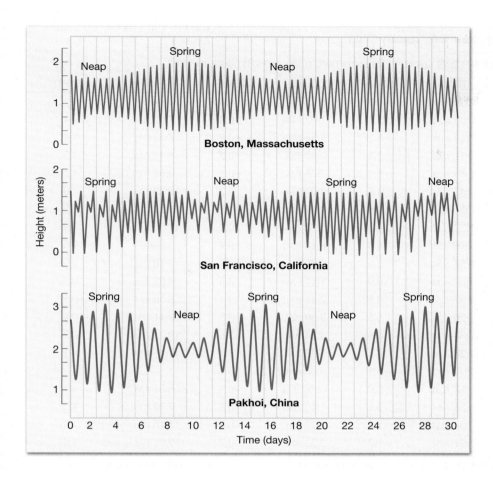

Figure 11.17 Tidal patterns for three locations, to accompany question 52.

54. On which day does the greatest tidal range occur? On which day does the smallest range occur?

Greatest tidal range: _____

Smallest tidal range: _____

55. Using Figure 11.15 as a reference, label the most likely lunar phases (e.g., full moon) associated with the tidal pattern above the appropriate days.

56. Assume that at 9:00 A.M. on January 5, a boat was anchored near a beach in 4 feet of water. When the owner returned at 3:30 P.M., the boat was resting on sand. Approximately how long did the owner have to wait before the boat was floating again?

The owner had to wait approximately _____.

Companion Website

The companion website provides numerous opportunities to explore and reinforce the topics of this lab exercise. To access this useful tool, follow these steps:

1. Go to www.mygeoscienceplace.com.

2. Click on "Books Available" at the top of the page.

3. Click on the cover of *Applications and Investigations in Earth Science, 7e.*

4. Select the chapter you want to access. Options are listed in the left column ("Introduction," "Web-based Activities," "Related Websites," and "Field Trips").

Table 11.1   Tidal Data for Long Beach, New York, January 2003

	TIMES ARE LISTED IN LOCAL STANDARD TIME (LST); HEIGHTS ARE IN FEET							
DAY	**TIME**	**HEIGHT**	**TIME**	**HEIGHT**	**TIME**	**HEIGHT**	**TIME**	**HEIGHT**
1	5:45 A.M.	5.5	12:16 P.M.	−0.7	6:12 P.M.	4.4	—	—
2	12:18 A.M.	−0.5	6:35 A.M.	5.6	1:07 P.M.	−0.8	7:03 P.M.	4.4
3	1:10 A.M.	−0.5	7:23 A.M.	5.5	1:56 P.M.	−0.8	7:53 P.M.	4.4
4	1:59 A.M.	−0.4	8:11 A.M.	5.4	2:42 P.M.	−0.7	8:42 P.M.	4.3
5	2:45 A.M.	−0.2	8:59 A.M.	5.1	3:25 P.M.	−0.5	9:32 P.M.	4.2
6	3:30 A.M.	0.0	9:47 A.M.	4.8	4:07 P.M.	−0.3	10:23 P.M.	4.0
7	4:14 A.M.	0.3	10:35 A.M.	4.6	4:49 P.M.	−0.1	11:12 P.M.	3.9
8	5:01 A.M.	0.6	11:22 A.M.	4.3	5:32 P.M.	0.2	11:59 P.M.	3.9
9	5:54 A.M.	0.8	12:09 P.M.	4.0	6:18 P.M.	0.4	—	—
10	12:45 A.M.	3.9	6:56 A.M.	0.9	12:57 P.M.	3.7	7:10 P.M.	0.5
11	1:31 A.M.	3.9	7:59 A.M.	0.9	1:47 P.M.	3.5	8:02 P.M.	0.5
12	2:19 A.M.	4.0	8:57 A.M.	0.8	2:41 P.M.	3.4	8:53 P.M.	0.5
13	3:10 A.M.	4.1	9:50 A.M.	0.6	3:39 P.M.	3.5	9:41 P.M.	0.4
14	4:02 A.M.	4.3	10:38 A.M.	0.3	4:34 P.M.	3.6	10:28 P.M.	0.2
15	4:51 A.M.	4.6	11:26 A.M.	0.1	5:23 P.M.	3.7	11:15 P.M.	0.1
16	5:36 A.M.	4.8	12:12 P.M.	−0.1	6:08 P.M.	3.9	—	—
17	12:02 A.M.	−0.1	6:17 A.M.	5.0	12:57 P.M.	−0.3	6:51 P.M.	4.1
18	12:49 A.M.	−0.2	6:58 A.M.	5.1	1:40 P.M.	−0.5	7:32 P.M.	4.2
19	1:35 A.M.	−0.4	7:38 A.M.	5.2	2:22 P.M.	−0.6	8:15 P.M.	4.3
20	2:20 A.M.	−0.4	8:21 A.M.	5.2	3:30 P.M.	−0.7	9:01 P.M.	4.4

Source: Center for Operational Oceanographic Products and Services, National Oceanographic and Atmospheric Association, National Ocean Service.

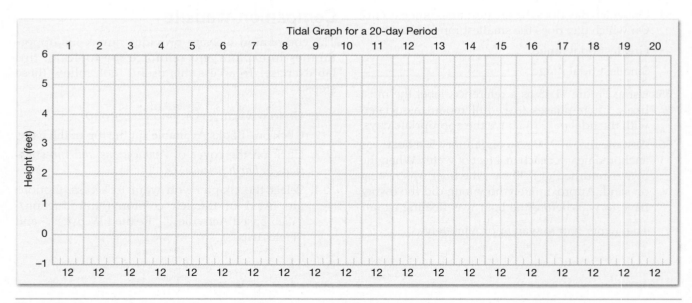

Figure 11.18   Tidal curve for Long Beach, New York. (*Note*: The number 12 along the bottom of the graph is 12 noon.)

Waves, Currents, and Tides

Name _____   **Course/Section** _____

Date _____   **Due Date** _____

1. On Figure 11.19, label the surf zone, a wave crest and a trough, the wavelength, and a deep-water wave. In addition, sketch the motion (orbit) of water particles at increasing depths for one deep-water wave and one shallow-water wave.

2. What happens to the shapes of waves as they approach shallow water?

   _____

3. Describe the formation and appearance of each of the following coastal features:

   Spit: _____

   _____

   Sea stack: _____

   _____

   Tombolo: _____

   _____

   Estuary: _____

   _____

   _____

4. Is the circulation of the surface currents in the South Atlantic Ocean (clockwise *or* counterclockwise)?

   The circulation is _____.

5. Compare a spring tide and a neap tide.

   _____

   _____

6. Explain what causes spring tides.

   _____

   _____

7. Neap tides are most likely to occur during which lunar phase(s)?

   _____

8. Which of the tidal patterns illustrated in Figure 11.17 exhibits the greatest tidal range?

   _____

9. Refer to Figure 11.16. Name the tidal patterns that occur at each of the following locations.

   Atlantic coast of the United States: _____

   Pacific coast of the United States: _____

   Gulf of Mexico: _____

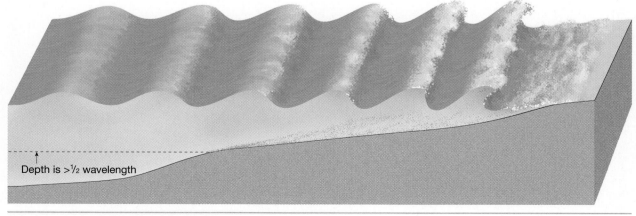

Depth is >½ wavelength

Figure 11.19 Features of waves to accompany summary question 1.

PART 3
Meteorology

EXERCISE 12

Earth–Sun Relationships

Objectives

Completion of this exercise will prepare you to:

A. Describe the effect that Sun angle has on the amount of solar radiation a place receives.

B. Explain why the intensity and duration of solar radiation vary with latitude.

C. Explain why the intensity and duration of solar radiation vary at any one place from season to season.

D. Describe the significance of these special parallels of latitude: Tropic of Cancer, Tropic of Capricorn, Arctic Circle, Antarctic Circle, and equator.

E. List the dates and characteristics of the solstices and equinoxes.

F. Determine the latitude where the overhead Sun is located on any day of the year.

G. Calculate the noon Sun angle for any place on Earth on any day.

H. Calculate the latitude of a place, using the noon Sun angle.

Materials

metric ruler	colored pencils
protractor	calculator
globe	large rubber band or string

Introduction

Weather is the state of the atmosphere at a particular place for a short period of time. In contrast, **climate** represents a long-term statistical perspective. Temperature is a critical part of any description of weather or climate. A very important factor influencing temperatures is the amount of solar energy a place receives. This, in turn, is strongly influenced by Earth–Sun relationships.

Solar Radiation and the Seasons

The amount of solar energy (radiation) striking the outer edge of the atmosphere is determined by the Sun's intensity and duration. **Intensity** depends on the angle at which the Sun's rays strike a surface (Figure 12.1). **Duration** refers to the length of daylight. Intensity and duration vary from place to place. In addition, every place on Earth experiences variations throughout the year.

Solar Radiation and Latitude

To illustrate how variations in sun angle and length of daylight influence the amount of solar radiation reaching Earth's surface, complete questions 1–11.

1. On Figure 12.2, extend the 1-cm-wide beam of sunlight from the Sun to Point A. Extend the second 1-cm-wide beam, beginning at the Sun, to the surface at Point B.

Notice in Figure 12.2 that the Sun is directly overhead at Point A and that the beam of sunlight strikes the surface at a 90° angle.

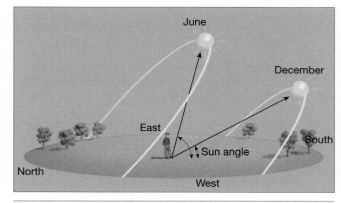

Figure 12.1 Daily paths of the Sun for June and December for an observer in the middle latitudes in the Northern Hemisphere. Notice that the angle of the Sun above the horizon is much greater in the summer than in the winter.

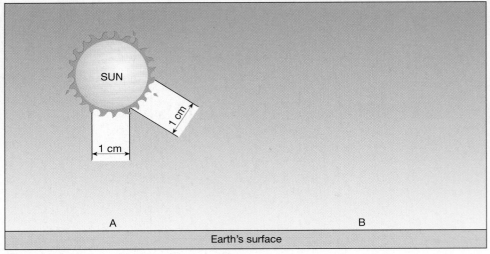

Figure 12.2 Vertical and oblique Sun beams.

Refer to Figure 12.2 to answer questions 2–5.

2. Using a protractor, measure the angle between the surface and the beam of sunlight coming from the Sun to Point B.

Angle of the Sun's rays: _____°

3. What are the lengths of the line segments on the surface covered by the Sun beam at Points A and B?

Point A: _____ mm

Point B: _____ mm

4. Which beam (A *or* B) is more spread out at the surface and therefore covers a larger area?

Beam _____ is more spread out.

5. Would more solar radiation be received by a square centimeter at Point A or at Point B?

Point _____ would receive more solar radiation.

Use Figure 12.3 to answer questions 6–11, which illustrates the amount of solar radiation intercepted by each 30° segment of latitude.

6. With a metric ruler, accurately measure the total width of incoming rays from Point x to Point y in Figure 12.3.

Total width: _____ cm (that is, _____ mm)

7. Assume that the length of line x–y represents 100% of the solar radiation that is intercepted by Earth. Each centimeter equals _____%, and each millimeter equals _____%.

8. What percentage of the total incoming radiation is concentrated in each of the following zones?

0°–30° = _____ mm = _____%

30°–60° = _____ mm = _____%

60°–90° = _____ mm = _____%

9. Use a protractor to measure the angle between the surface and Sun ray at each of the following locations. (Angle b is completed as an example.)

Angle a: _____°

Angle b: _____60°

Angle c: _____°

Angle d: _____°

10. Describe the relationship between the Sun angle and the amount of radiation received in each 30° segment.

_____

_____

11. What fact about Earth creates the unequal distribution of solar energy you noted in question 10?

_____

_____

_____

Yearly Variations in Solar Energy

The amount of solar radiation varies throughout the year at every location on Earth because of the following Earth–Sun relationships:

- Earth rotates on its axis (creating day and night) and also revolves around the Sun (with one revolution equaling one year).

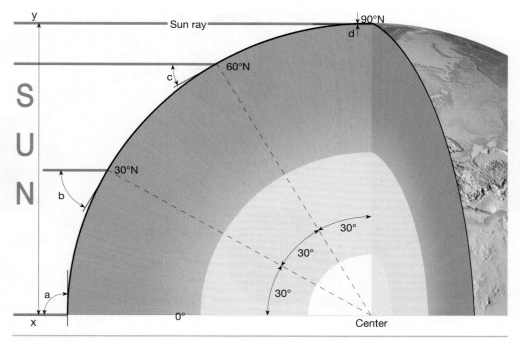

Figure 12.3 Distribution of solar radiation per 30° segment of latitude on Earth.

- The axis of Earth is inclined (tilted) 23.5° from the perpendicular to the plane of its orbit.
- Throughout the year, the axis of Earth points to the same place in the sky, which causes the overhead (vertical, or 90°) noon Sun to cross the **equator** (0° latitude) twice as it migrates from the **Tropic of Cancer** (23.5°N latitude) to the **Tropic of Capricorn** (23.5°S latitude) and back again.

Because the position of the vertical, or overhead, noon Sun shifts between the hemispheres, variations occur in the intensity and duration of solar radiation. These shifts cause seasonal temperature changes. To see how the intensity and duration of solar radiation vary throughout the year, answer questions 12–31.

12. Refer to a globe or world map to locate the important parallels of latitude listed below. Draw in any missing parallels with a colored pencil or a dashed line and label them on Figure 12.4.

   Equator

   Arctic Circle

   Tropic of Capricorn

   Antarctic Circle

   Tropic of Cancer

13. Figure 12.5 shows the position of Earth on the traditional first day of summer, fall, winter, and spring. September dates are provided.

   a. Write in the date represented by each of the other three positions.

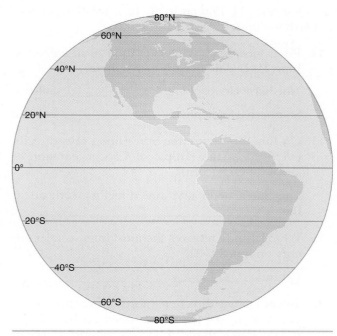

Figure 12.4 Locating important parallels of latitude, for use with question 12.

   b. Label the following for one equinox and one solstice:

   North Pole and South Pole

   Axis of Earth

   Arctic Circle and Antarctic Circle

   Circle of illumination (day–night line)

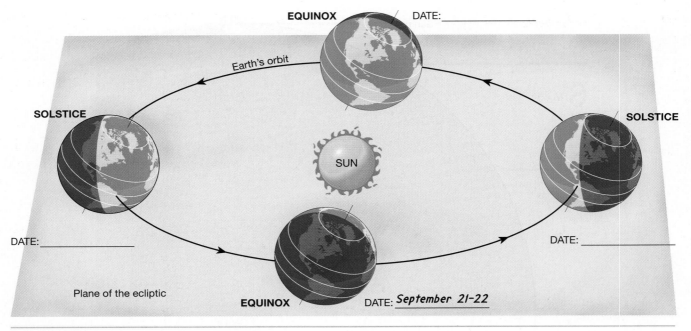

EQUINOX   DATE:_____

Earth's orbit

SOLSTICE

SUN

SOLSTICE

DATE:_____

DATE:_____

Plane of the ecliptic

EQUINOX   DATE: *September 21–22*

DATE:_____

Figure 12.5   Earth–Sun relationships.

Questions 14–19 refer to the June **solstice** position of Earth in Figure 12.5.

14. What term (e.g., winter) is used to describe the June 21–22 date in each hemisphere?

    Northern Hemisphere: _____ solstice

    Southern Hemisphere: _____ solstice

15. On June 21–22, are the Sun's noon rays directly overhead at the (Tropic of Cancer, equator, *or* Tropic of Capricorn)?

    The Sun's noon rays are directly overhead at the _____.

16. What latitude receives the most intense solar energy on June 21–22?

    Latitude: _____

17. Toward what direction (north *or* south) would you look to see the Sun at noon on June 21–22 if you lived at the following latitudes? (*Hint:* Figure 12.1 might be helpful.)

    40°N latitude: _____

    10°N latitude: _____

18. Position a rubber band, string, or pieces of tape on a globe corresponding to the *circle of illumination* on June 21–22. Then calculate the approximate length of daylight at the following latitudes. To do this, determine the number of degrees of longitude that a place located at each latitude spends in daylight as Earth rotates. Then divide that number by 15. (*Note:* Lines of longitude are called *meridians* and

extend from pole to pole. Earth rotates a total of 360° of longitude in 24 hours. Therefore, each 15° of longitude is equivalent to one hour 360 ÷ 24.)

70°N latitude: _____ hours, _____ minutes

40°N latitude: _____ hours, _____ minutes

40°S latitude: _____ hours, _____ minutes

90°S latitude: _____ hours, _____ minutes

 0° latitude: _____ hours, _____ minutes

19. On June 21–22, what is the length of daylight at latitudes north of the Arctic Circle and at latitudes south of the Antarctic Circle?

    North of the Arctic Circle: _____ hours

    South of the Antarctic Circle: _____ hours

Questions 20–24 refer to the December solstice position of Earth in Figure 12.5.

20. What name (season) is used to describe the December 21–22 date in each hemisphere?

    Northern Hemisphere: _____ solstice

    Southern Hemisphere: _____ solstice

21. At which parallel of latitude (the Tropic of Cancer, the equator, *or* the Tropic of Capricorn) are the Sun's noon rays directly overhead on December 21–22?

    Sun's noon rays overhead at _____.

Table 12.1 **Length of Daylight**

LATITUDE (DEGREES)	SUMMER SOLSTICE	WINTER SOLSTICE	EQUINOXES
0	12 h	12 h	12 h
10	12 h 35 min	11 h 25 min	12
20	13 12	10 48	12
30	13 56	10 04	12
40	14 52	9 08	12
50	16 18	7 42	12
60	18 27	5 33	12
66.5	24 h	0 00	12
70	24 h (for 2 mo)	0 00	12
80	24 h (for 4 mo)	0 00	12
90	24 h (for 6 mo)	0 00	12

22. On December 21–22, which hemisphere (Northern *or* Southern) receives the most intense solar energy?

_____ Hemisphere

23. If you lived at the equator on December 21–22, would you look toward the (north *or* south) to see the Sun at noon?

Toward the _____

24. Refer to Table 12.1. What is the length of daylight at each of the following latitudes on December 21–22?

90°N latitude: _____ hours, _____ minutes

40°S latitude: _____ hours, _____ minutes

40°N latitude: _____ hours, _____ minutes

90°S latitude: _____ hours, _____ minutes

0° latitude: _____ hours, _____ minutes

Questions 25–31 refer to the March and September **equinox** positions of Earth in Figure 12.5.

25. For those living in the Northern Hemisphere, what terms (seasons) are used to describe the following dates?

March 21: _____ equinox

September 22: _____ equinox

26. For those living in the Southern Hemisphere, what terms (seasons) are used to describe the following dates?

March 21: _____ equinox

September 22: _____ equinox

27. On March 21 and September 22, at what latitude are the Sun's noon rays directly overhead?

Latitude: _____

28. What latitude receives the most intense solar energy on March 21 and September 22?

Latitude: _____

29. If you lived at 20°S latitude, would you look toward the (north *or* south) to see the Sun at noon on March 21 and September 22?

Toward the _____

30. What is the relationship between the North Pole and South Pole and the circle of illumination on March 21 and September 22?

_____

31. Briefly describe the length of daylight everyplace on Earth on March 21 and September 22.

_____

As you have seen, the latitude where the noon Sun is directly overhead is easily determined for the solstices and equinoxes.

Figure 12.6 is a graph, called an **analemma**, that can be used to determine the latitude where the overhead

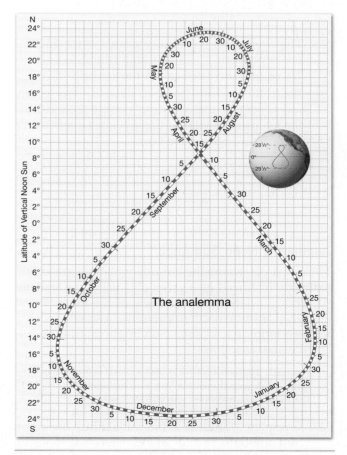

Figure 12.6 The analemma, a graph illustrating the latitude of the overhead (vertical) noon Sun throughout the year.

noon Sun is located for any date. To determine the latitude of the overhead noon Sun from the analemma, find the desired date on the graph and read the coinciding latitude along the left axis. Remember to indicate *North* (N) or *South* (S).

32. Using a colored pencil, draw and label lines that correspond to the equator, Tropic of Cancer, and Tropic of Capricorn on Figure 12.6.

33. Using the analemma (Figure 12.6), determine where the Sun is overhead at noon on the following dates: (An example is provided)

    December 10: _____23°S latitude_____

    March 21: _____

    May 5: _____

    June 22: _____

    August 10: _____

    October 15: _____

34. The position of the overhead noon Sun is always located on or between which two parallels of latitude?

    _____°N (the Tropic of _____)

    _____°S (the Tropic of _____)

35. The overhead noon Sun is located at the equator on September _____ and March _____. These two days are called the _____.

36. Refer to Figure 12.6 and describe the yearly movement of the overhead noon Sun and how the intensity of solar radiation varies over Earth's surface throughout the year.

    _____

    _____

    _____

    _____

Calculating Noon Sun Angle

If you first determine the latitude where the noon Sun is directly overhead on a given date using the analemma, you can determine the angle of the noon Sun at any other latitude on that same day. The relationship between latitude and **noon Sun angle** (Figure 12.7) is stated below as follows:

> For each degree of latitude that a place is away from the latitude where the noon Sun is overhead, the angle of the noon Sun becomes one degree *lower*.

37. Complete Table 12.2 by calculating the noon Sun angle for each of the indicated latitudes on

Table 12.2 Noon Sun Angle Calculations

LATITUDE OF OVERHEAD NOON SUN	MARCH 21 (_____)	APRIL 11 (_____)	JUNE 21 (_____)	DECEMBER 22 (_____)
	Noon Sun Angle			
90°N	_____	_____	_____	0°
40°N	50°	_____	_____	26½°
0°	_____	_____	66½°	_____
20°S	_____	62°	_____	_____

the dates given. Some have been completed for reference.

38. Which latitude in Table 12.2 has the highest *average* noon Sun angle?

    Latitude: _____

39. Calculate the noon Sun angle for the latitude where you live for today's date.

    Date: _____

    Latitude of noon Sun: _____

    Your latitude: _____

    Your noon Sun angle: _____

40. Calculate the yearly maximum and minimum noon Sun angles for your latitude.

MAXIMUM NOON SUN ANGLE	MINIMUM NOON SUN ANGLE
Date: _____	Date: _____
Angle: _____°	Angle: _____°

41. Calculate the yearly average noon Sun angle (maximum plus minimum, divided by 2) and the range of the noon Sun angle (maximum minus minimum) for your location.

    Average noon Sun angle: _____°

    Range of the noon Sun angle: _____°

Using the Noon Sun Angle

One practical use of knowing the noon Sun angle is in navigation. Specifically, if you know the date and angle of the noon Sun at your location, you can determine your latitude using the following procedure:

Step 1: Determine whether you are north or south of the latitude where the noon Sun is directly overhead. (*Hint:* If you look toward the *north*

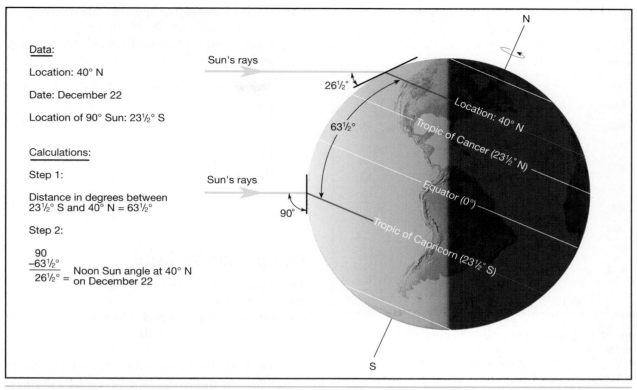

Figure 12.7 Calculating the noon Sun angle. Remember that on any given day, only one latitude receives vertical (90°) rays of the Sun. A place located 1° away (either north or south) receives an 89° noon Sun angle; a place 2° away, an 88° angle, and so forth. To calculate the noon Sun angle, simply find the number of degrees of latitude separating the location you want to know about from the latitude that is receiving the vertical rays of the Sun. Then subtract that value from 90°. The example in this figure illustrates how to calculate the noon Sun angle for a city located at 40° N latitude on December 22 (the winter solstice).

to see the noon Sun, you are *south* of the latitude receiving the Sun's most direct rays.)

Step 2: Subtract the noon Sun angle that was measured for your unknown latitude from 90° to determine how many degrees you are from the latitude where the Sun is directly overhead.

Step 3: Count that many degrees *in the proper direction* to find your latitude. Remember to indicate *North* (N) or *South* (S) with latitude.

42. What is your latitude if, on March 21, you observe the noon Sun to the north at an angle of 18°?

Latitude: _____

43. What is your latitude if, on October 16, you observe the noon Sun to the south at an angle of 39°?

Latitude: _____

Solar Radiation at the Outer Edge of the Atmosphere

Table 12.3 shows the average daily radiation received at the outer edge of the atmosphere at select latitudes for different months.

To help visualize the pattern, plot the data from Table 12.3 on the graph in Figure 12.8. Using a different color for each latitude, draw lines through the monthly values to obtain yearly curves. Then answer questions 44–47.

44. Why are there two periods of maximum solar radiation at the equator?

_____

45. In June, why does the outer edge of the atmosphere at the equator receive less solar radiation than both the North Pole and 40°N latitude?

_____

_____

Table 12.3 Solar Radiation at the Outer Edge of the Atmosphere (Langleys/day) at Various Latitudes During Select Months*

LATITUDE	MARCH	JUNE	SEPTEMBER	DECEMBER
90°N	50	1050	50	0
40°N	700	950	720	325
0°	890	780	880	840

*Langleys are a common measure of solar intensity. The higher the value, the more intense the solar radiation.

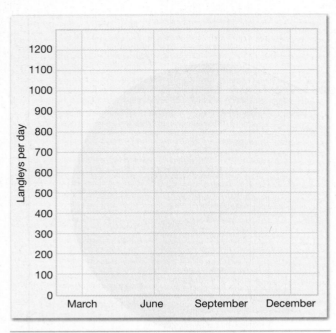

Figure 12.8 Graph of solar radiation received at the outer edge of the atmosphere.

46. Why does the outer edge of the atmosphere at the North Pole receive no solar radiation in December?

_____

47. What would be the approximate monthly values for solar radiation at the outer edge of the atmosphere at 40°S latitude? Explain how you arrived at the values.

March: _____

June: _____

September: _____

December: _____

Explanation: _____

_____

_____

Companion Website

The companion website provides numerous opportunities to explore and reinforce the topics of this lab exercise. To access this useful tool, follow these steps:

1. Go to www.mygeoscienceplace.com.
2. Click on "Books Available" at the top of the page.
3. Click on the cover of *Applications and Investigations in Earth Science, 7e.*
4. Select the chapter you want to access. Options are listed in the left column ("Introduction," "Web-based Activities," "Related Websites," and "Field Trips").

Earth–Sun Relationships

Name _____   **Course/Section** _____

Date _____   **Due Date** _____

1. From Figure 12.3, what was the calculated percentage of solar radiation that is intercepted by each of the following 30° segments of latitude?

   0°–30°: _____ %

   30°–60°: _____ %

   60°–90°: _____ %

2. How many hours of daylight occur at the following locations on the specified dates?

LATITUDE	MARCH 22	DECEMBER 22
40°N	_____ hours	_____ hours
0°	_____ hours	_____ hours
90°S	_____ hours	_____ hours

3. What is the noon Sun angle at these latitudes on April 11?

   40°N: _____ °

   0°: _____ °

4. What is the relationship between the angle of the noon Sun and the intensity of solar radiation received at the outer edge of the atmosphere?

   _____

   _____

5. Complete Figure 12.9, showing Earth's relation to the Sun on June 22. On the Earth, sketch and label the following:

   Axis

   Equator

   Tropic of Cancer

   Tropic of Capricorn

   Antarctic Circle

   Arctic Circle

   Circle of illumination

   Location of the overhead noon Sun

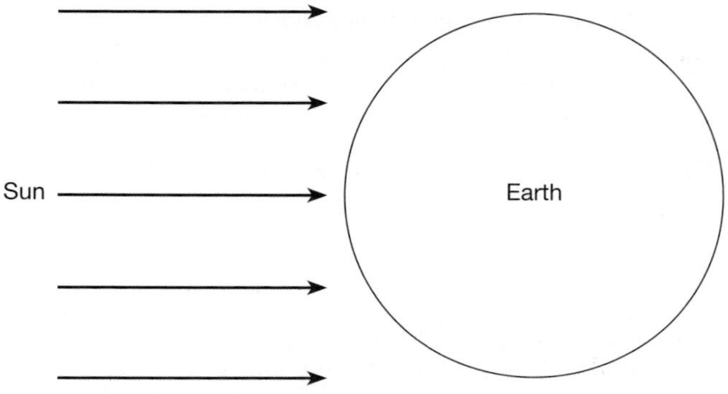

Figure 12.9   Earth's relation to the Sun on June 22.

6. What causes the intensity and duration of solar radiation received at any place to vary throughout the year? That is, what causes the seasons?

   _____

   _____

   _____

7. What is the date illustrated by the diagram in Figure 12.10? Calculate the noon Sun angle at 30°N latitude on this date and write a statement describing the distribution of solar radiation over Earth on this date.

   _____

   _____

   _____

   _____

8. What are the maximum and minimum noon Sun angles at your latitude, and what are the dates when they occur?

   Maximum noon Sun angle: _____ °

   on _____

   Minimum noon Sun angle: _____ °

   on _____

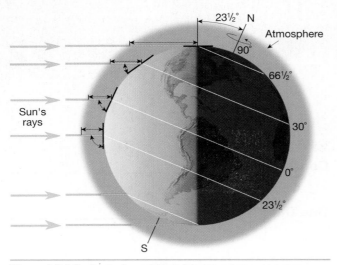

Figure 12.10 Earth–Sun relation diagram.

9. What are the maximum and minimum durations of daylight at your latitude?

   Maximum duration of daylight: _____ hours

   Minimum duration of daylight: _____ hours

10. Briefly describe how the intensity and duration of solar radiation change at your location throughout the year.

    _____

    _____

    _____

    _____

    _____

11. The day is March 22. You view the noon Sun to the south at an angle of 35°. What is your latitude?

    Latitude: _____

Heating the Atmosphere

Questions: 1-11
20-48

Objectives

Completion of this exercise will prepare you to:

A. Explain how Earth's atmosphere is heated.

B. Describe how absorption, scattering, and reflection influence incoming solar radiation.

C. List the gases in the atmosphere that are responsible for absorbing radiation emitted by Earth's surface.

D. Explain how albedo affects the heating of Earth's surface.

E. Describe the differences in the heating and cooling of land and water.

F. Summarize the global pattern of surface temperatures for January and July.

G. Determine windchill-equivalent temperatures.

Materials

calculator	colored pencils
light source	wood splints
black and silver containers	beaker of sand
beaker of water	two thermometers

Introduction

Temperature is a basic element of weather and climate. In this exercise, you will gain a better understanding of how Earth's atmosphere is heated. In addition, you will examine some of the factors that cause temperatures to vary from time to time and from place to place.

Radiation and the Electromagnetic Spectrum

From everyday experience, we know that the Sun emits light and heat as well as the ultraviolet rays that cause sunburn. Although these forms of energy comprise a major portion of the total energy that radiates

from the Sun, they are only part of a large array of energy called **radiation** or **electromagnetic radiation**. This array or spectrum is shown in Figure 13.1.

Visible light is the only portion of the spectrum we can see. It is often referred to as "white" light because it appears white in color. However, as shown in Figure 13.1B, visible light is really a mixture of colors—the colors of the rainbow.

Refer to Figure 13.1 to answer questions 1–3.

1. Which part (color) of visible light has the longest wavelength?

Red _____Violet_____ has the longest wavelength.

2. Which part (color) of visible light has the shortest wavelength?

Blue _____Yellow_____ has the shortest wavelength.

3. Which part of the electromagnetic spectrum (infrared *or* ultraviolet) has shorter wavelengths?

_____Ultraviolet_____ has shorter wavelengths.

What Happens to Incoming Solar Radiation?

When radiation strikes an object, there are usually three different results. (1) Some energy is *absorbed*, causing the temperature to increase. (2) Substances such as air and water are transparent to certain wavelengths and simply *transmit* the radiant energy. Radiation that is transmitted *does not* contribute energy to the object. (3) Some radiation may "bounce off" the object without being absorbed or transmitted. *Reflection* and *scattering* are responsible for redirecting incoming solar radiation.

Figure 13.2 shows what happens to incoming solar radiation averaged over the entire globe. Refer to this diagram when answering questions 4–7.

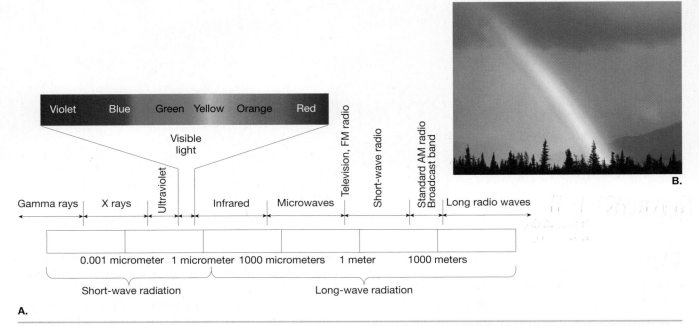

Figure 13.1 **A**. The electromagnetic spectrum, illustrating the wavelengths and names of various types of radiation. **B**. Visible light consists of an array of colors we commonly call the "colors of the rainbow." (Photo by Michael Giannechini/Photo Researchers, Inc.)

4. What percentage of incoming solar radiation is reflected and scattered back to space?

30 _____ %

5. What percentage of incoming solar radiation is absorbed by clouds and gases in the atmosphere?

_____20____ %

6. What percentage of incoming solar radiation passes through the atmosphere and is absorbed at Earth's surface?

____50____ %

7. Excluding the radiation that is reflected back to space, we can conclude that Earth's atmosphere is largely (transparent *or* not transparent) to the

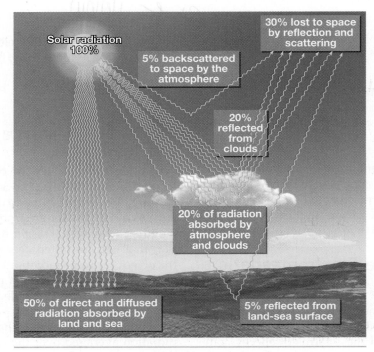

Figure 13.2 Solar radiation budget of the atmosphere and Earth.

remaining incoming solar radiation because most of this energy is (transmitted *or* absorbed) by the atmosphere.

The atmosphere is: _____transparent_____

Energy is: _____absorbed_____

How Is Earth's Atmosphere Heated?

Answers to the preceding questions showed that the atmosphere is largely transparent to solar radiation and that much of this energy is absorbed at Earth's surface. What happens to the energy absorbed at Earth's surface? In addition to getting hotter, the surface emits radiation into the atmosphere. Is Earth's radiation different from solar radiation? Yes, because the wavelengths emitted by an object depend on its temperature: *The hotter the object, the shorter the wavelengths it emits.*

To answer questions 8–11, refer to Figure 13.3.

8. Does the radiation emitted by Earth's surface have (longer *or* shorter) wavelengths compared to solar radiation?

   Earth's radiation has _____shorter_____ wavelengths.

9. Earth's surface emits (ultraviolet, visible, *or* infrared) radiation.

   Earth emits_____infrared_____ radiation.

10. Nitrogen, the most abundant constituent of the atmosphere (78 percent) is a relatively (good *or* poor) absorber of incoming solar radiation?

    Nitrogen is a _____good poor_____ absorber.

11. Which two gases are most effective in absorbing radiation emitted by Earth?

    Gases: _____Nitrogen_____ and _____oxygen and ozone_____
    CO₂   water vapor

   Questions 1–11 focus on how Earth's atmosphere is heated. Figure 13.4 summarizes this process, termed the **greenhouse effect**.

The Nature of Earth's Surface

Since the atmosphere is heated chiefly by radiation emitted by Earth's surface, differences in the heating properties of various surfaces must influence air temperatures.

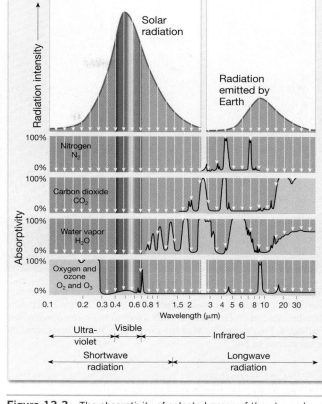

Figure 13.3 The absorptivity of selected gases of the atmosphere and the atmosphere as a whole.

Albedo refers to the reflectivity of a surface and is usually expressed as the percentage of radiation that is reflected. Since surfaces with high albedos are *not* efficient absorbers of radiation, they cannot return much long-wave radiation to the atmosphere for heating.

Experiment: The Influence of Color on Albedo

To better understand the effect of color on albedo, conduct the following experiment:

Step 1: Write a brief hypothesis that relates albedo to the heating and cooling of light-colored and dark-colored surfaces.

_____

_____

_____

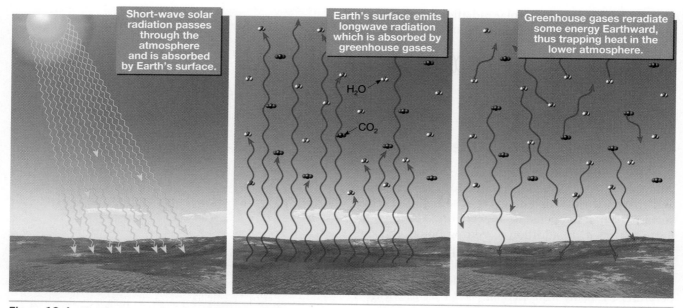

Figure 13.4 The heating of the atmosphere. Most of the short-wavelength radiation from the Sun passes through the atmosphere and is absorbed by Earth's land–sea surface. This energy is then emitted from the surface as longer-wavelength radiation, much of which is absorbed by carbon dioxide and water vapor in the lower atmosphere. Some of the energy absorbed by the atmosphere will be reradiated Earthward. This process, called the *greenhouse effect*, is responsible for keeping Earth's surface and lower atmosphere much warmer than it would otherwise be.

Step 2: Place the black and silver containers (with lids and thermometers) about six inches away from the light source (see Figure 13.5). Place both containers the same distance from the light, but *do not* have them touch.

Step 3: Using the albedo experiment data table, Table 13.1, record the starting temperature of both containers.

Step 4: Turn on the light and record the temperature of both containers on the data table at about 30-second intervals for 5 minutes.

Step 5: Turn off the light and continue to record the temperatures at 30-second intervals for another 5 minutes.

Step 6: Plot the temperatures from the data table on the albedo experiment graph, Figure 13.6. Use a different-color line to connect the points for each container.

12. For each container, calculate the *rate of heating* (change in temperature divided by the time the light was on) and the *rate of cooling* (change in

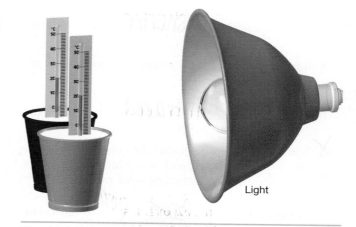

Figure 13.5 Albedo experiment lab equipment.

temperature divided by the time the light was off).

	RATE OF HEATING	RATE OF COOLING
Silver can	_____	_____
Black can	_____	_____

Table 13.1 Albedo Experiment Data Table

	STARTING TEMPERATURE	30 SEC	1 MIN	1.5 MIN	2 MIN	2.5 MIN	3 MIN	3.5 MIN	4 MIN	4.5 MIN	5 MIN	5.5 MIN	6 MIN	6.5 MIN	7 MIN	7.5 MIN	8 MIN	8.5 MIN	9 MIN	9.5 MIN	10 MIN
Black container																					
Silver container																						

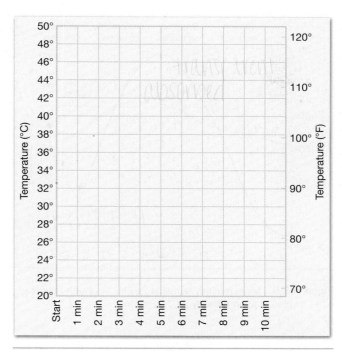

Figure 13.6 Albedo experiment graph.

13. Write a statement that summarizes and explains the results of the albedo experiment.

_____

_____

_____

_____

14. List at least two Earth surfaces that have high albedos and two that have low albedos.

   High albedos: _____

   Low albedos: _____

15. Given equal amounts of radiation reaching the surface, should the air over a snow-covered surface be (warmer _or_ colder) than air above a dark-colored, barren field? Explain your answer.

   _____

   _____

16. If you lived in an area with long, cold winters, would a (light-colored _or_ dark-colored) roof be the best choice for your house?

   _____

Figure 13.7 Land and water heating experiment lab equipment.

Experiment: Differential Heating of Land and Water

Investigate the differential heating of land and water by conducting the following experiment:

Step 1: Fill a beaker three-quarters full with dry sand and fill a second beaker three-quarters full with room-temperature water.

Step 2: Suspend a thermometer in each beaker so that the bulbs are _just below_ the surfaces of the sand and water, as shown in Figure 13.7.

Step 3: Hang a light from a stand so that both beakers receive the same amount of light.

Step 4: Record the starting temperatures for both the dry sand and water on the land and water heating data table, Table 13.2.

Table 13.2　**Land and Water Heating Data Table**

	STARTING TEMPERATURE	1 MIN	2 MIN	3 MIN	4 MIN	5 MIN	6 MIN	7 MIN	8 MIN	9 MIN	10 MIN
Water											
Dry sand											
Damp sand											

Step 5: Turn on the light and record temperatures on the data table at about 1-minute intervals for 10 minutes.

Step 6: Turn off the light for several minutes. Dampen the sand with water and record the starting temperature of the damp sand on the data table. Turn on the light and record the temperature of the damp sand on the data table at about 1-minute intervals for 10 minutes.

Step 7: Plot the temperatures for the water, dry sand, and damp sand from the data table on the graph in Figure 13.8. Use a different-color line to connect and label the points for each material.

17. Questions 17a and 17b refer to the land and water heating experiment.

   a. How does the temperature differ for dry sand and water when they are exposed to equal amounts of light energy?

   _____

   _____

   b. How does the temperature differ for dry sand and damp sand when they are exposed to equal amounts of light?

   _____

   _____

18. Suggest a reason for the differential heating of land and water.

   _____

   _____

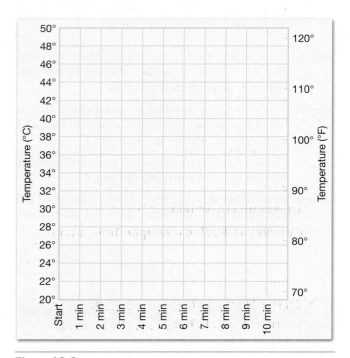

Figure 13.8 Land and water heating graph.

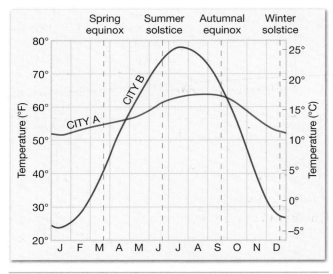

Figure 13.9 Mean monthly temperatures for two North American cities located at approximately 37°N latitude.

19. Suggest a reason why the damp sand heated differently than the dry sand.

   _____

   _____

   Figure 13.9 presents the annual temperature curves for two cities, A and B, that are located in North America at approximately 37°N latitude. On any date, both cities receive the same intensity and duration of solar radiation. One city is in the center of the continent, whereas the other is on the West Coast. Refer to Figure 13.9 to complete the following.

20. Which city has the highest monthly mean temperature?

   City: _____CITY ~~CITY~~ B_____

21. Which city has the lowest monthly mean temperature?

   City: _____CITY ~~AREA~~ B_____

22. Which city has the greatest *annual temperature range* (difference between highest and lowest monthly mean temperatures)?

   City: _____CITY B_____

23. Which city reaches its maximum monthly mean temperature at an earlier date?

   City: _____CITY B_____

24. Which city maintains more uniform temperatures throughout the year?

   City: _____CITY A_____

25. Which city is along the West Coast, and which city is in the center of the continent, far from the influence of the ocean? Explain you answer.

West Coast city: _City A_

Center continent city: _City B_

Explanation: _Because the west coast doesn't see as much of an annual climate change as a center continent city would._

Air Temperature Data

Daily Temperatures

Questions 26–33 refer to the daily temperature graph in Figure 13.10.

26. When does the coolest temperature of the day occur?
_4 AM_

27. When does the warmest temperature of the day occur?
_5 PM_   _~4 pm_

28. What is the *daily temperature range* (difference between maximum and minimum temperatures for the day)?

_79 20_
_-58 -14_
_21 12_

Daily temperature range: _21_ °F (_12_ °C).

29. What is the *daily mean temperature* (average of the maximum and minimum temperatures)?

Daily mean temperature: _08.5_ °F (_20_ °C).

30. Why does the coolest temperature of the day occur about sunrise?
Because the sun has not risen yet

31. How would cloud cover influence the daily maximum temperature? Explain your answer.
If there was more clouds the temperature would decrease

32. How would cloud cover influence the daily minimum temperature? Explain your answer.
It would probably remain the same.

33. On Figure 13.10, sketch and label a colored line that represents a daily temperature graph for a typical cloudy day.

Global Pattern of Temperature

The primary reason for global variations in surface temperatures is the unequal distribution of radiation over Earth due to seasonal changes in sun angle and length of daylight. Among the most important secondary factors affecting the global pattern of temperature

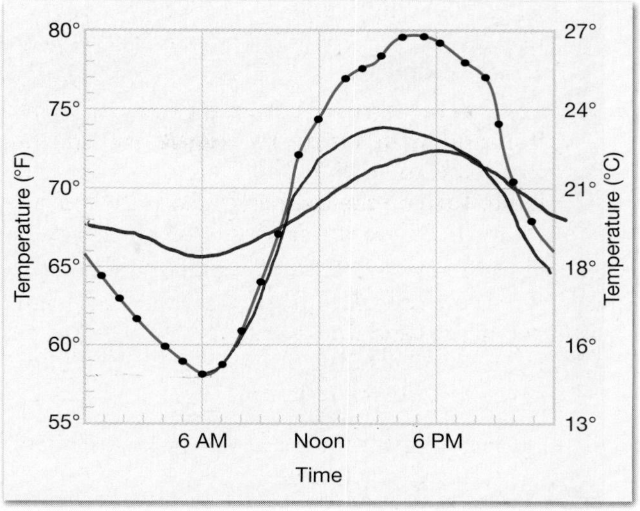

Figure 13.10 Typical daily temperature graph for a mid-latitude city during the summer.

are differential heating of land and water and the influence of cold and warm ocean currents.

Questions 34–46 refer to Figure 13.11. The lines on the maps, called **isotherms**, connect places of equal surface temperatures.

34. Is the general trend of the isotherms on the maps (north–south *or* east–west)?

The isotherms trend _East-West_.

35. In general, how do surface temperatures vary from the equator toward the poles? What is the likely cause of this variation? _Due to direct sun angle_
The temperatures are warmer near the equator and cooler near the poles.

36. Over which countries or oceans do the *warmest* and *coldest* temperatures occur?

Warmest global temperature: _Mexico, Africa, Middle East, Asia_

Coldest global temperature: _North Pole / Russia_

37. Are the locations of the warmest and coldest temperatures over (land *or* water)?

Warmest and coldest temperatures occur over _land_.

38. Calculate the *annual temperature range* at each of the following locations:

_-20 0_   Coastal Norway at 60°N: _20_ °C (_____ °F)   _~14_

_-20 15_ Siberia at 60°N, 120°E: _35_ °C (_____ °F)   _~45_

On the equator, over the center of the Atlantic Ocean: _0_ °C (_0_ °F)   _1°_

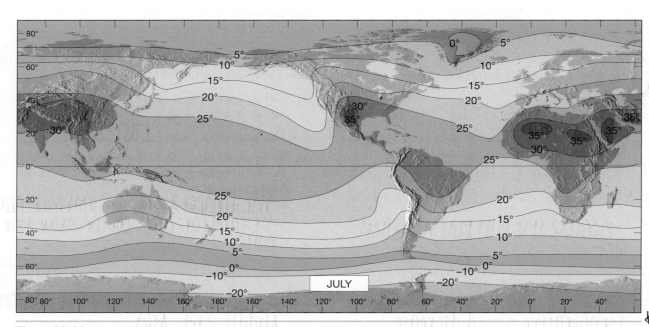

Figure 13.11 World distribution of mean surface temperatures (°C) for January and July.

in the middle of a continent

39. Explain the wide annual range of temperature in Siberia.

Siberia goes through seasonal changes where the sun is facing them or it is away from them.

ocean proximity

40. Why is the annual temperature range smaller along the coast of Norway than at the same latitude in Siberia?

Because it is located near the poles and doesn't see much light.

41. Why are temperatures relatively uniform throughout the year in the tropics?

Because temperatures increase towards the equator.

— due to sun angle + proximity to ocean

42. Using the two maps in Figure 13.11, calculate the approximate average annual temperature range for your location. How does your temperature range compare with the ranges in the tropics and Siberia?

Average annual temperature range: ____20____ °C
(_____°F)

Siberia fluctuates more but we see some of the same temperature changes

43. Trace the path of the 5°C isotherm over North America in January. Explain why the isotherm deviates from a true east–west trend where it crosses from the Pacific Ocean onto the continent.

Because it is over a large ocean which is going to be cooler with no land.

44. Trace the path of the 20°C isotherm over North America in July. Explain why the isotherm deviates from a true east–west trend where it crosses from the Pacific Ocean onto the continent.

Because of the flow of the water and its closeness to a continent.

45. Why do the isotherms in the Southern Hemisphere follow a true east–west trend more closely than those in the Northern Hemisphere?

Because there is less landmass.

Why does the entire pattern of isotherms shift northward from the January map to the July map?

Because they are changing directions in regards to the sun.

Windchill Equivalent Temperature

Windchill equivalent temperature is the term applied to the sensation of temperature that the human body feels, in contrast to the actual temperature of the air as recorded by a thermometer. Wind cools by evaporating perspiration and carrying heat away from the body. When temperatures are cool and the wind speed increases, the body reacts as if it were being subjected to increasingly lower temperatures—a phenomenon known as *windchill.*

47. Refer to the windchill equivalent temperature chart shown in Figure 13.12. What is the windchill equivalent temperature sensed by the human body in the following situations?

AIR TEMPERATURE (°F)	WIND SPEED (mph)	WINDCHILL EQUIVALENT TEMPERATURE (°F)
30°	10	21
−5°	20	−29
−20°	30	−53

48. Write a brief summary of the effect of wind speed on how long a person can be exposed before frostbite develops.

Wind speed increases cause an increase in the temperature humans sense. Therefore your body will be a lot colder than the actual temp.

Companion Website

The companion website provides numerous opportunities to explore and reinforce the topics of this lab exercise. To access this useful tool, follow these steps:

1. Go to www.mygeoscienceplace.com.
2. Click on "Books Available" at the top of the page.
3. Click on the cover of *Applications and Investigations in Earth Science, 7e.*
4. Select the chapter you want to access. Options are listed in the left column ("Introduction," "Web-based Activities," "Related Websites," and "Field Trips").

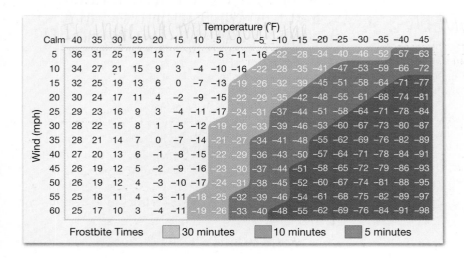

Figure 13.12 This windchill chart came into use in November 2001. Fahrenheit temperatures are used here because this is how the National Weather Service and the news media in the United States commonly report windchill information. The shaded areas on the chart indicate frostbite danger. Each shaded zone shows how long a person can be exposed before frostbite develops.

Notes and calculations.

Heating the Atmosphere

Name _____   **Course/Section** _____

Date _____   **Due Date** _____

1. What percentage of the solar radiation is absorbed by the atmosphere? What percentage is absorbed by Earth's surface?

   Atmospheric absorption: _____ %

   Absorption by Earth's surface: _____ %

2. What will be the atmospheric effect of each of the following?

   More carbon dioxide in the atmosphere:

   _____

   _____

   A surface with a high albedo:

   _____

   _____

3. Briefly explain how Earth's atmosphere is heated.

   _____

   _____

   _____

4. What are the two gases that absorb most of Earth's outgoing radiation? In general, what wavelength of radiation do they absorb?

   _____

   _____

5. What were the starting and five-minute temperatures you obtained for the black and silver containers in the albedo experiment?

	STARTING TEMPERATURE	FIVE-MINUTE TEMPERATURE
Black container	_____	_____
Silver container	_____	_____

6. Summarize the effect of color on the heating of an object.

   _____

   _____

7. What were the starting and ending temperatures you obtained for the water and dry sand in the land and water heating experiment?

	STARTING TEMPERATURE	ENDING TEMPERATURE
Water	_____	_____
Dry sand	_____	_____

8. Summarize the effects that equal amounts of radiation have on the heating of land and water.

   _____

   _____

9. Where on Earth are the highest and lowest monthly average temperature?

   Highest average monthly temperature:

   _____

   Lowest average monthly temperature:

   _____

10. Why does the Northern Hemisphere experience a greater annual temperature range than the Southern Hemisphere?

    _____

    _____

property, water freely leaves Earth's surface as a gas and returns again as a solid or liquid.

Whenever water changes state, heat is exchanged between water and its surroundings. For example, when ice melts, heat is absorbed. The heat absorbed or released during a change of state is termed **latent heat**.

1. On Figure 14.2, write the name of each process involved (choose from the following list) and indicate whether heat is absorbed by water or released by water.

PROCESSES

Freezing (liquid to solid) **Evaporation** (liquid to gas) **Deposition** (gas to solid)

Sublimation (solid to gas) **Melting** (solid to liquid) **Condensation** (gas to liquid)

2. When water evaporates, is heat (absorbed *or* released)?

   Heat is _____*absorbed*_____.

3. During condensation, do water molecules (absorb *or* release) heat?

   Water molecules _____*release*_____ heat.

4. Is the amount of latent heat associated with the process of deposition the (same as *or* less than) the total energy required to condense water vapor and then freeze the water?

   The amount of latent heat is _____*same as*_____ the total energy required.

Latent Heat Experiment

This experiment is intended to help you gain a better understanding of latent heat. You are going to heat a beaker that contains a mixture of ice and water (Figure 14.3). You will record temperature changes *as the ice melts* and continue to record the temperature changes *after the ice melts*. Conduct the experiment by completing the following steps:

Step 1: Write a brief hypothesis about how you expect the temperature of the ice–water mixture to change as heat is added.

_____

_____

_____

Step 2: Turn on the hot plate and set the temperature setting to about three-fourths the maximum (7 on a scale of 10).

CAUTION: Do not touch the heating surface of the hot plate.

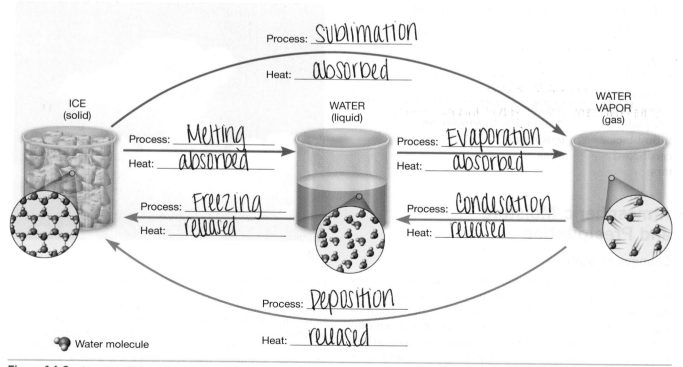

Figure 14.2 Changes of state of water always involve an exchange of heat.

Thermometer →

Figure 14.3 Setup for the latent heat experiment.

Step 3: Fill a 400-ml or larger beaker approximately half full with ice and add enough *cold* water to cover the ice.

Step 4: Gently stir the ice–water mixture for about 15 seconds with the thermometer and record the temperature in the "Starting" temperature space in Table 14.1.

Table 14.1 Latent Heat Data Table

TIME (MINUTES)	TEMPERATURE (———°)
Starting	____
1	____
2	____
3	____
4	____
5	____
6	____
7	____
8	____
9	____
10	____
11	____
12	____
13	____
14	____
15	____

Step 5: Place the beaker with the ice–water mixture and thermometer on the hot plate, and while *stirring the mixture frequently*, record the temperature of the mixture at *one-minute intervals* on the data table. Watch the ice closely as it melts. Circle the exact time (minute) on the data table when all the ice has melted.

Step 6: Continue stirring the mixture and recording its temperature for at least three or four minutes after all the ice has melted.

Step 7: Plot the temperatures from the data table on the graph in Figure 14.4.

Questions 5–8 refer to the latent heat experiment.

5. How did the temperature of the mixture change prior to and after the ice melted?

_____

_____

6. Calculate the *average temperature change per minute* of the ice–water mixture *prior to* the ice melting and the average rate *after* the ice melted.

Average rate prior to melting: _____

Average rate after melting: _____

7. With your answers to questions 5 and 6 in mind, write a statement comparing your results to your hypothesis in Step 1.

_____

_____

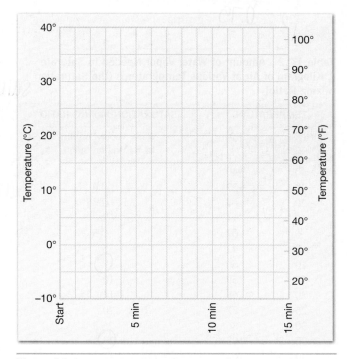

Figure 14.4 Latent heat experiment graph.

8. Explain how the absorption of latent (hidden) heat accounts for the temperature changes that occured before the ice melted as compared to after the ice melted.

_____

_____

_____

_____

Water Vapor Content

Any measure of water vapor in the air is referred to as *humidity*. The amount of water vapor required for saturation is directly related to temperature.

The mass of water vapor in a unit of air compared to the remaining mass of dry air is referred to as the *mixing ratio*. Table 14.2 presents the mixing ratios for saturated air at various temperatures. Use the table to answer questions 9–12.

9. To illustrate the relationship between the amount of water vapor needed for saturation and temperature, prepare a graph by plotting the data from Table 14.2 on Figure 14.5.

10. Using Table 14.2 and/or Figure 14.5, determine how much water vapor is required to saturate 1 kilogram of air at the following temperatures:

40°C: __47__ grams

20°C: __14__ grams

0°C: __3.5__ grams

–20°C: __0.75__ grams

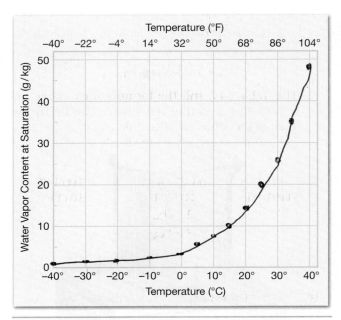

Figure 14.5 Graph of water vapor content at saturation of 1 kilogram of air versus temperature.

11. Refer to Table 14.2. How much more water vapor is contained in 1 kilogram of saturated air at 35°C compared to 1 kilogram of air at 10°C?

The difference is __28__ grams. Or, put another way saturated air at 35°C contains __4__ times more water vapor than saturated air at 10°C.

12. Using Table 14.2 and/or Figure 14.5 as a guide, describe how the amount of water vapor needed for saturation relates to temperature.

As the temperature increases the water vapor that is needed for saturated also increases. It increases at a more rapid rate over the increase of temperature.

Table 14.2 **Amount of Water Vapor Needed to Saturate 1 Kilogram of Air at Various Temperatures (the Saturation Mixing Ratio)**

TEMPERATURE		SATURATION MIXING RATIO
(°C)	(°F)	Water Vapor Content at Saturation (g/kg)
–40	–40	0.1
–30	–22	0.3
–20	–4	0.75
–10	14	2
0	32	3.5
5	41	5
10	50	7
15	59	10
20	68	14
25	77	20
30	86	26.5
35	95	35
40	104	47

Measuring Humidity

Relative humidity is the most common measurement used to describe the amount of water vapor in the air. It expresses how close the air is to being saturated. Relative humidity is a *ratio* of the air's actual water vapor content (amount actually in the air) compared with the amount of water vapor required for saturation at that temperature (saturation mixing ratio), expressed as a percentage. The general formula is:

$$\text{Relative humidity (\%)} = \frac{\text{Water vapor content}}{\text{Saturation mixing ratio}} \times 100$$

For example, from Table 14.2, the saturation mixing ratio of air at 25°C is 20 grams per kilogram. If the actual amount of water vapor in the air were 5 grams per kilogram (the water vapor content), the relative humidity of the air would be calculated as follows:

$$\text{Relative humidity (\%)} = \frac{5\,\text{g/kg}}{20\,\text{g/kg}} \times 100 = 25\%$$

13. Use Table 14.2 and the formula for relative humidity to determine the relative humidity for each of the following situations. Notice that temperature remains unchanged.

AIR TEMPERATURE	WATER VAPOR CONTENT	RELATIVE HUMIDITY
15°C	2 g/kg	*20* % *2/10*
15°C	5 g/kg	*50* %
15°C	7 g/kg	*10* %

14. Based on question 13, answer the following questions. When air temperature remains constant, will adding water vapor (increase *or* decrease) relative humidity? How does relative humidity change when water vapor content is reduced?

Adding water vapor will ___*increase*___ relative humidity.

Reducing water vapor will ___*decrease*___ relative humdity.

15. Use Table 14.2 and the formula for relative humidity to determine the relative humidity for each of the following situations in which the *water vapor content remains unchanged*:

AIR TEMPERATURE	WATER VAPOR CONTENT	RELATIVE HUMIDITY
25°C	5 g/kg	*25* % *5/20*
15°C	5 g/kg	*50* % *5/10*
5°C	5 g/kg	*100* % *5/5*

16. Based on question 15, if the amount of water vapor in the air remains constant, will cooling the air (raise *or* lower) the relative humidity? What happens to relative humidity when the air temperature is increased?

Cooling will ___*decrease*___ the relative humidity.

The relative humidity increases

17. In the winter, air in homes is heated. What effect does heating have on the relative humidity inside the home? What can be done to lessen this effect?

The relative humidity should be increased as the temperature is increased. Add a dehumidifier/humidifier.

18. Explain why the air in a cool basement is relatively humid (damp) in the summer.

Because the basement is cooler causing a lower relative humidity.

19. Briefly describe the two ways relative humidity can be changed in nature.

a. *Temperature Change*

b. *Amt. of Precipitation*

One misconception concerning relative humidity is that it gives an accurate indication of the actual quantity of water vapor in the air. For example, on a winter day, if you hear on the radio that the relative humidity is 90%, can you conclude that the air contains more water vapor than on a summer day that records a relative humidity of 40%? Completing question 20 will help you find the answer.

20. Refer to Table 14.2 to determine the water vapor content for each of the following situations. As you do the calculations, keep in mind the definition of relative humidity.

SUMMER	WINTER
Air temperature = 77°F	Air temperature = 41°F
Relative humidity = 40%	Relative humidity = 90%
Content = *20* g/kg	Content = *5* g/kg

21. Explain why relative humidity does *not* give an accurate indication of the actual amount of water vapor in the air.

Because it shows that there is a higher content in the summer, yet precipitation is seen at almost a constant during the winter.

Dew-Point Temperature

The temperature at which saturation occurs is called the **dew-point temperature**. Put another way, the dew point is the temperature at which the relative humidity of the air is 100%.

In question 15, you determined that 1 kilogram of air at 25°C, containing 5 grams of water vapor, has a relative humidity of 25% and is not saturated. However, when the temperature is lowered to 5°C, the air has a relative humidity of 100% and is saturated. Therefore, 5°C is the dew-point temperature of the air in that example.

22. By referring to Table 14.2, determine the dew-point temperature of 1 kilogram of air that contains 7 grams of water vapor.

    Dew-point temperature = __10__ °C

23. What are the relative humidity and dew-point temperature of 1 kilogram of 25°C air that contains 10 grams of water vapor?

    Relative humidity = __100__ %

    Dew-point temperature = __15__ °C   10/10

Using a Psychrometer

One method of determining the relative humidity and dew-point temperature of air is by using a **psychrometer** (Figure 14.6). A psychrometer consists of two identical thermometers mounted side by side. One of the thermometers, the *dry-bulb* thermometer, measures the air temperature. The other thermometer, the *wet-bulb thermometer*, has a piece of wet cloth wrapped around its bulb. As the psychrometer is spun for approximately one minute, water on the wet-bulb thermometer evaporates, and cooling results. In dry air, the rate of evaporation will be high, and a low wet-bulb temperature will be recorded. After using the psychrometer and recording both the dry- and wet-bulb temperatures, you can determine the relative humidity

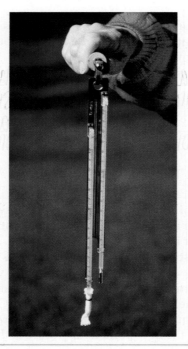

Figure 14.6 The sling psychrometer is one instrument that is used to determine relative humidity and dew-point temperature. (Photo by E. J. Tarbuck)

and dew-point temperatures by using Table 14.3 and Table 14.4 (page 202).

24. Use Table 14.3 to determine the relative humidity for each of the following psychrometer readings:

	READING 1	READING 2
Dry-bulb temperature	20°C	32°C
Wet-bulb temperature	18°C	25°C
Difference between dry- and wet-bulb temperatures	_____	_____
Relative humidity	_____%	_____%

25. Based on question 24, describe the relationship between the *difference* in the dry-bulb and wet-bulb temperatures compared to the relative humidity of the air.

    _____

    _____

26. Use Table 14.4 to determine the dew-point temperature for these two psychrometer readings:

	READING 1	READING 2
Dry-bulb temperature	8°C	30°C
Wet-bulb temperature	6°C	24°C
Difference between dry- and wet-bulb temperatures	_____	_____
Dew-point temperature	_____°C	_____°C

   If a psychrometer is available in the laboratory, your instructor will explain the procedure for using it. FOLLOW THE SPECIFIC DIRECTIONS OF YOUR INSTRUCTOR to complete question 27.

27. Use the psychrometer to determine the relative humidity and dew-point temperature of the air in the room and outside the building. Record your information in the following spaces:

	ROOM	OUTSIDE
Dry-bulb temperature	_____	_____
Wet-bulb temperature	_____	_____
Difference between dry- and wet-bulb temperatures	_____	_____
Relative humidity	_____%	_____%
Dew-point temperature	_____	_____

Table 14.3 Relative Humidity (percentage)*

DRY-BULB TEMPERA-TURE (°C)	DEPRESSION OF WET-BULB TEMPERATURE (°C) (DRY-BULB TEMPERATURE – WET-BULB TEMPERATURE = DEPRESSION OF THE WET BULB)																				
	1	2	3	4	5	6	7	8	9	10	11	12	13	14	15	16	17	18	19	20	21	22
−20	28																					
−18	40																					
−16	48	0																				
−14	55	11																				
−12	61	23																				
−10	66	33	0																			
−8	71	41	13																			
−6	73	48	20	0																		
−4	77	54	43	11																		
−2	79	58	37	20	1																	
0	81	63	45	28	11																	
2	83	67	51	36	20	6																
4	85	70	56	42	27	14																
6	86	72	59	46	35	22	10	0														
8	87	74	62	51	39	28	17	6														
10	88	76	65	54	43	33	24	13	4													
12	88	78	67	57	48	38	28	19	10	2												
14	89	79	69	60	50	41	33	25	16	8	1											
16	90	80	71	62	54	45	37	29	21	14	7	1										
18	91	81	72	64	56	48	40	33	26	19	12	6	0									
20	91	82	74	66	58	51	44	36	30	23	17	11	5	0								
22	92	83	75	68	60	53	46	40	33	27	21	15	10	4	0							
24	92	84	76	69	62	55	49	42	36	30	25	20	14	9	4	0						
26	92	85	77	70	64	57	51	45	39	34	28	23	18	13	9	5						
28	93	86	78	71	65	59	53	47	42	36	31	26	21	17	12	8	2					
30	93	86	79	72	66	61	55	49	44	39	34	29	25	20	16	12	8	4				
32	93	86	80	73	68	62	56	51	46	41	36	32	27	22	19	14	11	8	4			
34	93	86	81	74	69	63	58	52	48	43	38	34	30	26	22	18	14	11	8	5		
36	94	87	81	75	69	64	59	54	50	44	40	36	32	28	24	21	17	13	10	7	4	
38	94	87	82	76	70	66	60	55	51	46	42	38	34	30	26	23	20	16	13	10	7	
40	94	89	82	76	71	67	61	57	52	48	44	40	36	33	29	25	22	19	16	13	10	7

Relative Humidity Values

(Dry-bulb (Air) Temperature — vertical axis label)

*To determine the relative humidity and dew point, find the air (dry-bulb) temperature on the vertical axis (far left) and the depression of the wet bulb on the horizontal axis (top). Where the two meet, the relative humidity or dew point is found. For example, use a dry-bulb temperature of 20°C and a wet-bulb temperature of 14°C. From Table 14.3, the relative humidity is 51%, and from Table 14.4, the dew point is 10°C.

Condensation

If air is cooled below its dew-point temperature, water vapor will *condense* (change to liquid) if the dew point is above freezing. *Deposition* (gas to solid) occurs if the dew point temperature is below freezing. In the atmosphere, the tiny particles on which water vapor condenses are called **condensation nuclei**. Condensation (or deposition) may result in the formation of dew or frost on the ground and clouds or fog in the atmosphere.

28. Examine the process of condensation by filling a beaker approximately one-third full of water and then gradually adding ice. As you add the ice, stir the water–ice mixture gently with a thermometer. Note the temperature when water begins to condense on the outside surface of the beaker. After you complete your observations, answer questions 28a and 28b.

a. The temperature at which water began condensing on the outside surface of the beaker was _____.

b. How does the temperature at which water began to condense compare to the dew-point temperature of the air in the room, which you determined in question 27 using the psychrometer?

Table 14.4 Dew-Point Temperature (°C)*

DRY-BULB TEMPERATURE (°C)	DEPRESSION OF WET-BULB TEMPERATURE (°C) (DRY-BULB TEMPERATURE − WET-BULB TEMPERATURE = DEPRESSION OF THE WET BULB)																					
	1	2	3	4	5	6	7	8	9	10	11	12	13	14	15	16	17	18	19	20	21	22
−20	−33																					
	−28																					
	−24																					
	−21	−36																				
	−14	−22																				
	−14	−22																				
	−12	−18	−29																			
	−10	−14	−22																			
	−7	−22	−17	−29																		
	−5	−8	−13	−20																		
0	−3	−6	−9	−15	−24																	
2	−1	−3	−6	−11	−17																	
4	1	−1	−4	−7	−11	−19																
6	4	1	−1	−4	−7	−13	−21															
8	6	3	1	−2	−5	−9	−14															
10	8	6	4	1	−2	−5	−9	−14														
12	10	8	6	4	1	−2	−5	−9	−16													
14	12	11	9	6	4	1	−2	−5	−10	−17												
16	14	13	11	9	7	4	1	−1	−6	−10	−17											
18	16	15	13	11	9	7	4	2	−2	−5	−10	−19										
20	19	17	15	14	12	10	7	4	2	−2	−5	−10	+19									
22	21	19	17	16	14	12	10	8	5	3	−1	−5	−10	−19								
24	23	21	20	18	16	14	12	10	8	6	2	−1	−5	−10	−18							
26	25	23	22	20	18	17	15	13	11	9	6	3	0	−4	−9	−18						
28	27	25	24	22	21	19	17	16	14	11	9	7	4	1	−3	−9	−16					
30	29	27	26	24	23	21	19	18	16	14	12	10	8	5	1	−2	−8	−15				
32	31	29	28	27	25	24	22	21	19	17	15	13	11	8	5	2	−2	−7	−14			
34	33	31	30	29	27	26	24	23	21	20	18	16	14	12	9	6	3	−1	−5	−12	−29	
36	35	33	32	31	29	28	27	25	24	22	20	19	17	15	13	10	7	4	0	−4	−10	
38	37	35	34	33	32	30	29	28	26	25	23	21	19	17	15	13	11	8	5	1	−3	−9
40	39	37	36	35	34	32	31	30	28	27	25	24	22	20	18	16	14	12	9	6	2	−2

*See footnote to Table 14.3.

29. Refer to Table 14.2. How many grams of water vapor will condense on a surface if 1 kilogram of 50°F air with a relative humidity of 100% is cooled to 41°F?

_____ grams of water will condense.

30. Assume a situation in which 1 kilogram of 25°C air contains 10 grams of water vapor. Refer to Table 14.2 and determine how many grams of water will condense if the air temperature is lowered to each of the following temperatures.

5°C: _____ grams of condensed water

−10°C: _____ grams of condensed water

Daily Temperature and Relative Humidity

Figure 14.7 shows a graph of typical daily variations in air temperature, relative humidity, and dew-point temperature during two consecutive spring days in a middle-latitude city. Use the graph to answer questions 31–35.

31. On which day and at what time is relative humidity at its maximum?

Day: __1____, time: __U AM_____

32. On which day and at what time does the lowest temperature occur?

Day: ___1___, time: __U AM_____

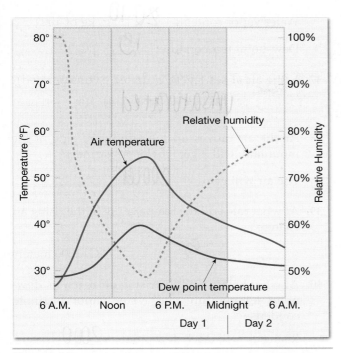

Figure 14.7 Typical variations in air temperatures, relative humidity, and dew-point temperature during two consecutive spring days in a middle-latitude city.

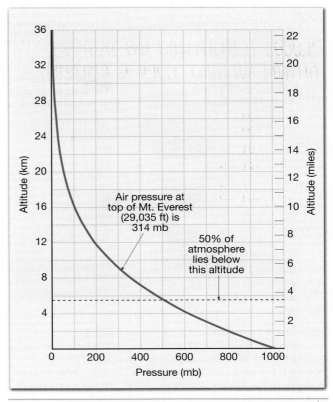

Figure 14.8 Atmospheric pressure variation with altitude. The rate of pressure decrease with an increase in altitude is not constant. Pressure decreases rapidly near Earth's surface and more gradually at greater heights.

33. On which day and at what time does the lowest relative humidity occur?

Day: _____1_____, time: ___4:00 PM___

34. Write a general statement describing the relationship between temperature and relative humidity throughout the time period shown in the figure.

When the temperature is higher the relative humidity is lower and vice versa.

35. Did dew or frost occur on either of the two days represented in the graph? If so, indicate the day and time and explain how you arrived at your answer.

6 am Yes because the dew point temperature was below 30°F in the morning

Adiabatic Processes

Cloud formation occurs above Earth's surface when air is cooled below its dew-point temperature. The process responsible for most cloud formation is easy to visualize. You have probably experienced the cooling effect of a propellant gas as you applied hair spray or spray deodorant. As the compressed gas in the can is released, it quickly expands and cools. This drop in temperature occurs even though heat is neither added nor subtracted. Such changes are known as **adiabatic temperature changes** and result when air is compressed or allowed to expand. *When air is allowed to expand, it cools, and when it is compressed, it warms.*

As Figure 14.8 shows, air pressure drops rapidly above Earth's surface. Thus, anytime air ascends, it passes through regions of successively lower pressure. Therefore, ascending air expands and cools adiabatically. Rising unsaturated air cools at the constant rate of 10°C per 1000 meters (1°C per 100 meters). This is called the **dry adiabatic rate**. The dry adiabatic rate also applies when unsaturated air descends and is compressed and warmed.

Once rising air reaches the dew-point temperature and condensation occurs, latent heat that was stored in the water vapor is liberated. The heat released by the condensing water slows down the rate of cooling of the air. Rising *saturated* air will continue to cool by expansion, but at a lesser rate of about 5°C per 1000 meters (0.5°C per 100 meters)—the **wet adiabatic rate**. Figure 14.9 summarizes this cloud-forming process.

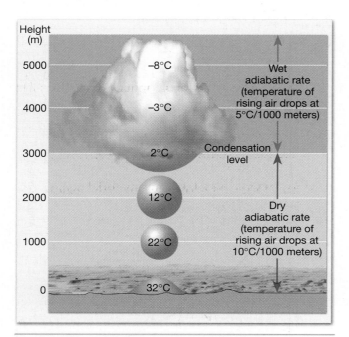

Figure 14.9 Adiabatic cooling and cloud formation. Rising air cools at the dry adiabatic rate of 10°C per 1000 meters until the air reaches the dew-point temperature and condensation (cloud formation) begins. As air continues to rise, the latent heat released by condensation reduces the rate of cooling. The wet adiabatic rate is therefore always less than the dry adiabatic rate.

Figure 14.10 shows air that begins with a temperature of 25° and a relative humidity of 50% flowing from the ocean over a coastal mountain range. Assume that the dew-point temperature remains constant in dry air (relative humidity less than 100%). If the air parcel becomes saturated, the dew-point temperature will cool at the wet adiabatic rate as it ascends, but will not change as the air parcel descends. Use this information to complete the following.

36. Use Table 14.2 to determine the saturation mixing ratio, water vapor content, and dew-point temperature of the air at sea level.

Saturation mixing ratio: _____20_____ g/kg of air

Water vapor content: ___~~20~~ 10___ g/kg of air

Dew-point temperature: ___15___ °C

37. Is the air at sea level (saturated *or* unsaturated)?
___unsaturated___

38. As the air moves up the windward side of the mountain, will it get (cooler *or* warmer)?

The air will get ___cooler___.

39. At what rate will the temperature of the rising air change?

___5°___ °C/1000 meters

40. At what height will the rising air reach its dew-point temperature and water vapor begin to condense?

Rising air will reach dew point at ___2000___ meters.

41. From the altitude where condensation begins until the air reaches the top of the mountain, the rising air will continue to expand and will cool at a rate of about ___5___ °C per 1000 meters. What term is applied to this rate of cooling?
___Wet Adiabatic Rate___

42. What will be the air temperature and dew-point temperature of the rising air at the summit of the mountain?

Temperature: ___10___ °C

Dew point: ___15___ °C

43. As the air descends the leeward side of the mountain will it (expand *or* be compressed)?

The air will ___expand___.

$$0.50 = \frac{X}{20}$$

$$Rel.\ Hum. = \frac{X}{14} \quad \frac{14}{20}$$

$$\frac{10}{20}$$

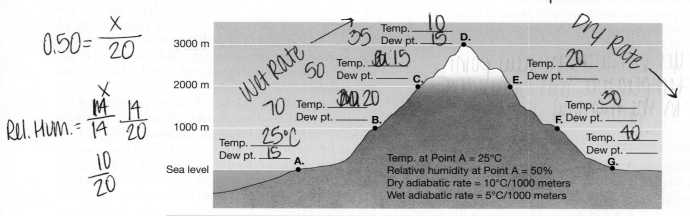

Figure 14.10 Adiabatic processes associated with a mountain barrier.

44. Will the temperature of the descending air (increase *or* decrease)?

    The temperature will ___increase___.

45. Assume that the relative humidity of the air is below 100% during its descent to the valley. At what rate will the temperature change?

    Temperature will change at ___10___ °C/1000 m.

46. As the air descends and warms on the leeward side of the mountain, will its relative humidity (increase *or* decrease)?

    Relative humidity will ___increase___

47. What will be the air temperature and dew point when it reaches the valley?

    Temperature: ___40___ °C

    Dew point: ___15___ °C

48. Explain why mountains might cause dry conditions on their leeward sides.

    ___Because they create rain shadows as a result of condensation and air being forced upward because of the mountain.___

Global Patterns of Precipitation

Use Figure 14.11 to answer questions 49–53.

49. List at least four of the areas of the world that receive the greatest average annual (over 200 cm/yr) precipitation.

    ___South America, Islands in Pacific Ocean, small areas in Africa Madagascar___

50. Are precipitation totals in equatorial regions generally (high *or* low)?

    Precipitation totals are generally ___low___.

51. What is the average annual precipitation at your location?

    ___60-90___ centimeters per year, which is equivalent to ___20-39___ inches per year

52. Describe the pattern of average annual precipitation in North America.

    ___It varies by region. The coastal states appear to get more precipitation and there is a dry region near the western coast.___

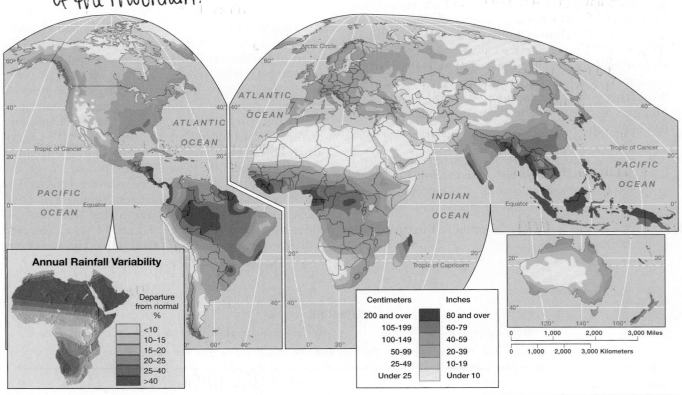

Figure 14.11 Average annual precipitation, in centimeters and inches.

53. The inset (left) in Figure 14.11 illustrates the annual variability, or reliability, of annual precipitation in Africa. After you compare the inset to the annual precipitation map for Africa, summarize the relationship between the amount of precipitation an area receives and the variability or reliability of that precipitation.

There is more variability in dry regions and less variability in "wetter" areas that get more precipitation.

Air Pressure and Wind

Atmospheric pressure and wind are two elements of weather that are closely interrelated. Although people usually overlook pressure in weather reports, the pressure differences in the atmosphere drive the winds that often bring changes in temperature and moisture.

Atmospheric Pressure

Atmospheric pressure is the force exerted by the weight of the atmosphere. The instrument used to determine atmospheric pressure is the **barometer** (Figure 14.12). Two units that are frequently used to express air pressure are *inches of mercury* and *millibars.* Inches of mercury refers to the height to which a column of mercury will rise in a glass tube that has been inverted into a reservoir of mercury. The millibar is a unit that expresses the actual force of the atmosphere pushing down on a surface. Standard pressure at sea level is 29.92 inches of mercury, or 1013.2 millibars (Figure 14.13). Since surface elevations vary, barometric readings are usually adjusted to indicate what the pressure would be if the barometer were located at sea level. This provides a common standard for mapping pressure distribution and comparing places, regardless of elevation.

54. When calculating the barometric pressure for a city located 200 meters above sea level, would you (add *or* subtract) units to its barometric reading in order to correct its pressure to sea level?

You would ___*add*___ units to the barometric reading.

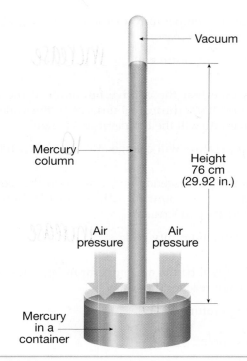

Figure 14.12 Simple mercury barometer. The weight of the column of mercury is balanced by the pressure exerted on the dish of mercury by the air above. If the pressure decreases, the column of mercury falls; if the pressure increases, the column rises.

Figure 14.14 is a sketch of a smaller, more portable, instrument for measuring air pressure—the **aneroid barometer**. The face of an aneroid barometer is inscribed with words such as *fair, change, rain,* and *stormy.* Notice that "fair weather" corresponds with high-pressure readings, and "rain" is associated with low pressures. Although current barometer readings may indicate the present weather, this is not always the case. If you want to "predict" the weather in a local area, *the change in air pressure* over the past few hours (known as the *pressure tendency*) is much more useful than the current pressure reading. Most aneroid barometers have a dial that you can move (shown in yellow on Figure 14.14) to coincide with the current reading. Later, when you check the

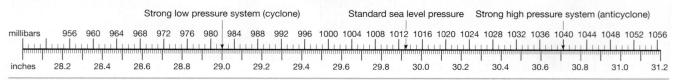

PRESSURE

Figure 14.13 Scale for comparing pressure readings in millibars and inches of mercury. (After NOAA)

Figure 14.14 An aneroid barometer has a partially evacuated chamber that changes shape, compressing as atmospheric pressure increases and expanding as pressure decreases.

instrument, you can see if the pressure has gone up or down. Falling pressure is often associated with increasing cloudiness and the possibility of precipitation, whereas rising air pressure generally indicates clearing conditions.

55. Refer to Figure 14.14.

   a. What is the current reading (black dial) on this instrument?

   _30.19_ inches of mercury, or _1022.5_ millibars

   b. Is the pressure tendency rising or falling?

   Falling

Factors Affecting Wind: The Pressure Gradient

Wind is simply horizontal movement of air that results from horizontal differences in air pressure. *Air flows from areas of higher pressure toward areas of lower pressure. The greater the pressure differences, the stronger the wind speed.*

Over Earth's surface, variations in air pressure are determined from measurements taken at hundreds of weather stations. These data are plotted on maps using **isobars**—lines that connect places of equal air pressure (Figure 14.15). The spacing of isobars indicates the amount of pressure change occurring over a given distance and is termed the **pressure gradient**. *Closely spaced isobars indicate a steep pressure gradient and strong winds, whereas widely spaced isobars indicate a weak pressure gradient and light winds.*

Figure 14.15 is a weather map that shows only the distribution of air pressure. Notice that isobars

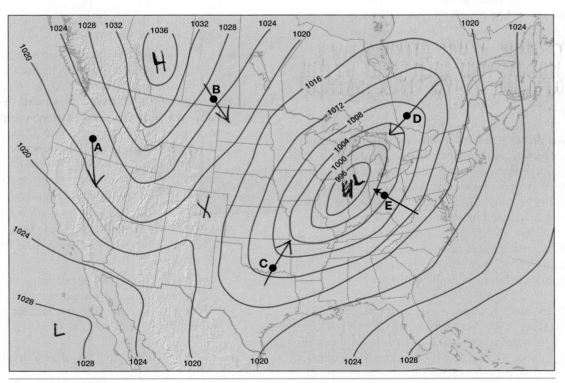

Figure 14.15 Weather map with isobars, for use with questions 56–60.

are seldom straight but usually form broad curves. Concentric rings of isobars are also common. If each succeeding isobar *decreases* in value going toward the center, the isobars are depicting a *low-pressure center*, or **cyclone**, usually labeled at the center with a large letter *L*. If the value of isobars *increases* toward the center, the concentric isobars are depicting a *high-pressure center*, or **anticyclone**, labeled at the center with a large letter *H*.

Refer to Figure 14.15 to complete questions 56–60.

56. At what interval are isobars drawn on this map? That is, what is the difference in millibars between each succeeding isobar?

The interval is ___4___ millibars.

57. The map has two large areas of concentric isobars. Label the center of each as either a low-pressure center (*L*) or a high-pressure center (*H*).

58. Assume for the moment that the pressure gradient is the *only* factor affecting winds. Draw arrows through Points A, B, C, and D that represent the wind direction. An arrow has already been drawn at Point E.

59. Which of the five stations (A–E) likely has the strongest winds? How were you able to figure this out?

___D - it is the closest to the___
___highest pressure center.___

60. Pick a spot on the map where winds should be weak. Mark the place with an *X*. Why did you select that place?

___Because there aren't any___
___isobars near each other in this___
___area like in the other examples___

Factors Affecting Wind: The Coriolis Effect

In the preceding section, you learned that the pressure gradient creates winds and strongly affects wind speed and direction. Another factor affecting wind is the **Coriolis effect**—the deflective effect of Earth's rotation. *All free-moving objects, including wind, are deflected to the right of their path of motion in the Northern Hemisphere and to the left in the Southern Hemisphere.*

To better understand how the Coriolis effect influences the motion of objects as they move across Earth's surface, conduct the following experiment:

Step 1: Working in groups of two or more, construct a rotating "table" that represents Earth's surface by first taping a thumbtack upside down on the table top (or inserting it through a piece of cardboard). Next, center a sheet of heavy-weight

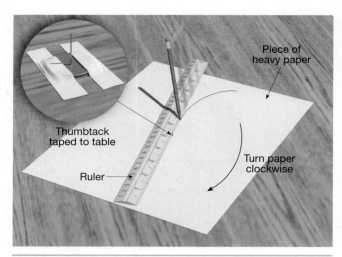

Figure 14.16 Coriolis effect experiment setup—Southern Hemisphere.

paper or thin cardboard over the point of the tack and push the sharp end of the tack through the paper (Figure 14.16). Turning the paper about the thumbtack represents the rotating Earth viewed from above the pole.

Step 2: Place a straight edge (or a 12-inch ruler) across the paper, resting it on the sharp point of the thumbtack.

Step 3: Using the straight edge as a guide and beginning at the edge nearest you, draw a straight line across the entire piece of paper. Mark the beginning of the line you drew with an arrow pointing in the direction the pencil moved. Label the line "no rotation."

Step 4: Have one person hold the straight edge and another turn the paper *counterclockwise* (the direction the Earth rotates when viewed from above the North Pole) about the thumbtack. Next, while spinning the paper at a slow, constant rate, draw a second line along the ruler, using a different-color pencil. Mark the beginning of the line you drew with an arrow pointing in the direction the pencil moved. Label the resulting line "Northern Hemisphere."

Step 5: Repeat Step 4, but this time rotate the paper *clockwise* (representing the Southern Hemisphere). Label the resulting line "Southern Hemisphere."

Step 6: Repeat Step 4 three more times, varying the speed of rotation of the paper with each trial. Identify each new line by labeling it either "slow," "fast," or "very fast."

Questions 61–63 refer to the Coriolis effect experiment.

61. Describe the apparent path of a free-moving object over Earth's surface on a "nonrotating" Earth, in the Northern Hemisphere (counterclockwise rotation), and in the Southern Hemisphere (clockwise rotation).

_____

_____

_____

62. In the Northern Hemisphere, is the deflection of objects to the (right *or* left)? In the Southern Hemisphere, is the deflection to the (right *or* left)?

Northern Hemisphere: _____

Southern Hemisphere: _____

63. Summarize your observations regarding the rate of rotation and the magnitude of the Coriolis effect you observed in Step 6.

_____

_____

_____

_____

From your observations in Step 6, you would suspect that the deflection in the path of a free-moving object would be maximized near the equator, where Earth's rotational velocity is greatest. However, this is *not* the case. Vector motion on the surface of a sphere is complex; however, the Coriolis effect is controlled primarily by rotation about a *vertical* axis. In your experiment, all points on the paper were rotating about a vertical axis, the thumbtack. However, the flat paper does not represent the "real" spherical Earth. On Earth, the vertical axis of rotation is a line connecting the geographic poles that is vertical to Earth's surface only at each pole. On the Equator, the axis is *parallel* to the surface; therefore, there is no rotation about a vertical axis. As a consequence, *the Coriolis effect is strongest at the poles and weakens equatorward, becoming nonexistent at the equator.* (To help visualize this changing orientation, obtain a globe and hold a pencil parallel to Earth's axis at the North Pole. Notice that the axis is vertical to the surface, just as in the experiment. Now, *while keeping the pencil parallel to Earth's axis of rotation,* slowly move the pencil over the surface to the equator. Notice how the orientation of the pencil changes relative to the surface as it is moved. At the equator, the pencil is now parallel to the surface, and rotation about the pencil

is directed toward and away from Earth's center, not over the surface.)

64. Considering what you have learned about the Coriolis effect, briefly describe the Coriolis effect on the atmosphere of the planet Venus, which is about the same size as Earth but has a period of rotation of 244 Earth days. What would be the Coriolis effect on Jupiter, a planet much larger than Earth, with a 10-hour period of rotation?

_____

_____

_____

_____

65. To summarize, examine surface wind directions associated with high- and low-pressure cells in both hemispheres by combining your understanding of pressure gradient and Coriolis effect. Refer to Figure 14.17. The concentric circles represent isobars. Add arrows to show the winds. The diagram for a low-pressure cell in the Northern Hemisphere is already completed. Complete the

Northern Hemisphere

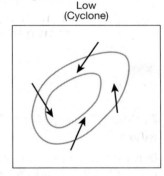

Southern Hemisphere

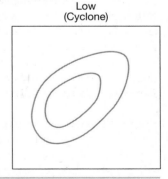

Figure 14.17 Northern and Southern Hemisphere pressure cells.

other three pressure cells with arrows that show the wind pattern.

66. In the following spaces, indicate the movements of air in high- and low-pressure cells for each hemisphere. Write one of the two choices given in italics for each blank.

NORTHERN HEMISPHERE

	HIGH	LOW
Surface air moves *into* or *out of*	_____	_____
Surface air motion is *clockwise* or *counterclockwise*	_____	_____

SOUTHERN HEMISPHERE

	HIGH	LOW
Surface air moves *into* or *out of*	_____	_____
Surface air motion is *clockwise* or *counterclockwise*	_____	_____

67. Briefly describe the difference in surface air movement between an anticyclone in the Northern Hemisphere and an anticyclone in the Southern Hemisphere.

_____

_____

_____

Companion Website

The companion website provides numerous opportunities to explore and reinforce the topics of this lab exercise. To access this useful tool, follow these steps:

1. Go to www.mygeoscienceplace.com.
2. Click on "Books Available" at the top of the page.
3. Click on the cover of *Applications and Investigations in Earth Science, 7e.*
4. Select the chapter you want to access. Options are listed in the left column ("Introduction," "Web-based Activities," "Related Websites," and "Field Trips").

Atmospheric Moisture, Pressure, and Wind

Name _____   **Course/Section** _____

Date _____   **Due Date** _____

1. Answer the following by circling the correct response.
   a. Liquid water changes to water vapor through a process called (condensation, evaporation, *or* deposition).
   b. (Warm *or* Cold) air has the greatest saturation mixing ratio.
   c. Lowering the air temperature will (increase *or* decrease) the relative humidity.
   d. At the dew-point temperature, the relative humidity is (25%, 50%, 75%, *or* 100%).
   e. When condensation occurs, heat is (absorbed *or* released) by water vapor.
   f. Rising air (warms *or* cools) by (expansion *or* compression).
   g. In the early morning hours, when the daily air temperature is often coolest, relative humidity is generally at its (lowest *or* highest).

2. What is the dew-point temperature of air when a psychrometer measures an 8°C dry-bulb temperature and a 6°C wet-bulb reading?

   Dew-point temperature = _____°C

3. Explain the principle that governs the operation of a psychrometer for determining relative humidity.

   _____
   _____
   _____

4. Describe the adiabatic process and how it is responsible for causing clouds to form.

   _____
   _____
   _____

5. Refer to question 41 in the section "Adiabatic Processes." What was the altitude where condensation occurred as the air was rising over the mountain?

   _____ meters

6. Number the following statements about development of clouds in proper sequence (1 for the first).

   _____: Dew-point temperature reached.

   _____: Air begins to rise.

   _____: Condensation occurs.

   _____: Adiabatic cooling.

7. Assume a parcel of air on the surface has a temperature of 29°C and a relative humidity of 50%. If the parcel rises, at what altitude should clouds begin to form?

   _____

8. Describe the Coriolis effect and its influence in both the Northern and Southern Hemispheres.

   _____
   _____
   _____

9. On Figure 14.18, use arrows to indicate the winds associated with each pressure cell.

10. Complete Figure 14.19 by using the indicated surface barometric pressures to assist in drawing appropriate isobars. Begin with the 988 mb isobar and use a 4 mb interval between successive isobars.

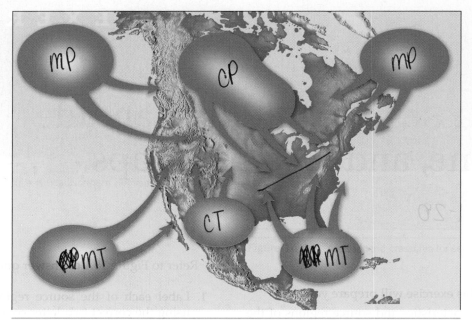

Figure 15.1 Source regions of North American air masses.

an active cold front overtakes a warm front, the advancing cold air wedges the warm front upward, creating an **occluded front**.

5. On Figure 15.1, draw a line where the cP and mT air masses are likely to collide and as a result where fronts develop. Where does this boundary occur?

   ____Midwest____

6. In the central United States, east of the Rocky Mountains, a (cP *or* mT) air mass will most likely be found north of a front, and a (cP *or* mT) mass to the south.

   Air mass north of front: ____CP____

   Air mass south of front: ____CT____

Figure 15.2 shows profiles of typical cold and warm fronts. Refer to the diagrams in this figure to answer questions 7–14.

7. Which diagram (A *or* B) represents a *warm* front? Which diagram (A *or* B) represents a *cold* front? Write you answer here and label the fronts on the diagrams.

   Warm front: ____B____

   Cold front: ____A____

8. Sketch and label the symbols for cold and warm fronts in the space below. (Hint: See Figure 15.2)

   cold   warm

9. The cold air is the aggressive, or "pushing," air along the (cold *or* warm) front.

   Cold air is aggressive along a ____cold____ front.

10. Along a (cold *or* warm) front, warm air rises at the steepest angle.

    Warm air rises at the steepest angle along a ____~~warm~~ cold____ front.

11. Along which front (cold *or* warm) are extensive areas of stratus clouds and periods of prolonged precipitation most likely? Explain why you expect longer periods of precipitation to be associated with this type of front.

    Front: A warm front because the warm air is rising more gradually where in a cold front it rises rapidly.

12. Assume that the fronts are moving from left to right in Figure 15.2. A drop in temperature is most likely to occur with the passing of a (cold *or* warm) front.

    Front: ____warm____

13. Clouds of vertical development and perhaps thunderstorms are most likely along a (cold *or* warm) front.

    Front: ____cold____

14. As a (cold *or* warm) front approaches, clouds gradually become lower, thicker, and cover more of the sky.

    Front: ____cold____

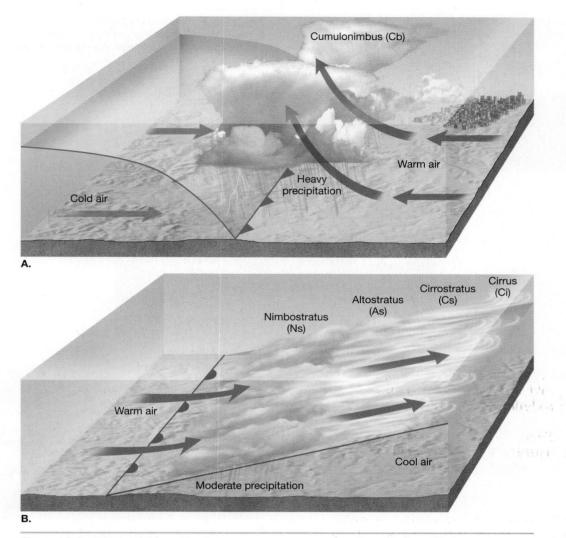

Figure 15.2 Profiles of different fronts.

15. Figure 15.3 is a cross-section (side view) showing the development of an occluded front. Based on this diagram, the (cold or warm) front is advancing most rapidly.

    Front: _Cold_

16. The symbol for an occluded front is a *combination* of the cold and warm front symbols. Sketch the occluded front symbol below.

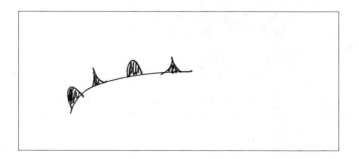

Middle-Latitude Cyclones

To understand day-to-day weather patterns in many parts of North America, it is useful to have an understanding of **middle-latitude cyclones**, also called **mid-latitude cyclones**—large centers of low pressure that generally travel from west to east and last from a few days to more than a week.

Figure 15.4 depicts an idealized mature middle-latitude cyclone. Refer to this figure to complete questions 17–26.

17. On Figure 15.4:

    a. Label the cold front, warm front, and occluded front.

    b. Draw arrows showing the surface wind directions at Points A, C, E, F, and G.

    c. Label the areas most likely experiencing precipitation with the word *precipitation*.

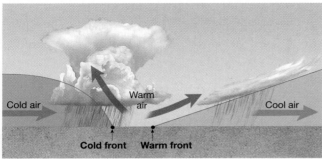

A.

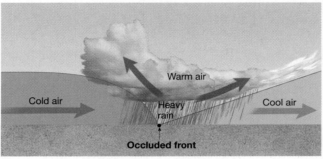

B.

C.

Figure 15.3 Stages in the formation of an occluded front. In this example, the air behind the cold front is colder (denser) than the cool air ahead of the warm front. As a result, the cold front wedges the warm front aloft.

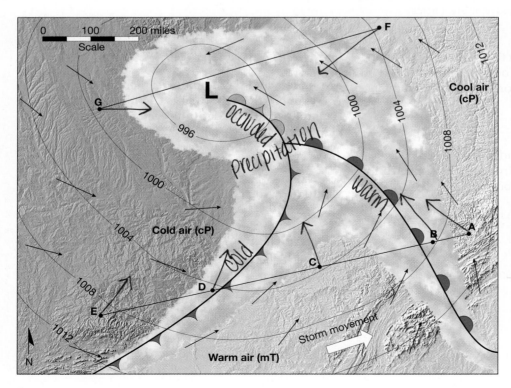

Figure 15.4 Idealized drawing of a mature, middle-latitude cyclone showing, isobars, wind direction, and fronts.

18. Are the surface winds in the cyclone (converging *or* diverging)?

The surface winds are ___converging___.

19. Is the air in the center of the cyclone (subsiding *or* rising)? What effect will this have on the potential for condensation and precipitation? Explain your answer.

The air is ___subsiding because it is___ ___a low pressure system.___

20. As the cold front passes Point A, will the barometric pressure be (rising *or* falling)?

The pressure will be ___falling___.

21. As the warm front passes, in what direction will the wind shift at Point B?

Wind will shift from ___North East___ to ___North West___.

22. Describe the changes in wind direction and barometric pressure that will likely occur at Point D after the cold front passes.

___The pressure will rise and the wind___ ___direction will go Northeast.___

23. Considering the air mass types and their locations in a middle-latitude cyclone, will the amount of water vapor in the air most likely (increase *or* decrease) at Point A after the warm front passes?

Water vapor will ___decrease___.

24. Will the water vapor content of the air at Point D most likely (increase *or* decrease) after the cold front passes?

Water vapor content will ___increase___.

25. Use Figure 15.4 to describe the sequence of weather conditions—temperature, pressure tendency, wind direction, clouds, and precipitation—expected for a city as a middle-latitude cyclone moves and the city's position relative to the cyclone changes from location A to B, and then to C, D, and E. Fill in Table 15.1 below.

26. What type of front is found near the center of the low pressure?

___Occluded___ front

a. What happens to the warm (mT) air in this type of front? (*Hint*: See Figure 15.3.)

___The warm air rises___

b. Why is there a good chance for precipitation with this type of front?

___Because you have both a cold___ ___front and warm front.___

Weather Station Analysis and Forecasting

Weather Station Data

To manage the great quantity of data needed for accurate weather maps, meteorologists developed a system for coding weather data. Figure 15.5 shows the system and many of the symbols that are used to record data for a weather station. (*Note:* When plotting barometric pressure in millibars for a weather station, to conserve space, the initial number 9 or 10 is omitted and the last digit is tenths of a millibar. For example, on a map, a barometric pressure of 216 for a station would be read as 1021.6 mb. For numbers that begin with (0–4) add 10, for (5–9) add 9.)

Figure 15.6 is a coded weather station, shown as it would appear on a simplified surface weather map.

27. Using the specimen station model and explanations shown in Figure 15.5 as a guide, interpret the weather conditions reported at the station illustrated in Figure 15.6.

Percentage of sky cover: ___50___ %

Wind direction: ___From NE___

Wind speed: ___9-14___ mph

Temperature: ___50___ °F

Dew-point temperature: ___42___ °F

Table 15.1   **Sequence of Weather**

STATION	TEMPERATURE (WARM OR COLD)	PRESSURE TENDENCY (RISING, FALLING, STEADY)	WIND DIRECTION	CLOUD TYPE	PRECIPITATION (YES OR NO)
A	cold	steady	NW	none	no
B	warm	falling	NW	stratus	yes
C	warm	falling	N	both	yes
D	cold	falling	NE	nimbus	yes
E	cold	steady	NE	none	no

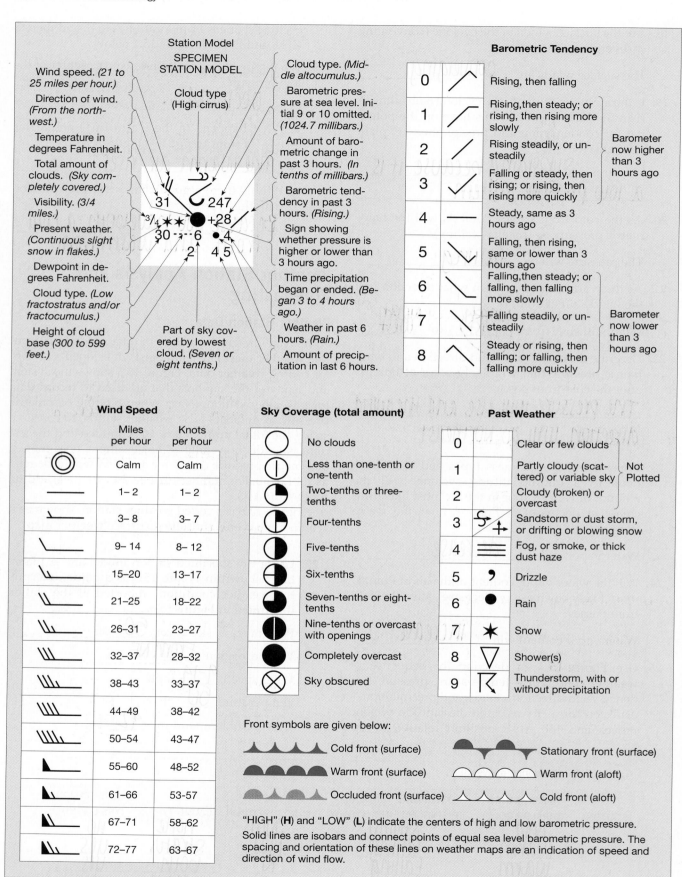

Figure 15.5 Specimen weather station model and standard symbols. (Source: National Weather Service)

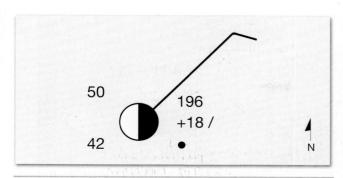

Figure 15.6   Coded weather station (abbreviated).

Barometric pressure: __196__ millibars

Barometric change in past three hours: __+18__ mb

Weather during the past six hours: __Rain__

28. Encode and plot the weather conditions for the following weather station on the station symbol shown below. Use Figures 15.5 and 15.6 as guides.

The sky is six-tenths covered by clouds. Air temperature is 82°F, with a dew point of 50°F. Wind is from the south, at 22 miles per hour. Barometric pressure is 1022.4 millibars and has fallen from 1023.0 millibars during the past three hours. There has been no precipitation during the past six hours.

Preparing a Weather Map and Forecast

Table 15.2 contains weather data for several cities in the central and eastern United States on a December day. Data for several of the cities have been plotted on the map in Figure 15.7. Use Table 15.2 and Figure 15.7 to complete questions 29 and 30.

29. Refer to stations where the data have already been plotted and the station model in Figure 15.5. Plot the data for the remaining stations (shown in

boldface) on the map. Then complete the following steps:

Step 1: Beginning with 988 millibars, draw and label isobars as accurately as possible at 4-mb intervals (992 mb, 996 mb, 1000 mb, etc.). (*Note:* You will have to estimate pressures between cities to determine the locations of isobars. Also, it may be a good idea to first lightly sketch the isobars in pencil.)

Step 2: Observe the data plotted on the map and determine the locations of the cold and warm fronts. Using the proper symbols, label the cold and warm fronts on the map.

Step 3: Label the air mass types that are likely located to the northwest and to the southeast of the cold front.

Step 4: Indicate areas of precipitation by lightly shading the map with a pencil.

30. Assume that the middle-latitude cyclone shown on the map is moving northeastward. Indicate your forecast (temperature, wind direction, probability of precipitation, cloud cover, and barometric pressure) for the next 12–24 hours at the following locations:

Chattanooga, Tennessee: _____

_____

_____

Little Rock, Arkansas: _____

_____

_____

Jackson, Mississippi: _____

_____

_____

Roanoke, Virginia: _____

_____

_____

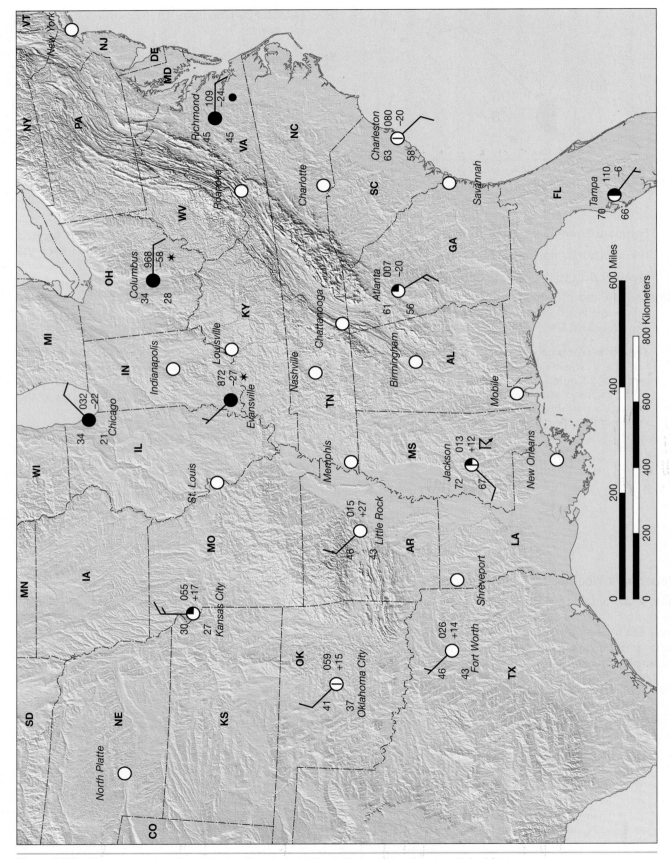

Figure 15.7 Weather map for a December day for selected cities in the central and eastern United States.

Table 15.2 December Surface Weather Data for Selected Cities in the Central and Eastern United States

STATION	% CLOUD COVER	WIND DIRECTION	WIND SPEED (MPH)	TEMP.	DEW PT. TEMP.	PRESSURE (MB)	PRESSURE LAST 3 HOURS (MB)	PRECIP.
Atlanta, GA	20	SE	18	61	56	1000.7	−2.0	
Birmingham, AL	80	SW	15	70	64	998.1	−1.4	
Charleston, SC	10	SE	12	63	58	1008.0	−2.0	
Charlotte, NC	70	SW	14	54	49	1003.4	−4.4	Drizzle
Chattanooga, TN	60	SW	12	66	60	995.3	−2.5	Drizzle
Chicago, IL	100	NE	13	34	21	1003.2	−2.2	
Columbus, OH	100	E	10	34	28	996.8	−5.8	Snow
Evansville, IN	100	NW	7	45	43	987.2	−2.7	Snow
Fort Worth, TX	0	NW	5	46	43	1002.6	+1.4	
Indianapolis, IN	100	NE	30	34	32	993.2	−5.6	**Snow**
Jackson, MS	40	SW	10	72	67	1001.3	+1.2	Thunderstorm
Kansas City, MO	30	N	18	30	27	1005.5	+1.7	
Little Rock, AR	0	NW	10	46	43	1001.5	+2.7	
Louisville, KY	100	E	12	34	34	993.0	−4.2	**Snow**
Memphis, TN	80	NW	12	50	45	996.7	+5.8	
Mobile, AL	60	SW	10	72	68	1004.3	−0.3	
Nashville, TN	100	SW	18	56	55	991.5	−0.1	Rain
New Orleans, LA	20	SW	11	75	70	1003.9	−0.1	
New York, NY	100	NE	23	36	18	1016.9	−2.1	
North Platte, NB	30	N	9	9	1	1017.4	+3.2	
Oklahoma City, OK	10	NW	13	41	37	1005.9	+1.5	
Richmond, VA	100	E	10	45	45	1010.9	−2.4	Rain
Roanoke, VA	100	SE	10	39	39	1007.5	−3.6	Snow
Savannah, GA	30	SE	7	61	55	1007.6	−2.0	
Shreveport, LA	0	NW	8	46	43	1002.6	+1.4	
St. Louis, MO	100	NW	10	32	32	999.7	+1.7	Showers
Tampa, FL	50	SE	8	70	66	1011.0	−0.6	

Companion Website

The companion website provides numerous opportunities to explore and reinforce the topics of this lab exercise. To access this useful tool, follow these steps:

1. Go to www.mygeoscienceplace.com.
2. Click on "Books Available" at the top of the page.

3. Click on the cover of *Applications and Investigations in Earth Science, 7e.*
4. Select the chapter you want to access. Options are listed in the left column ("Introduction," "Web-based Activities," "Related Websites," and "Field Trips").

Air Masses, the Middle-Latitude Cyclone, and Weather Maps

Name _____   **Course/Section** _____

Date _____   **Due Date** _____

1. List the source region(s) and winter temperature/moisture characteristics of each of the following North American air masses:

   cP: _____

   mT: _____

   mP: _____

2. In the following space, sketch a profile (side view) of a cold front. Label the cold air and warm air and sketch the probable cloud type at the appropriate location. Draw an arrow on the diagram to show the direction the front is moving.

   Cold Front Profile

3. Indicate the type of front (cold *or* warm) associated with each of the following statements.

   a. Steep wall of cold air: _____

   b. Warm air replaces cool air: _____

   c. Thunderstorms: _____

   d. Drop in temperature: _____

   e. After the front passes, winds come from a southerly direction: _____

   f. Narrow belt of precipitation: _____

   g. Gradual rise of warm air over cool air: _____

4. Describe the sequence of weather events that a city would experience as an idealized, mature, middle-latitude cyclone that has not developed an occluded front passes over it. Assume that the center of the cyclone is moving west to east and is 150 miles to the north of the city. Sketching a diagram may be helpful.

   _____

   _____

   _____

   _____

5. Based on the weather map that you constructed at the end of the exercise, from question 30, what was your forecast for Jackson, Mississippi?

   _____

   _____

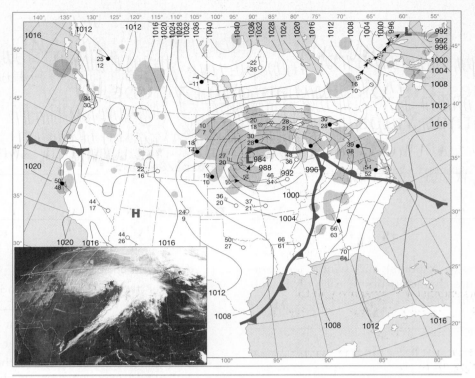

Figure 15.8 Surface weather map for a March day with a satellite image showing the cloud patterns on that day. The gray areas on the map are areas where precipitation is occurring. (Courtesy of NOAA/Seattle)

6. Use Figure 15.8 to a brief analysis of the weather in the following areas: The Great Plains, Upper Midwest, and the Southeast.

Great Plains: _____

_____

_____

Upper Midwest: _____

_____

_____

_____

Southeastern U.S.: _____

_____

_____

Global Climates

Questions 1-11
15-47

Objectives

Completion of this exercise will prepare you to:

A. Understand the nature of classification systems.

B. Read and prepare a climograph.

C. List the criteria used to define the five principal climate groups in the Köppen system of classification.

D. Describe the general location of the principal climate groups.

Materials

ruler   calculator
world map or atlas

Climate

The broad diversity of climates around the globe is the primary focus of this exercise. **Climate** is the long-term aggregate of weather. Average values of climate elements based on many years of data are an important part of a climate description. However, in order to accurately portray the character of a place or an area, variations and extremes are also considered. Figure 16.1 is a graph that shows daily temperature data for New York City. In addition to showing the average daily maximum and minimum temperatures for each month, it also shows extremes.

Refer to Figure 16.1 to answer questions 1–2.

1. **a.** What is the record high daily maximum for the entire year?

    _42_ °C (_107_ °F)

   b. How does the record high compare to the average high temperature for the same month?

    _29_ °C (_83_ °F) higher

2. What are the values of the record high temperature and the record low temperature for the month of January?

   Record January high: _22_ °C (_73_ °F)

   Record January low: _-19_ °C (_-2_ °F)

Examining Global Temperatures

Temperature is one of the essential elements in any description of climate. Complete questions 3–5, using the temperature data in Table 16.1.

3. Stations 1–3 in Table 16.1 are representative of three North American cities at approximately 40°N latitude. Choose which station represents each of the following locations and give the reason for your selection.

   Interior of the continent: _Station 1 because the temperatures are similar to the Midwest and they vary a lot seasonally, but more than we vary._

   West coast of the continent: _Station 2 because the temperatures are very consistent_

   East coast of the continent: _Station 3 because the temperatures vary but not by a huge amount._

4. Selecting from Stations 4–10, answer questions 4a–c.

   a. Station _8_ is in the Southern Hemisphere. How were you able to figure this out?

   Because the seasons are reversed.

   b. Identify two stations that must be very near the equator? How did you know?

   Station 7 and 8 because they have consistently high, hot temperatures, which are characteristics of countries on the equator.

Table 16.1   Monthly and Annual Temperature Data (°F)

STATION	J	F	M	A	M	J	J	A	S	O	N	D	ANNUAL MEAN
1	25	28	36	48	61	72	75	74	66	55	39	28	50
2	48	48	49	50	54	55	57	57	57	54	52	49	52
3	31	31	38	49	60	69	74	73	69	59	44	35	52
4	25	26	35	45	58	70	75	72	65	55	45	35	50
5	20	34	38	46	51	54	60	53	49	40	32	25	40
6	42	48	50	55	56	58	59	60	59	56	48	46	51
7	76	76	76	76	78	78	77	77	76	75	74	74	76
8	90	88	84	79	77	73	70	71	76	79	82	87	80
9	−2	11	25	36	48	54	70	60	50	36	18	10	34
10	59	58	59	59	60	58	60	59	60	58	59	60	59
11	70	69	69	66	61	57	56	60	66	70	71	70	66
12	−46	−35	−10	16	41	59	66	60	42	16	−21	−41	12
13	22	25	37	51	62	72	77	75	66	54	39	27	50
14	82	82	82	83	83	82	82	83	83	83	82	83	83
15	27	30	34	41	48	54	57	55	50	43	36	31	42
16	40	44	48	54	61	69	77	76	67	57	47	41	57

(handwritten annotations at left: "varies" by row 1, "Mod" by row 2, "High" by row 3)

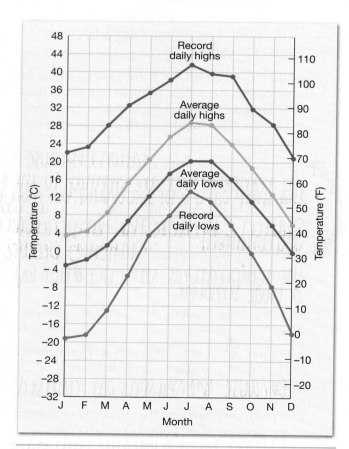

Figure 16.1 Graph showing daily temperature data for New York City. In addition to showing the average daily maximum and minimum temperatures for each month, the graph also shows extremes.

c. Which station must be a high-altitude location? Explain your answer. *it is [handwritten] Station 9 because they are still cold during July and it is in the northern hemisphere*

5. Selecting from Stations 11–16, answer questions 5a and 5b.

a. Which of these stations is located near the Arctic Circle in the interior of a large continent?

Station: _12_

b. Which of these stations is located about 40°S latitude near the coast?

Station: _14_

Using a Climograph

Temperature and precipitation are presented on a **climograph** such as the one shown in Figure 16.2. Monthly mean temperatures are connected with a single line and read from the temperature scale on the left side of the graph. Average precipitation for each month is presented as a bar graph and read from the precipitation scale on the right side of the graph.

Refer to the climograph in Figure 16.2 to answer questions 6–11.

6. What month receives the greatest amount of precipitation?

June

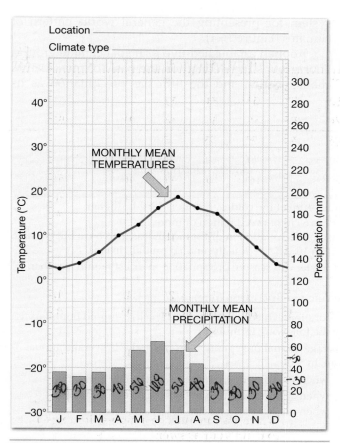

Location _____
Climate type _____

Figure 16.2 Typical climograph. Letters along the bottom represent the months. Average monthly temperatures (°C) are plotted with points, using the scale on the left axis. Precipitation for each month (in mm) is plotted with a bar, using the right axis.

7. What month has the lowest monthly mean temperature occur?
~~November~~ January

8. What month has the highest average temperature?
July

9. What is the *total* annual precipitation?
Annual precipitation is about ___517___ mm.

10. Is the place represented by the climograph in the (Northern Hemisphere *or* Southern Hemisphere)?
Hemisphere: ___Northern___

11. During what three-month span does this location receive the greatest amount of precipitation?
May – July

The Nature of Classification

Because weather is highly variable, no two places separated by even a short distance have precisely the same climate. To manage such variety, scientists have devised ways to classify the vast array of meteorological data available. By establishing groups of items that have common characteristics, order is introduced—which aids comprehension and understanding and facilitates analysis and explanation.

It is important to remember that the classification of climates (or anything else) is not a natural phenomenon but the product of human ingenuity.

Complete questions 12–14, using the climographs in Figure 16.3 found on page 235.

12. Working in groups of three to five students, develop the best possible classification scheme for the stations represented by the climographs in Figure 16.3 by arranging them into groups with similar characteristics. When you have finished, describe your classification system, listing the criteria you established. (*Note:* Cut Figure 16.3 into individual stations. This makes it easier to arrange them into groups.)

_____
_____
_____
_____
_____
_____
_____

13. Why is the classification your group devised better than other possible systems you considered?

_____
_____
_____
_____
_____
_____
_____

14. Compare the classification your group devised with those developed by two other groups. Which is the best classification scheme? Why?

_____
_____
_____
_____
_____
_____
_____
_____
_____
_____
_____

Köppen System of Climate Classification

Table 16.2 presents the climate classification system devised by Wladimir Köppen. Since its introduction, the **Köppen system**, with some modifications, has become the best-known and most-used classification system for presenting the general world pattern of climates (Figure 16.4).

Köppen established five principal climate groups. Four groups are defined on the basis of temperature characteristics, and the fifth has precipitation as its

Table 16.2 Köppen System of Climatic Classification*

NAME	LETTER SYMBOL			CHARACTERISTICS
	FIRST	SECOND	THIRD	
	A			Average temperature of the coldest month is 18°C or higher.
Humid Tropical (type A) Climates		f		Every month has 6 cm of precipitation or more.
		m		Short dry season; precipitation in driest month less than 6 cm but equal to or greater than $10 - R/25$ (R is the annual rainfall, in cm).
		w		Well-defined winter dry season; precipitation in driest month less than $10 - R/25$.
		s		Well-defined summer dry season (rare).
	B			Potential evaporation exceeds precipitation. The dry–humid boundary is defined by the following formulas: (*Note:* R is average annual precipitation, in cm, and T is average annual temperature, in °C) $R < 2T + 28$ when 70% or more of rain falls in warmer 6 months; $R < 2T$ when 70% or more of rain falls in cooler 6 months; $R < 2T + 14$ when neither half year has 70% or more of rain
Dry (type B) Climates		S		Steppe ⎤ The BS–BW boundary is 1/2 the dry–humid boundary.
		W		Desert ⎦
			h	Average annual temperature is 18°C or greater.
			k	Average annual temperature is less than 18°C.
	C			Average temperature of the coldest month is under 18°C and above −3°C.
Humid Middle-Latitude with Mild Winters (type C) Climates		w		At least 10 times as much precipitation in a summer month as in the driest winter month.
		s		At least three times as much precipitation in a winter month as in the driest summer month; precipitation in driest summer month less than 4 cm.
		f		Criteria for w and s cannot be met.
			a	Warmest month is over 22°C; at least four months over 10°C.
			b	No month above 22°C; at least four months over 10°C.
			c	One to three months above 10°C.
	D			Average temperature of coldest month is −3°C or below; average temperature of warmest month is greater than 10°C.
Humid Middle-Latitude with Severe Winters (type D) Climates		s		Same as under C climates.
		w		Same as under C climates.
		f		Same as under C climates.
			a	Same as under C climates.
			b	Same as under C climates.
			c	Same as under C climates.
			d	Average temperature of the coldest month is −38°C or below.
	E			Average temperature of the warmest month is below 10°C.
Polar (type E) Climates		T		Average temperature of the warmest month is greater than 0°C and less than 10°C.
		F		Average temperature of the warmest month is 0°C or below.

*When classifying climate data using this table, you should first determine whether the data meet the criteria for the E climates. If the station is not a polar climate, proceed to the criteria for B climates. If your data do not fit either the E or B groups, check the data against the criteria for A, C, and D climates, in that order.

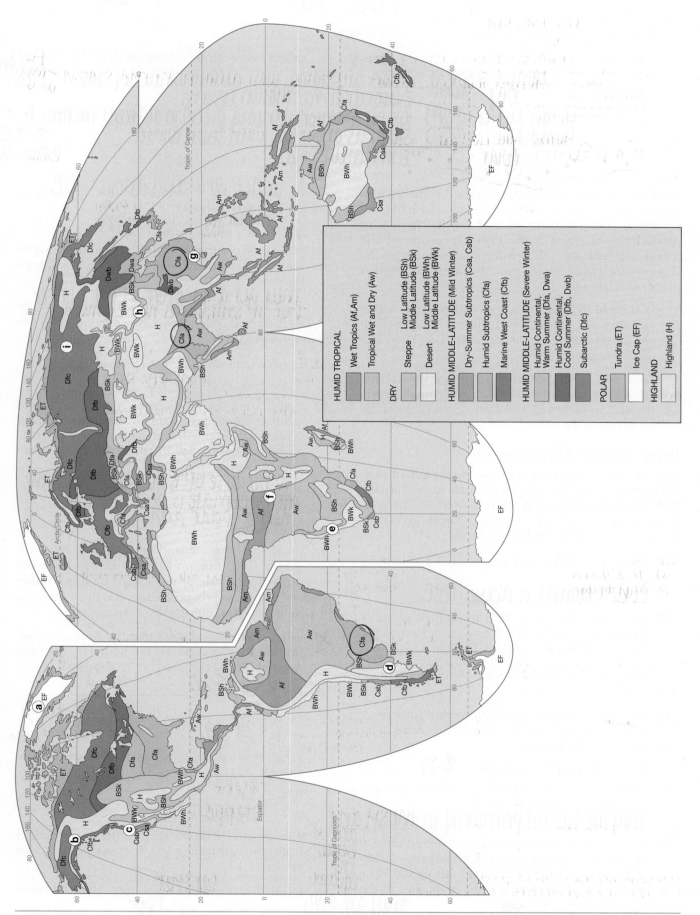

Figure 16.4 Climates of the world (Köppen system). (Adapted from E. Willard Miller, *Physical Geography*, Columbis, Ohio, Macmillan/Merrill, 1985. Plate 2)

$$\frac{2619 \text{ mm}}{1} \times \frac{0.1 \text{ cm}}{1 \text{ mm}} \times \frac{0.39 \text{ in}}{1 \text{ cm}} = 102.141 \quad \frac{102 \text{ in}}{1} \times \frac{1 \text{ ft}}{12 \text{ in}} = 8.5$$

Table 16.3 Characteristics of the Principal Climate Groups

CLIMATE GROUP	NAME	TEMPERATURE AND/OR PRECIPITATION CHARACTERISTICS	
A:	Humid Tropical	Short dry season, well defined winter dry season	Every month has at least 6 cm
B:	Dry	Evaporation > Precipitation, Deserts,	
C:	Humid Mid-Lat (mild)	10x precipitation varying precip in different months	18→-3
D:	Humid Mid-Lat (sev)	Same as above but with severe winter	-3→10
E:	Polar	Extremely cold	Below 10°C

primary criterion. Further division of the groups into climate types allows for more detailed climate descriptions. Köppen believed that the distribution of natural vegetation was the best expression of the totality of climate. Therefore, the boundaries he chose were based largely on the limits of certain plant associations.

15. On Table 16.3, list the names and general characteristics of each principal climate group next to its designated classification letter. Use Table 16.2 as a reference.

Table 16.4 contains climate data for several stations that are representative of Köppen climate types. Use the data in Table 16.2 and Table 16.4 to answer the following questions.

Humid Tropical (A) Climates

The constantly high temperatures and abundant rainfall in regions near the equator combine to produce the lush vegetation of the *tropical rain forest*. In the zone poleward of the wet tropics, the rain forests gives way to a tropical grass land called a *savanna*.

16. What temperature criterion is used for defining an A climate?
    Coldest month is above 18°C

17. Use Figure 16.5 (page 232) to plot the monthly temperature and precipitation data for Iquitos, Peru, an A climate, given in Table 16.4.

Use the Iquitos, Peru, climograph to answer questions 18–21.

18. What is the *annual temperature range* (difference between highest and lowest monthly mean temperatures) for Iquitos?
    Annual temperature range is __2.3__ °C.

19. Why are month-to-month temperature changes so small for A climates?
    They are located primarily in the SH and are below the equator. They are also very humid.

20. Notice that Iquitos receives an average of 2619 mm of precipitation per year. How many inches of precipitation per year would this equal? How many feet? (*Hint:* You may have to refer to the

conversion tables located on the inside back cover of this manual.)
2,619 mm = __102__ inches = __8.5__ feet

21. Is the precipitation at Iquitos concentrated in one season or distributed fairly evenly throughout the year?
    They get a lot more in the winter but it still varies not by much.

Use the world climate map shown in Figure 16.4 to answer questions 22–23.

22. In what latitude belt are A climates located?
    A climates are in __0°-20° SH__.

23. Considering the locations of A climates, would weather fronts produced by the interaction of continental polar (cP) and maritime tropical (mT) air be responsible for the precipitation? Explain your answer.
    No because cP is not near where the A climate is because it is not near a polar area.

Dry (B) Climates

Dry climates cover more land area than any other climate group. Classification as a dry climate does not necessarily imply little or no precipitation but rather indicates that the yearly precipitation is not as great as the potential loss of moisture by evaporation.

24. What are three variables that the Köppen classification uses to establish the boundary between dry and humid climates?
    a. $R < 2T + 28$ annual temp
    b. $R < 2T$ annual precip
    c. $R < 2T + 14$ seasonal precip

25. What names are applied to the following climate types?
    BW: Desert
    BS: Steppe

26. Are deserts (BW) located (closer to *or* farther from) the wet tropics (A) than steppes (BS) (see Figure 16.4)?
    Located: __farther__ A climates.

Table 16.4 Climate Data for Representative Stations

	J	F	M	A	M	J	J	A	S	O	N	D	YEAR
IQUITOS, PERU (Af); LAT. 3°39′ S; 115 M													
Temp. (°C)	25.6	25.6	24.4	25.0	24.4	23.3	23.3	24.4	24.4	25.0	25.6	25.6	24.7
Precip. (mm)	259	249	310	165	254	188	168	117	221	183	213	292	2619
RIO DE JANEIRO, BRAZIL (Aw); LAT. 22°50′ S; 26 M													
Temp. (°C)	25.9	26.1	25.2	23.9	22.3	21.3	20.8	21.1	21.5	22.3	23.1	24.4	23.2
Precip. (mm)	137	137	143	116	73	43	43	43	53	74	97	127	1086
FAYA, CHAD (BWh); LAT. 18°00′ N; 251 M													
Temp. (°C)	20.4	22.7	27.0	30.6	33.8	34.2	33.6	32.7	32.6	30.5	25.5	21.3	28.7
Precip. (mm)	0	0	0	0	0	2	1	11	2	0	0	0	16
SALT LAKE CITY, UTAH (BSk); LAT. 40°46′ N; 1288 M													
Temp. (°C)	−2.1	0.9	4.7	9.9	14.7	19.4	24.7	23.6	18.3	11.5	3.4	−0.2	10.7
Precip. (mm)	34	30	40	45	36	25	15	22	13	29	33	31	353
WASHINGTON, D.C. (Cfa); LAT. 38°50′ N; 20 M													
Temp. (°C)	2.7	3.2	7.1	13.2	18.8	23.4	25.7	24.7	20.9	15.0	8.7	3.4	13.9
Precip. (mm)	77	63	82	80	105	82	105	124	97	78	72	71	1036
BREST, FRANCE (Cfb); LAT. 48°24′ N; 103 M													
Temp. (°C)	6.1	5.8	7.8	9.2	11.6	14.4	15.6	16.0	14.7	12.0	9.0	7.0	10.8
Precip. (mm)	133	96	83	69	68	56	62	80	87	104	138	150	1126
ROME, ITALY (Csa); LAT. 41°52′ N; 3 M													
Temp. (°C)	8.0	9.0	10.9	13.7	17.5	21.6	24.4	24.2	21.5	17.2	12.7	9.5	15.9
Precip. (mm)	83	73	52	50	48	18	9	18	70	110	113	105	749
PEORIA, ILLINOIS (Dfa); LAT. 40°52′ N; 180 M													
Temp. (°C)	−4.4	−2.2	4.4	10.6	16.7	21.7	23.9	22.7	18.3	11.7	3.8	−2.2	10.4
Precip. (mm)	46	51	69	84	99	97	97	81	97	61	61	51	894
VERKHOYANSK, RUSSIA (Dfd); LAT. 67°33′ N; 137 M													
Temp. (°C)	−46.8	−43.1	−30.2	−13.5	2.7	12.9	15.7	11.4	2.7	−14.3	−35.7	−44.5	−15.2
Precip. (mm)	7	5	5	4	5	25	33	30	13	11	10	7	155
IVIGTUT, GREENLAND (ET); LAT. 61°12′ N; 129 M													
Temp. (°C)	−7.2	−7.2	−4.4	−0.6	4.4	8.3	10.0	8.3	5.0	1.1	−3.3	−6.1	0.7
Precip. (mm)	84	66	86	62	89	81	79	94	150	145	117	79	1132
MCMURDO STATION, ANTARCTICA (EF); LAT. 77°53′ S; 2 M													
Temp. (°C)	−4.4	−8.9	−15.5	−22.8	−23.9	−24.4	−26.1	−26.1	−24.4	−18.8	−10.0	−3.9	−17.4
Precip. (mm)	13	18	10	10	10	8	5	8	10	5	5	8	110

27. Find at least two desert locations that are classified BWk. What does the letter "k" in BWk tell you about this type of desert climate?

Average annual temperature is less than 18°C

To answer questions 28 and 29, refer to the world climate map shown in Figure 16.4.

28. At what latitudes (north and south) are the most extensive arid areas located?

20-40° N and S

29. The Sahara Desert in North Africa is the largest area in the world with a BWh climate. What are some other regions that have this climate?

Middle East, Australia

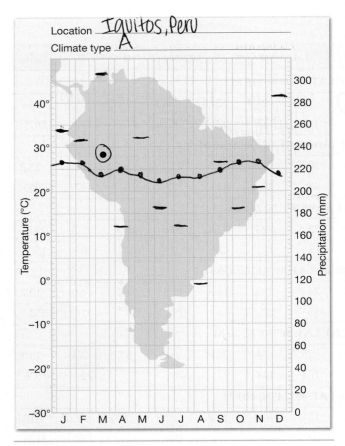

Location __Iquitos, Peru__

Climate type __A__

Figure 16.5 Climograph for Iquitos, Peru.

Humid Middle-Latitude Climates with Mild Winters (C)

A large percentage of the world's population is located in areas with C climates. This climate is characterized by weather contrasts brought about by changing seasons. The regions with C climates are frequently influenced by contrasting air masses and middle-latitude cyclones.

30. What temperature criterion is used to define the boundary of C climates?
    under 18°C and above -3°C

31. Prepare climographs for Washington, D.C., and Rome, Italy, by plotting the data in Table 16.4 on the graphs in Figure 16.6. Then answer questions 32–35.

32. In what manner are the temperature curves for each of the two cities similar?
    They both peak during May-Oct.

33. How does the annual distribution of precipitation vary between the two cities?
    Rome gets less and they are almost inverted

34. What is the difference between a Cf (Washington, D.C.) climate and a Cs (Rome, Italy) climate?
    The temp. in Washington DC vary more and correlate with precip. In Rome there isn't a similar correlation.

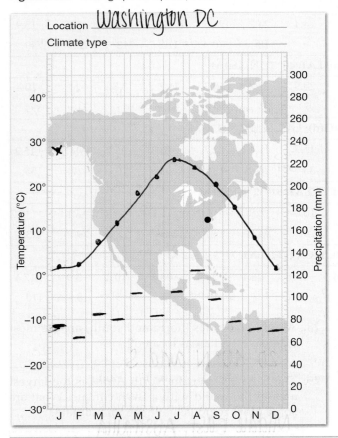

Location __Washington DC__

Climate type _____

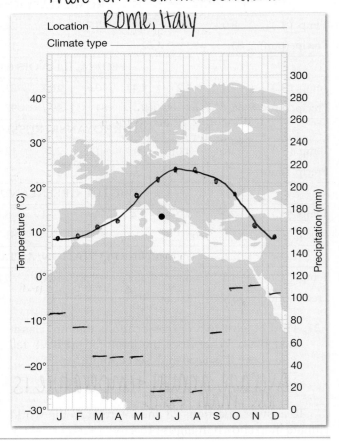

Location __Rome, Italy__

Climate type _____

Figure 16.6 Climographs for Washington, D.C., and Rome, Italy.

Use the world climate map shown in Figure 16.4 to answer questions 35–37.

35. What countries in Asia have areas of climate similar to that of Washington, D.C.?

India (circled on map)

36. What Southern Hemisphere countries have climates similar to that of Washington, D.C.?

circled on map

37. Which U.S. state has a climate similar to that of Rome, Italy?

Oregon and washington

Humid Middle-Latitude Climates with Severe Winters (D)

The harsh winters and relatively short growing season restrict agricultural activity in much of the area of D climates. The northern portions of D climate regions are covered by coniferous forests, with lumbering being a significant economic activity.

38. What criteria are used to define a D climate?

Avg. temp. of coldest month is -3°C or below. Avg. temp. of warmest month is greater than 10°C.

39. Use the data from Table 16.4 to plot a climograph for Peoria, Illinois, on Figure 16.7. Then use this climograph to answer questions 40–42.

40. What is the annual range of temperature in Peoria, Illinois?

_28.3°C_

41. How does the annual range of temperature in Peoria, Illinois, compare to the annual temperature ranges in Iquitos, Peru, and Rome, Italy?

It changes monthly (has a difference between high and low temp) so it is similar in that way to Rome, but not Peru. However, Illinois has much larger temperature changes than Rome does because they have severe winters.

42. During what season does Peoria receive its greatest precipitation? How does this compare with the seasonal distribution of precipitation for Rome, Italy?

It appears they recieve the most precip. from May–August, so summer. Rome gets the majority of their precip from October–February, so winter. Therefore the seasons they get the most precip during are opposite.

Use the world climate map shown in Figure 16.4 to answer questions 43–45.

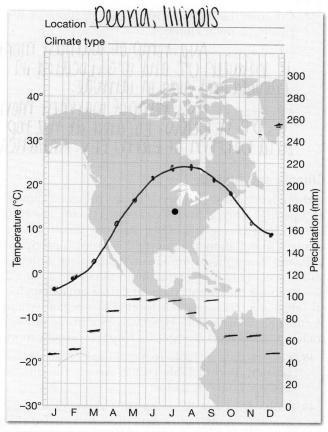

Location _Peoria, Illinois_

Climate type _____

Figure 16.7 Climograph for Peoria, Illinois.

43. Which continent has the greatest continuous expanse of D climates?

Europe / North America

44. D climates are located in only one hemisphere. Which one?

Hemisphere: _Northern_

45. Suggest a reason why D climates are located in only one hemisphere.

They are probably colder because they are located near the polar air masses. They are also near polar climates in the Northern Hemisphere where in the Southern Hemisphere there is more ocean too.

Polar (Type E) Climates

The polar climates, found at high latitudes and scattered high altitudes in mountains, are regions of cold temperatures and sparse population. Low evaporation rates allow these areas to be classified as humid, even though the annual precipitation is modest.

46. What criterion is used to define E climates?

The average temperature of the warmest month is below 10°C.

47. Contrast the characteristics and locations of the two basic polar climates.

ET climates: *Avg. temp. of warmest month is between 0°C and 10°C. Located in Canada and high in latitude.*

EF climates: *Avg. temp of warmest month is 0°C or below. Located at the top and bottom of Earth (ex. Antartica)*

Climate and Altitude

Although often not included with the principal climate groups, *high-altitude*, or *highland* (type H), climates exist in all climatic regions. They are the result of the changes in radiation, temperature, humidity, precipitation, and atmospheric pressure that take place with elevation and orientation of mountain slopes.

Examine the climate data for Cruz Loma, Ecuador, in Table 16.5. This station is located in a region where A climates are expected. However, its altitude of 3888 meters (12,753 feet) changes its Köppen classification.

48. Use the data in Table 16.5 and the Köppen classification chart in Table 16.2 to classify the climate of Cruz Loma, Ecuador: _____

49. Although Cruz Loma has this classification, the monthly temperature data provide a strong clue to the fact that this station is very near the equator. Explain.

_____

_____

50. Why would you expect the vegetation in the area of Cruz Loma to be different from that found at Guayaquil, Ecuador, a city located at the same latitude, but on the coast?

_____

_____

51. Examine Figure 16.4. Where is the greatest continuous expanse of high-altitude (highland) climate located?

_____

_____

Companion Website

The companion website provides numerous opportunities to explore and reinforce the topics of this lab exercise. To access this useful tool, follow these steps:

1. Go to www.mygeoscienceplace.com.
2. Click on "Books Available" at the top of the page.
3. Click on the cover of *Applications and Investigations in Earth Science, 7e.*
4. Select the chapter you want to access. Options are listed in the left column ("Introduction," "Web-based Activities," "Related Websites," and "Field Trips").

Table 16.5   Climate Data for CRUZ LOMA, Ecuador

CRUZ LOMA, ECUADOR; LAT. 0°, LONG. 79°W; 3888 M												
	J	F	M	A	M	J	J	A	S	O	N	D	YEAR
Temp. (°C)	6.1	6.6	6.6	6.6	6.6	6.1	6.1	6.1	6.1	6.1	6.6	6.6	6.4
Precip. (mm)	198	185	241	236	221	122	36	23	86	147	124	160	1779

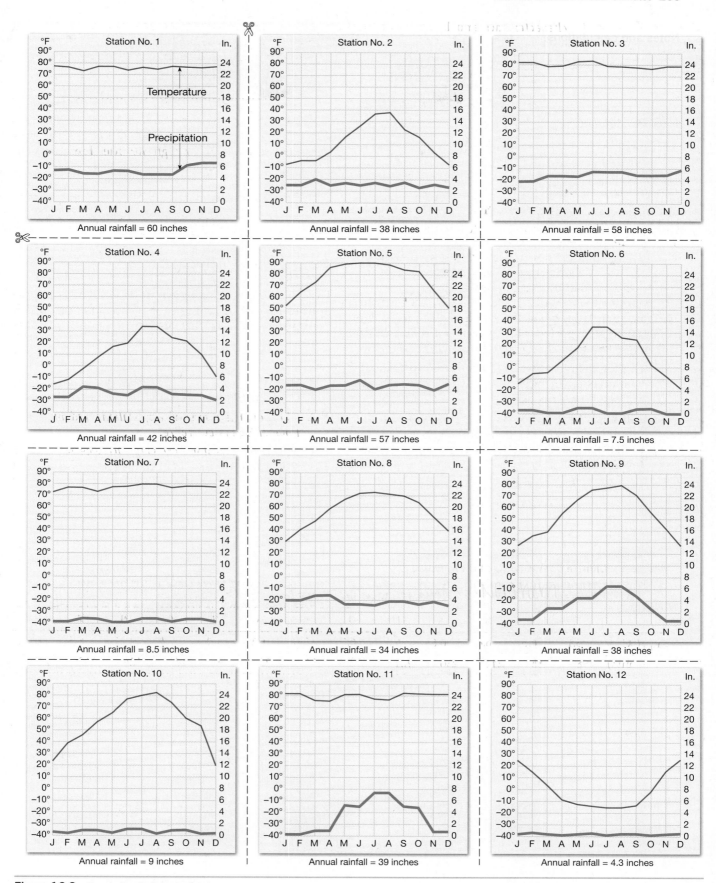

Figure 16.3 Generalized climographs.

Notes and calculations.

Global Climates

Name _____   **Course/Section** _____

Date _____   **Due Date** _____

1. Explain how a climograph is constructed.

   _____

   _____

   _____

   _____

   _____

2. In your own words, provide some key words to describe the characteristics of each of the following Köppen climate groups.

   A climates: _____

   _____

   B climates: _____

   _____

   C climates: _____

   _____

   D climates: _____

   _____

   E climates: _____

   _____

   H climates: _____

   _____

   _____

3. Indicate the name of the Köppen climate group best described by each of the following statements.

   Vast areas of northern coniferous forests: _____

   _____

   Smallest annual range of temperature: _____

   _____

   Highest annual precipitation: _____

   _____

   Mean temperature of the warmest month is below 10°C: _____

   _____

   The result of high elevation and mountain slope orientation: _____

   _____

   Potential evaporation exceeds precipitation: ___

   _____

   Very little change in the monthly precipitation and temperature throughout the year: _____

   _____

PART 4
Astronomy

David Nunuk/Photo Researchers, Inc.

Astronomical Observations

Objectives

Completion of this exercise will prepare you to:

  A. Describe the changing position of the Sun as it rises and sets.

  B. Describe how the Sun angle changes daily.

  C. Explain how the position of the Moon changes over a period of several weeks.

  D. Explain how the positions of stars in the night sky relate to Earth's rotational motion.

Materials

meterstick (or yardstick)	star chart
small weight	ruler
protractor	calculator
string	
atlas	

Introduction

Scientific inquiry often starts with the systematic collection of data from which hypotheses—and occasionally theories—are developed. The study of astronomy, an observational science, begins with careful observation and recording of the changing positions of the Sun, Moon, planets, and stars, using both the unaided eye and a telescope.

Sun Observations

Many people are unaware that the Sun rises and sets at different locations on the horizon each day. As Earth **revolves** about the Sun, the orientation of its axis to the Sun continually changes. As a result, the location of the rising and setting Sun, as well as the **altitude** (angle above the horizon) of the Sun at noon, changes throughout the year.

Sunset (or Sunrise) Observations

Use the following procedure to observe and record the Sun's location on the horizon at sunset or sunrise:

Step 1: Several minutes prior to sunset, *estimate* where the Sun will set on the horizon. Draw the prominent features (buildings, trees, etc.) to the north and south of the Sun's approximate setting position on the sunset data sheets in Figure 17.1.

> **CAUTION:** Never look directly at the Sun; eye damage may result.

Step 2: As the Sun sets, draw its position on the data sheet relative to the fixed features on the horizon.

Step 3: Note the date and time of your observation on the data sheet.

Step 4: Return to the same location several days later. Repeat your observation and record the results on a new data sheet.

  1. Following Steps 1–4, record at least four separate observations of the setting (or rising) Sun. (*Note*: Although the directions are for sunset, some minor adjustments will allow their use for sunrise.) Gather the data over a period of several weeks; *wait four or five days between observations*.

  2. After completing your observations, describe the changing location of the Sun at sunset (or sunrise).

  _____

  _____

  _____

  _____

Measuring the Noon Sun Angle

Observe the method for measuring the altitude (angle) of the noon Sun above the horizon illustrated in Figure 17.2.

SUNSET (Sunrise) DATA SHEETS

South   HORIZON   North

Date of observation _____   Time of observation _____

South   HORIZON   North

Date of observation _____   Time of observation _____

South   HORIZON   North

Date of observation _____   Time of observation _____

South   HORIZON   North

Date of observation _____   Time of observation _____

Figure 17.1 Sunset (sunrise) data sheets.

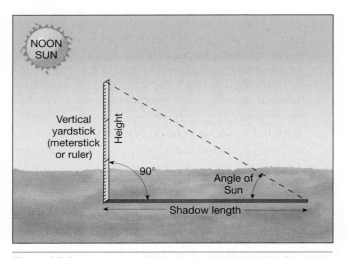

Figure 17.2 Illustration of the method for measuring the angle of the Sun above the horizon at noon. With each observation, be certain that the yardstick, meterstick, or ruler is perpendicular (at a 90° angle) to the ground or table top.

Then, following the steps below, determine the altitude of the Sun at noon on several days.

To determine the altitude (angle) of the Sun at noon:

Step 1: Place a yardstick (or meterstick or ruler) perpendicular to the ground or a tabletop.

Step 2: When the Sun is at its highest position in the sky (noon standard time or 1 P.M. daylight saving time), measure the length of the shadow.

Step 3: Divide the height of the stick by the length of the shadow.

Step 4: Determine the Sun angle by referring to Table 17.1. Locate the number on the table that comes closest to your answer from Step 3 and then record the angle.

Step 5: Repeat the measurement at the same time on four days, spaced over four or five weeks.

Table 17.1 Data Table for Determining the Noon Sun Angle*

IF $\dfrac{\text{HEIGHT OF STICK}}{\text{LENGTH OF SHADOW}}$	THEN SUN ANGLE IS	IF $\dfrac{\text{HEIGHT OF STICK}}{\text{LENGTH OF SHADOW}}$	THEN SUN ANGLE IS
0.2679	15°	1.235	51°
0.2867	16°	1.280	52°
0.3057	17°	1.327	53°
0.3249	18°	1.376	54°
0.3443	19°	1.428	55°
0.3640	20°	1.483	56°
0.3839	21°	1.540	57°
0.4040	22°	1.600	58°
0.4245	23°	1.664	59°
0.4452	24°	1.732	60°
0.4663	25°	1.804	61°
0.4877	26°	1.881	62°
0.5095	27°	1.963	63°
0.5317	28°	2.050	64°
0.5543	29°	2.145	65°
0.5774	30°	2.246	66°
0.6009	31°	2.356	67°
0.6249	32°	2.475	68°
0.6494	33°	2.605	69°
0.6745	34°	2.748	70°
0.7002	35°	2.904	71°
0.7265	36°	3.078	72°
0.7536	37°	3.271	73°
0.7813	38°	3.487	74°
0.8098	39°	3.732	75°
0.8391	40°	4.011	76°
0.8693	41°	4.332	77°
0.9004	42°	4.705	78°
0.9325	43°	5.145	79°
0.9657	44°	5.671	80°
1.0000	45°	6.314	81°
1.0360	46°	7.115	82°
1.0720	47°	8.144	83°
1.1110	48°	9.514	84°
1.1500	49°	11.430	85°
1.1920	50°		

*Select the number nearest to the quotient determined by dividing the height of the stick by the length of the shadow. Read the corresponding Sun angle.

3. Record the dates and results of the measurements in the following spaces.

   Date: _____ Noon Sun angle: _____
   Date: _____ Noon Sun angle: _____
   Date: _____ Noon Sun angle: _____
   Date: _____ Noon Sun angle: _____

4. Did the altitude of the noon Sun (increase *or* decrease) over the period of the measurements?

   The altitude (angle) _____.

5. How many degrees did the noon Sun angle change over the period of your observations?

   _____

6. What is the approximate average change of the noon Sun angle per day?

   Change: _____° per day

7. Based on your answer to question 6, how many degrees will the noon Sun angle change over a six-month period?

   Change: _____° over a six-month period

Moon Observations

Most people have noticed that the shape of the illuminated portion of the Moon, as observed from Earth, changes regularly. However, very people few take the time to systematically record and explain why these changes occur.

8. Record at least four observations of the Moon by completing the following steps at a two- or three-day interval.

Step 1: On a Moon observation data sheet provided in Figure 17.3, indicate the approximate east–west position of the Moon in the sky by drawing a circle at the appropriate location.

Step 2: Shade the circle to indicate the shape of the illuminated portion of the Moon you observe.

Step 3: Record the date and time of your observation.

Step 4: Keep in mind that the approximate time between moonrise on the eastern horizon and moonset on the western horizon is 12 hours.

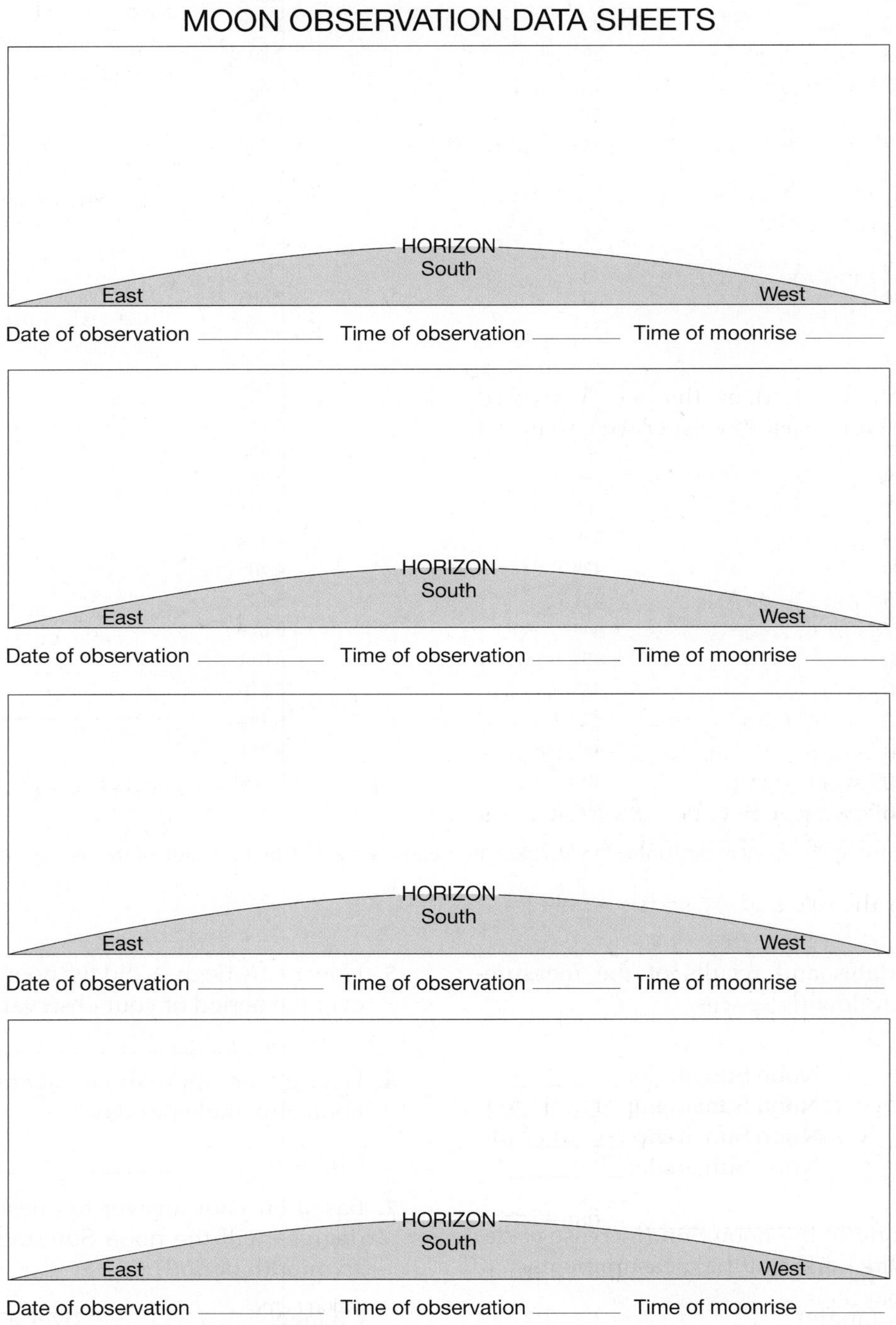

MOON OBSERVATION DATA SHEETS

HORIZON
South
East   West
Date of observation _____ Time of observation _____ Time of moonrise _____

HORIZON
South
East   West
Date of observation _____ Time of observation _____ Time of moonrise _____

HORIZON
South
East   West
Date of observation _____ Time of observation _____ Time of moonrise _____

HORIZON
South
East   West
Date of observation _____ Time of observation _____ Time of moonrise _____

Figure 17.3 Moon observation data sheets.

Estimate when the Moon may have risen and when it may set. Write your estimates on the data sheet.

Step 5: Repeat your observation several days later using a new data sheet.

Answer questions 9–12 after you have completed all your observations of the Moon.

9. What happened to the size and shape of the illuminated portion of the Moon over the period of your observations?

_____

_____

10. Did the Moon move farther (eastward *or* westward) in the sky with each successive observation?

The Moon moved further _____.

11. Did moonrise and moonset occur (earlier *or* later) with each successive observation?

Moonrise and moonset occurred _____ with each observation.

12. Based on your observations and your answers to questions 10 and 11, does the Moon revolve around Earth from (east to west *or* west to east)?

Moon revolves around Earth from _____.

Star Observations

Throughout history, people have been recording the positions and nightly movements of stars that result from Earth's **rotation**, as well as the seasonal changes in the constellations as Earth revolves around the Sun. Ancient observers offered many explanations for the changes before the nature of planetary motions was understood in the 17th century.

Complete the following activity on a clear, moonless night, preferably in an area away from bright city lights.

13. Make a list of the *colors* of the stars you observe.

_____

Select a bright star that is about 45 degrees above the southern horizon and observe its movement over a period of one hour.

14. With your arm extended, measure approximately how many widths of your fist the position of the star has changed.

Approximately _____ fist widths

15. Does the star appear to have moved (eastward *or* westward) over a period of one hour?

Star appears to have moved _____.

16. How is the movement of the star you observed in question 15 related to the direction of Earth's rotation?

_____

_____

Use a star chart to locate several constellations and the North Star (Polaris).

17. Refer to Figure 17.4. Sketch the pattern of stars for two constellations you were able to locate. List the name of each constellation next to the sketch you drew.

18. Use Figure 17.5 to construct a simple **astrolabe** and measure the angle of the North Star (Polaris) above the horizon as accurately as possible.

_____° above the horizon

Constellation Star Pattern

Constellation name: _____

Constellation Star Pattern

Constellation name: _____

Figure 17.4 Constellation sketches.

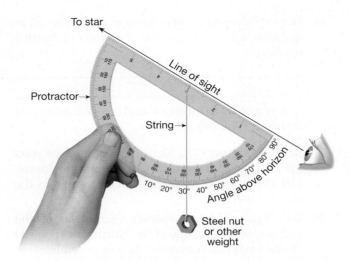

Figure 17.5 Simple astrolabe, an instrument used to measure the angle of an object above the horizon. The angle is read where the string crosses the outer edge of the protractor—32° in the example illustrated here. Notice that the angle above the horizon is the difference between 90° and the angle printed on the protractor.

Over a period of several hours, observe the motion of the stars in the vicinity of Polaris.

19. Write a brief summary of the motion of the stars in the vicinity of Polaris.

_____

_____

If possible, return to the same location several weeks later, at the same time, to answer questions 20–22.

20. Did the same star you observed overhead in question 14 (remain overhead, move eastward, *or* move westward)?

The star _____.

21. How is the change in position of the star you observed overhead several weeks earlier related to the revolution of Earth?

_____

_____

_____

22. Using the astrolabe you constructed in question 20, repeat your measurement of the angle of the North Star (Polaris) above the horizon. List your new measurement and compare it to the measurement you obtained several weeks earlier. Explain your result(s).

_____

_____

_____

_____

Companion Website

The companion website provides numerous opportunities to explore and reinforce the topics of this lab exercise. To access this useful tool, follow these steps:

1. Go to www.mygeoscienceplace.com.

2. Click on "Books Available" at the top of the page.

3. Click on the cover of *Applications and Investigations in Earth Science, 7e.*

4. Select the chapter you want to access. Options are listed in the left column ("Introduction," "Web-based Activities," "Related Websites," and "Field Trips").

Astronomical Observations

Name _____   **Course/Section** _____

Date _____   **Due Date** _____

1. On Figure 17.6, prepare a single sketch illustrating your observed positions of the setting (or rising) Sun during a span of several weeks. Show the reference features on the horizon that you used. Label each position of the Sun with the date of the observation. Write a brief summary of your observations below the diagram.

2. From question 3, in the exercise, list the noon Sun angle that you calculated for the first day and last day of your measurements.

   Date of observation: _____
   Noon Sun angle on the first day: _____°

   Date of observation: _____
   Noon Sun angle on the last day: _____°

3. Draw two sketches of the Moon—the first illustrating the Moon as you saw it on your first lunar observation and the second as you saw it on your last observation. Label the date and time of each.

FIRST MOON OBSERVATION	LAST MOON OBSERVATION
Date: _____	Date: _____
Time: _____	Time: _____

4. Did the Moon rise (earlier *or* later) each night you observed it?

   _____

5. List the different colors of stars you observed.

   _____

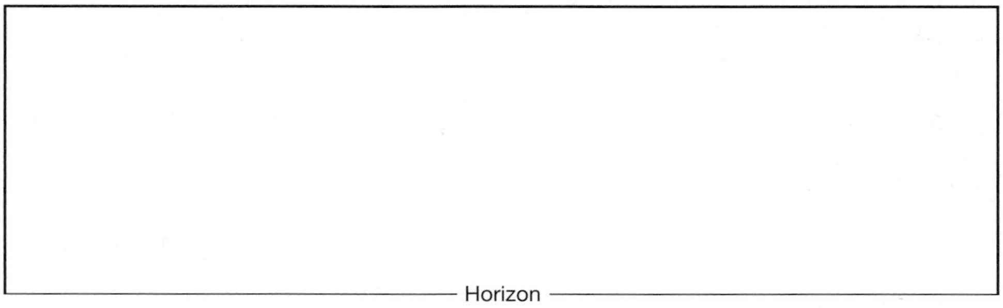

Summary: _____

_____

Figure 17.6 Sunset (sunrise) observations.

6. Approximately how many widths of your fist, with your arm extended, will a star appear to move in one hour? Toward which direction do the stars appear to move throughout the night? What is the reason for the apparent motion?

_____

_____

_____

7. What was your measured angle of the North Star (Polaris) above the horizon at your location? Did the angle change over a several-week period? Explain.

_____

_____

_____

8. Refer to Figure 17.7. Sketch and name the pattern of stars for any constellation you have been able to locate in the sky.

Constellation Star Pattern

Constellation name: _____

Figure 17.7 Constellation sketch.

Patterns in the Solar System

Objectives

Completion of this exercise will prepare you to:

A. Describe and distinguish characteristics of the terrestrial and Jovian planets.

B. Describe the motions of the planets.

C. Explain temperature differences among planetary surfaces.

Materials

ruler
calculator
black containers
  with covers and
  thermometers

colored pencils
4-meter length of adding
  machine paper
meterstick
light source (150-watt bulb)

Introduction

The **nebular theory**, which explains the formation of the solar system, states that the Sun and planets formed about 4.6 billion years ago, from a rotating cloud of interstellar gases and dust called the **solar nebula**. As the solar nebula contracted, most of the material collected in the center to form the Sun. The remaining materials formed a thick, flattened, rotating disk, within which matter gradually cooled and condensed into grains and clumps of icy, rocky, and metallic material. Repeated collisions resulted in most of the material accreting to form the planets and innumerable smaller bodies in the solar system.

Located closest to the Sun, the planets Mercury, Venus, Earth, and Mars evolved under nearly the same conditions and consequently exhibit similar characteristics. Because these planets are rocky objects with solid surfaces, they are collectively called the **terrestrial (Earth-like) planets**.

The outer planets, Jupiter, Saturn, Uranus, and Neptune, being farther from the Sun than the terrestrial planets, formed under much colder conditions and are gaseous objects with central cores of ices and rock. Since the four outer planets are similar, they are grouped together and called the **Jovian (Jupiter-like) planets**.

Table 18.1 lists many of the characteristics of the planets in the solar system.

1. Draw lines on both the upper and lower parts of Table 18.1 to separate the terrestrial planets from the Jovian planets. Label the lines *Belt of Asteroids*.

2. On both parts of the table write the word *terrestrial* next to Mercury, Venus, Earth, and Mars and the word *Jovian* next to Jupiter, Saturn, Uranus, and Neptune.

Inclination of Orbits

The orbits of the planets all lie in nearly the same plane, called the **plane of the ecliptic** (Figure 18.1). The column labeled "Inclination of Orbit" in Table 18.1 lists how many degrees the orbit of each planet is inclined from the plane of the ecliptic.

3. Other than Mercury, whose orbit is inclined (4, 7, *or* 10) degrees, the orbits of the remaining planets are all inclined less than (4, 7, *or* 10) degrees.

   Mercury is inclined _____ degrees.

   The other planets are inclined less than _____ degrees.

4. The orbit of the dwarf planet Pluto is inclined 17°. When compared to the eight planets, is Pluto's inclination (smaller, average, *or* larger)?

   Pluto's inclination is _____.

5. Considering the nebular origin of the solar system, suggest a reason why the orbits of the planets are nearly all in the same plane.

   _____

   _____

Table 18.1 Planetary Data

| PLANET | SYMBOL | MEAN DISTANCE FROM SUN | | | PERIOD OF REVOLUTION | INCLINATION OF ORBIT | ORBITAL VELOCITY | |
		AU	MILLIONS OF MILES	MILLIONS OF KILOMETERS			mi/s	km/s
Mercury	☿	0.387	36	58	88^d	7°00′	29.5	47.5
Venus	♀	0.723	67	108	224^d	3°24′	21.8	35.0
Earth	⊕	1.000	93	150	365.25^d	0°00′	18.5	29.8
Mars	♂	1.524	142	228	687^d	1°51′	14.9	24.1
Jupiter	♃	5.203	483	778	11.86^{yr}	1°18′	8.1	13.1
Saturn	♄	9.539	886	1427	29.46^{yr}	2°29′	6.0	9.6
Uranus	♅	19.180	1783	2870	84^{yr}	0°46′	4.2	6.8
Neptune	♆	30.060	2794	4497	165^{yr}	1°46′	3.3	5.3

| PLANET | PERIOD OF ROTATION | DIAMETER | | RELATIVE MASS EARTH = 1 | AVERAGE DENSITY g/cm^3 | POLAR FLATTENING (%) | MEAN TEMPERATURE (°C) | NUMBER OF KNOWN SATELLITES |
		MILES	KILOMETERS					
Mercury	59^d	3015	4854	0.056	5.4	0.0	167	0
Venus	244^d	7526	12,112	0.82	5.2	0.0	464	0
Earth	23^{h}56^{m}04^s	7920	12,751	1.00	5.5	0.3	15	1
Mars	24^{h}37^{m}23^s	4216	6788	0.108	3.9	0.5	−65	2
Jupiter	9^{h}50^m	88,700	143,000	317.87	1.3	6.7	−110	63
Saturn	10^{h}14^m	75,000	121,000	95.14	0.7	10.4	−140	61
Uranus	17^{h}14^m	29,000	47,000	14.56	1.2	2.3	−195	27
Neptune	16^{h}03^m	28,900	46,529	17.21	1.7	1.8	−200	13

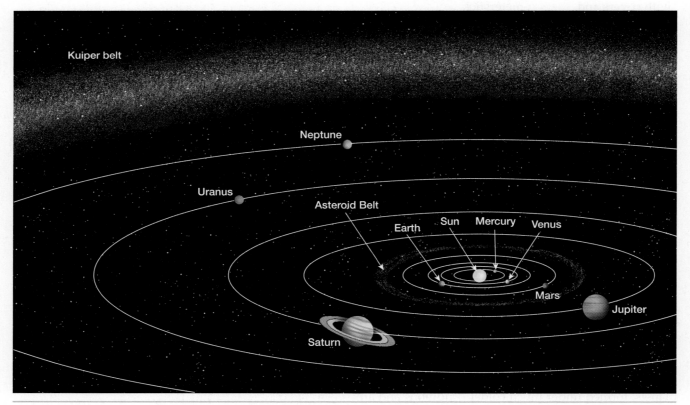

Figure 18.1 The solar system, showing the orbits of the planets.

Spacing of the Planets

Although many ancient astronomers tried to establish a pattern for the distribution of the planets, it was not until the mid-1700s that astronomers found a simple mathematical relationship that described the arrangement of the known planets.

A Scale Model of Planetary Distances

6. Prepare a scale model of the solar system, using the following steps:

    Step 1: Obtain a 4-meter length of adding machine paper and a meterstick from your instructor.

    Step 2: Place an *X* about 10 centimeters from one end of the adding machine paper and label it *Sun*.

    Step 3: Using the mean distances of the planets from the Sun, in miles, found in Table 18.1 and the following scale, draw a small circle for each planet at its scale mile distance from the Sun. Use a different colored pencil for the terrestrial planets than for the Jovian planets and write the name of each planet next to its position.

    SCALE
    1 millimeter = 1 million miles
    1 centimeter = 10 million miles
    1 meter = 1,000 million miles

    Step 4: Label the area 258 million miles from the Sun *Asteroid belt*.

7. What feature of the solar system separates the terrestrial planets from the Jovian planets?

    _____

8. Observe your scale model diagram and summarize the spacing for each of the two groups of planets.

    Terrestrial planets spacing: _____

    Jovian planets spacing: _____

9. Briefly describe the spacing of the planets in the solar system.

    _____

    _____

10. Which Jovian planet varies the most from the general pattern of spacing?

    _____

Comparing the Terrestrial and Jovian Planets

Size of the Planets

11. To visually compare the relative sizes of the planets and the Sun, complete the following steps, using the unmarked side of your 4-meter long adding machine paper.

    Step 1: Determine the radius of each planet, in kilometers, by dividing its diameter (in kilometers) by 2. List your answers in the "Radius" column of Table 18.2.

    Step 2: Use a scale of 1 cm = 2,000 km. Determine the scale model radius of each planet and list your answer in the "Scale Model Radius" column of Table 18.2.

    Step 3: Place an *X* about 10 cm from one end of the adding machine paper and label it *Starting point*.

    Step 4: Using the scale model radius in Table 18.2, begin at the starting point and mark the radius of each planet with a line on the paper. Use a different colored pencil for the terrestrial planets than for the Jovian planets. Label each line with the corresponding planet's name.

    Step 5: The diameter of the Sun is approximately 1,350,000 kilometers. Using the same scale as you used for the planets (1 cm = 2000 km), determine the scale model radius of the Sun. Mark the Sun's radius on the adding machine paper, using a different colored pencil than you used for the two planet groups. Label the line *Sun*.

Answer questions 12–17, using Table 18.1.

12. Which is the largest terrestrial planet, and what is its diameter?

    Planet: _____, diameter: _____ km

Table 18.2   Planetary Radii with Scale Model Equivalents

PLANET	RADIUS	SCALE MODEL RADIUS
Mercury	_____ km	_____ cm
Venus	_____ km	_____ cm
Earth	_____ km	_____ cm
Mars	_____ km	_____ cm
Jupiter	_____ km	_____ cm
Saturn	_____ km	_____ cm
Uranus	_____ km	_____ cm
Neptune	_____ km	_____ cm

13. Which is the smallest Jovian planet, and what is its diameter?

Planet: _____, diameter: _____ km

14. How many times larger than the largest terrestrial planet is the smallest Jovian planet?

The largest Jovian planet is _____ times larger than the smallest terrestrial planet.

15. Write a general statement that compares the sizes of the terrestrial planets to the sizes of the Jovian planets.

_____

_____

16. How many times larger is the Sun than Earth? How many times larger is the Sun than Jupiter?

The Sun's diameter is _____ times larger than Earth's diameter.

The Sun's diameter is _____ times larger than Jupiter's diameter.

17. The diameter of the dwarf planet Pluto is approximately 2300 km, which is about (one-third, one-half, *or* twice) the diameter of Mercury, the smallest planet.

The diameter of Pluto is about _____ that of Mercury.

Mass and Density of the Planets

Mass is a measure of the quantity of matter an object contains. In Table 18.1, the masses of the planets are given in relation to the mass of Earth. For example, the mass of Mars is 0.108, which means that it is only about 11 percent as massive as Earth. On the other hand, the Jovian planets are all many times more massive than Earth.

Density is the mass per unit volume of a substance. In Table 18.1, the average densities of the planets are expressed in grams per cubic centimeter . As a reference, the density of water is approximately 1 gram per cubic centimeter.

Using the masses of the planets given in Table 18.1, complete the following.

18. What is the most massive planet in the solar system? How many times more massive is this planet than Earth?

Most massive planet: _____

This planet is _____ times more massive than Earth.

19. What is the least massive planet? What is its mass compared to that of Earth?

Least massive planet: _____

This planet is only _____ percent as massive as Earth.

The gravitational attraction of a planet is directly related to its mass.

20. Which planet exerts the greatest gravitational attraction? Which planet exerts the least attraction?

Greatest gravitational attraction: _____

Least gravitational attraction: _____

The mass of an object is measured by the amount of material it contains. However, the **weight** of an object when measured on Earth depends on the gravitational attraction between the object and Earth.

21. The surface gravities of Mars and Jupiter are respectively about 0.4 and 2.5 times that of Earth. What would be the approximate weight of a 200-pound person on these planets?

This person would weigh _____ pounds on Mars.

This person would weigh _____ pounds on Jupiter.

22. Which of the two groups of planets are more likely to attract and hold low-density gaseous material such as hydrogen and helium?

_____

Diameter versus Density

Use the data in Table 18.1 to complete the graph in Figure 18.2, using the procedure below.

23. Plot a point on the graph in Figure 18.2 for each planet where its diameter and density intersect. Label each point with the planet's name. Use a different colored pencil for the terrestrial than for the Jovian planets.

24. What relationship exists between a planet's size (diameter) and its density?

_____

25. The average density of Earth's lithosphere is similar to that of the igneous rock basalt—about 3.0 g/cm^3. Is the average density of the terrestrial planets (greater *or* less) than the density of Earth's lithosphere?

The average density is _____ than the density of Earth's lithosphere.

26. Does the term (rocky *or* gaseous) best describe the terrestrial planets?

The terrestrial planets are _____.

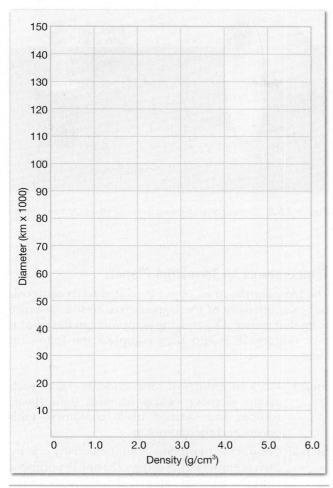

Figure 18.2 Diameter verses density graph.

27. The average density of Earth is about 5.5 g/cm³. Since the density of Earth's crustal rocks is only 3.0 g/cm³, what does this suggest about the density of Earth's interior?

_____

_____

28. Which of the planets has a density less than that of water and therefore would "float"?

_____

_____

29. Briefly compare the densities of the Jovian planets to the density of water.

_____

_____

30. Explain why Jupiter is such a massive object and yet has such a low density. (*Hint*: Remember that density is the relationship between mass and volume.)

_____

_____

31. Write a general statement comparing the densities of the terrestrial planets to the densities of the Jovian planets.

_____

_____

32. Why are the densities of the terrestrial and Jovian planets so different?

_____

_____

Number of Moons

The column labeled "Number of Known Satellites" in Table 18.1 indicates the number of known moons orbiting each planet.

33. Write a brief statement comparing the number of known moons of the terrestrial planets to the number orbiting the Jovian planets.

_____

_____

34. What is the relationship between the planet's number of moons compared to its mass? Suggest a reason for your answer.

_____

_____

_____

Rotation and Revolution of the Planets

All the planets, with the exception of Venus, rotate about their axis in a counterclockwise direction. Venus exhibits a very slow clockwise rotation. The time that it takes for a planet to complete one 360° rotation on its axis is called the **period of rotation**. The units used to measure a planet's period of rotation are Earth hours and days.

Revolution is the motion of a planet around the Sun. The time that it takes a planet to complete one revolution about the Sun is the length of its year, called the *period of revolution*. The units used to measure a planet's period of revolution are Earth days and years. Without exception, the direction of revolution of the planets is counterclockwise around the Sun.

Use the planetary data in Table 18.1 to answer the following questions.

35. If you lived on Venus, approximately how long would you have to wait between sunrises?

On Venus, sunrise occurs every _____ days.

36. If you lived on Jupiter, approximately how long would you wait between sunrises?

On Jupiter, sunrise occurs every _____ hours.

37. Compare the periods of rotation of the terrestrial planets to those of the Jovian planets.

_____

_____

The giant planet Jupiter rotates once on its axis approximately every 10 hours. If an object were on the equator of Jupiter and rotating with it, it would travel approximately 280,000 miles (the equatorial circumference or distance around the equator) in about 10 hours.

38. Calculate the equatorial rotational velocity of Jupiter, using the following formula:

$$\text{Velocity} = \frac{\text{Distance}}{\text{Time}} = \frac{\text{miles}}{\text{hour}}$$

39. The equatorial circumference of Earth is about 24,000 miles. What is the approximate equatorial rotational velocity of Earth?

_____ miles/hour

40. How many times faster is Jupiter's equatorial rotational velocity than Earth's?

_____ times faster

41. In one Earth year, how many revolutions will the planet Mercury complete? What fraction of a revolution will Neptune accomplish?

Mercury: _____ revolutions in one Earth year

Neptune: _____ of a revolution in one Earth year

42. On Venus, how many days (sunrises) would there be in each of its years?

Venus: _____ day(s) per year

43. How many days (rotations) will Mercury complete in one of its years?

Mercury: _____ day(s) in one Mercury year

44. Do the terrestrial planets have, on average (longer *or* shorter) days and (longer *or* shorter) years as compared to the Jovian planets?

Terrestrial planet days are _____.

Terrestrial planet years are _____.

In the early 1600s, Johannes Kepler set forth three laws of planetary motion. According to Kepler's third law, the period of revolution of a planet, measured in Earth years, is related to its distance from the Sun in astronomical units. One **astronomical unit (AU)** is defined as the average distance from the Sun to Earth— 93 million miles, or 150 million kilometers. The law states that a planet's orbital period squared is equal to its mean solar distance cubed $p^2 = d^3$.

45. Applying Kepler's third law, what would be the period of revolution of a hypothetical planet that is 4 AUs from the Sun? Show your calculation in the following space.

_____

_____

_____

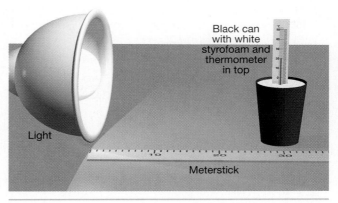

Figure 18.3 Terrestrial planet temperatures lab equipment setup.

Temperatures on Terrestrial Planets

The temperature of an object is related to its composition, the intensity of the heat source, and its distance from that source. Observe the equipment in the laboratory (Figure 18.3) and then complete the following steps:

Step 1: Work in groups of four students. Have each group member obtain *identical* light (heat) sources and *identical* black containers with covers and thermometers.

Step 2: Conduct experiments simultaneously, one by each member of the group. Do one experiment with the covered can and thermometer 15 cm from the light source, another with the can 30 cm from the light source, the third 45 cm, and the fourth 60 cm.

Step 3: Record the starting temperature for each container on Table 18.3; the temperatures should all be the same.

Step 4: For each of the four setups, turn on the light. After 10 minutes record the temperature of the container in Table 18.3.

Step 5: Plot the temperatures from Table 18.3 on the graph in Figure 18.4. Connect the points and label the graph "Temperature change with distance."

Table 18.3 Temperature Data

DISTANCE FROM LIGHT SOURCE (cm)	STARTING TEMPERATURE (°C)	10-MINUTE TEMPERATURE
15		
30		
45		
60		

46. Briefly relate temperature to distance, using the "Temperature change with distance" graph you constructed in Figure 18.4.

_____

_____

_____

Refer to the "Mean Temperature" column in Table 18.1. Plot the mean temperatures of the terrestrial planets at their appropriate distances from the Sun on the graph in Figure 18.4. Use a scale of 40 cm = 1 AU in addition to the temperature scale on the right axis of the graph. Label each point with the planet's name. Connect the points and label the graph "Mean terrestrial planet temperatures."

47. The mean terrestrial planet temperatures graph represents the actual mean temperatures of the planets. Compare this graph to the theoretical temperature change with distance graph. How are the graphs similar? How are they different?

_____

_____

_____

_____

48. Briefly suggest the reason(s) for the difference(s) between the two graphs you noted in question 47.

_____

_____

_____

_____

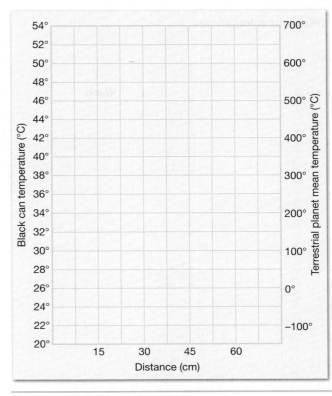

Figure 18.4 Terrestrial planet temperatures graph.

Companion Website

The companion website provides numerous opportunities to explore and reinforce the topics of this lab exercise. To access this useful tool, follow these steps:

1. Go to www.mygeoscienceplace.com.

2. Click on "Books Available" at the top of the page.

3. Click on the cover of _Applications and Investigations in Earth Science, 7e._

4. Select the chapter you want to access. Options are listed in the left column ("Introduction," "Web-based Activities," "Related Websites," and "Field Trips").

Notes and calculations.

Patterns in the Solar System

Name _____   **Course/Section** _____

Date _____   **Due Date** _____

1. On Figure 18.5, prepare a sketch illustrating the planets Mercury, Venus, Earth, and Mars at their approximate distances from the Sun. Draw arrows around each planet to illustrate its direction of rotation and arrows that show its direction of revolution.

2. Briefly describe the spacing of the planets in the solar system.

   _____

   _____

   _____

   _____

3. Use the nebular theory to explain why the planets revolve around the Sun in the same direction.

   _____

   _____

4. Briefly compare the terrestrial to the Jovian planets based on the following characteristics:

   Diameter: _____

   _____

   _____

   Density: _____

   _____

   _____

Period of rotation: _____

   _____

   _____

Number of moons: _____

   _____

   _____

Mass: _____

   _____

5. If you knew the distance of a planet from the Sun, explain how you would calculate its period of revolution.

   _____

   _____

   _____

6. How does a planet's distance from the Sun affect the solar radiation the planet receives?

   _____

   _____

   _____

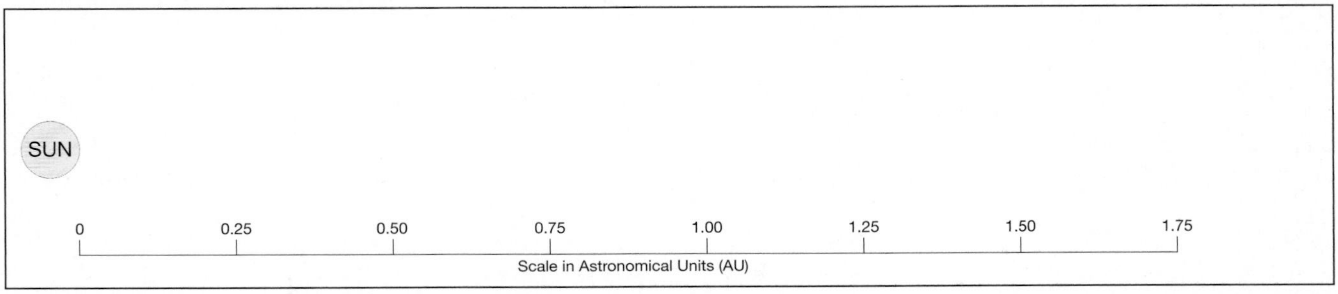

Figure 18.5 Spacing and motion of the terrestrial planets.

Locating the Planets

Objectives

Completion of this exercise will prepare you to:

A. Explain the motions of the planets when viewed from Earth.

B. Prepare a scale model of the inner solar system for a specified date.

C. Use a diagram of the solar system to estimate the times when a planet will rise and set.

D. Determine whether a planet can be seen from Earth on a specified date.

Materials

ruler   colored pencils
calculator

A Model of the Inner Solar System

The ability to recognize patterns in the night sky is one of the first skills that astronomers develop (Figure 19.1). It begins with an understanding of the motions of celestial objects and the proficient use of astronomical charts. This exercise examines a "working" model of the inner solar system. Using this model, you will investigate the movements of the planets that are visible to the unaided eye and whose names are used for the days of our week. The information in this exercise can also be used to locate the positions of the outer planets.

Figure 19.2 illustrates the orbits of Mercury, Venus, Earth, and Mars, as viewed from above Earth's Northern Hemisphere. Earth's orbit has been subdivided into months and select days. (The red marks indicate the first day of each month.) Surrounding the orbits is a large circle, marked in degrees, which is used to reference the locations of the planets. Since the planets and Sun lie in nearly the same plane, they move along

Figure 19.1 Mars, the "red planet," as viewed from Earth at night with the unaided eye. (Courtesy of NASA)

the same region of the sky, called the *zodiac* ("zone of animals"). Also shown on the outer reference circle are positions of the 12 **constellations of the zodiac**, which form the background of stars.

Revolution of the Planets

The **period of revolution** of a planet is directly related to its distance from the Sun. The direction of revolution around the Sun is counterclockwise (looking down from on Earth's northern hemisphere) for all planets.

On Figure 19.2, draw an arrow on the orbit of each planet to show its direction of revolution around the Sun.

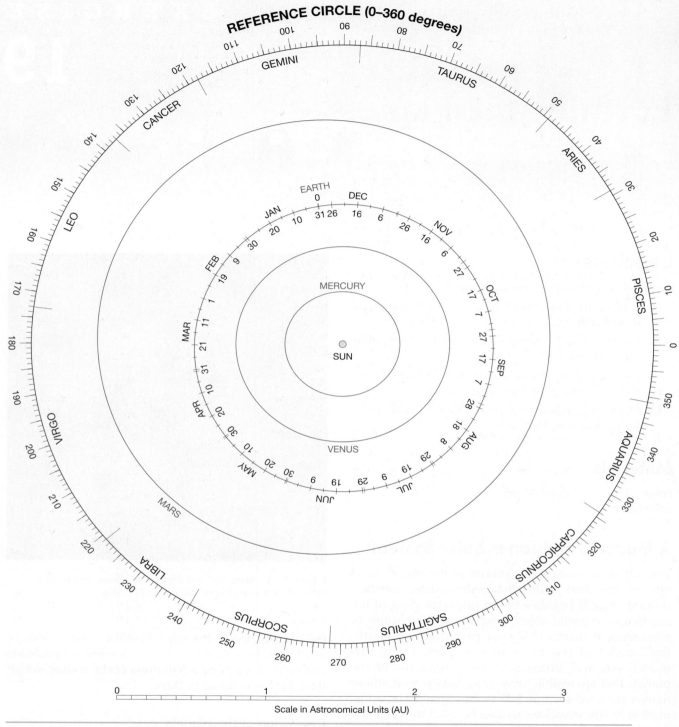

Figure 19.2 The orbits of Mercury, Venus, Earth, and Mars to scale, as viewed from above the Northern Hemisphere.

1. Refer to Table 18.1, "Planetary Data," in Exercise 18 (page 250), and record the distance from the Sun in astronomical units (average distance from Earth to Sun) and the period of revolution for the planets listed below.

PLANET	DISTANCE (AU)	PERIOD OF REVOLUTION
Mercury	_____	_____
Venus	_____	_____
Earth	_____	_____
Mars	_____	_____
Jupiter	_____	_____
Neptune	_____	_____

2. Using the information in question 1, write a brief statement comparing the period of revolution of a planet to its distance from the Sun.

_____

_____

Positioning a Planet Using the Reference Circle

Table 19.1 shows the locations (bearings) of Mercury, Venus, and Mars for the years 2011–2014 on the first day of each month. The locations are expressed in degrees and correspond to the equivalent degree marks found in the reference circle in Figure 19.2. For example, the position of Mercury on the first day of January 2011 is 156 degrees.

Use the following steps to determine a planet's position on Figure 19.2.

Step 1: Use Table 19.1 to determine the location of a planet on a specified date. (*Note:* Simple interpolation of the data can be used to estimate a planet's position on dates other than the first of each month.)

Step 2: Mark the location you determine in Step 1 on the reference circle on Figure 19.2.

Step 3: Use a ruler, or any other straight edge, to connect the Sun with the marked position from Step 2.

Step 4: Place a dot representing the planet's position where the ruler crosses the planet's orbit. Label the dot with the date.

3. Using Table 19.1, determine the location, in degrees, of the following planets for the dates indicated. (One location is provided for reference.)

PLANET	MAY 1, 2012	JUNE 1, 2014
Mercury	304°	_____
Venus	_____	_____
Mars	_____	_____

4. Using the data in question 3, plot the May 1, 2012, locations of Mercury, Venus, Earth, and Mars on Figure 19.2. (*Note:* Earth's position can be read directly from its orbit.)

Planet Location Among the Constellations

A planet is often referred to as being "in" a constellation. When viewing a planet from Earth, this refers to the constellation that is positioned directly behind the planet. The following steps are used to determine in which constellation a planet is located:

Step 1: Using Figure 19.2, mark the position of Earth and the planet to be observed for the same date.

Table 19.1 Planetary Heliocentric Longitudes (in degrees)

2011	MERCURY	VENUS	MARS
January	156°	139°	294°
February	259	189	313
March	352	234	331
April	165	283	350
May	262	331	9
June	11	20	28
July	178	68	45
August	273	118	62
September	33	168	79
October	193	217	94
November	285	266	108
December	51	314	122

2012	MERCURY	VENUS	MARS
January	208°	3°	136°
February	298	53	150
March	71	99	163
April	217	150	176
May	304	198	189
June	96	248	204
July	227	296	218
August	317	345	233
September	121	34	250
October	238	82	266
November	332	132	284
December	138	181	302

2013	MERCURY	VENUS	MARS
January	249°	231°	322°
February	349	280	341
March	148	325	359
April	255	14	18
May	354	62	36
June	167	112	53
July	263	160	70
August	14	211	85
September	184	260	101
October	274	308	115
November	36	357	129
December	195	45	142

2014	MERCURY	VENUS	MARS
January	286°	95°	156°
February	60	145	169
March	202	190	181
April	292	24	240
May	66	287	209
June	215	337	224
July	302	24	240
August	92	74	256
September	227	124	273
October	315	173	291
November	116	223	310
December	236	271	329

Source: Data from West Virginia University/Johns Hopkins/Tomchin Planetarium.

Step 2: Beginning at the position of Earth, draw a line that extends through the position of the planet until it intersects the outer reference circle.

Step 3: Note the constellation that lies in the planet's background.

5. Determine the constellation in which each of the following planets is observed on May 1, 2012:

PLANET	CONSTELLATION ON MAY 1, 2012
Mercury	_____
Venus	_____
Mars	_____

Relative Movement of the Planets

The planets that are farthest from the Sun take longer to complete one revolution than those that are nearest the Sun. Therefore, given the same time interval, planets that are closer to the Sun will move farther in their orbits relative to the background of stars than will planets that are farther from the Sun.

6. Using the information in question 1, how many revolutions will Mercury complete in one Earth year (365 days)? How does this motion influence how the position of Mercury changes throughout the year?

_____

_____

7. What fraction of a revolution will Mars complete in one Earth year? How does this rate of motion influence how the position of Mars changes throughout the year?

_____

_____

8. Using the data in Table 19.1, estimate how many degrees the following planets will move in their orbit over a period of one month:

PLANET	MOVEMENT IN ONE MONTH
Mercury	_____
Venus	_____
Earth	_____
Mars	_____

On Figure 19.2, plot the locations of Mercury, Venus, Earth, and Mars for July 1, 2012, and label each 7/1/12. These will be the positions of the planets two months after those you plotted in question 4.

9. If you had observed Mercury, Venus, and Mars each night over the two-month period May 1 to July 1, how would the position of each have changed in the sky relative to the background of stars? Example: Mercury would have traveled 283 degrees and moved from the constellation Pisces to Cancer

Venus would have traveled _____ degrees and moved from the constellation _____ to _____.

Mars would have traveled _____ degrees and moved from the constellation _____ to_____.

Retrograde Motion

The fact that Earth periodically overtakes the more distant planets makes the motion of the outer planets *appear* to move backward. The apparent backward, or westward, movement of a planet when viewed from Earth is called **retrograde motion**. A similar observation is made when one vehicle passes another going the same direction on a highway. For a period of time, relative to a fixed background, the vehicle being passed *appears* to move backward.

Figure 19.3 illustrates six positions of Earth and Mars over a period of approximately six months. Complete the following steps, using Figure 19.3.

Step 1: On Figure 19.3, draw a line connecting Earth to Mars at each of the six positions in the orbits. Extend each line from Mars to the background of stars shown on the figure.

Step 2: Mark the place where each line intersects the background of stars and label it with the number of the position. Positions 1 and 2 have been completed.

Step 3: Using a continuous line, connect the six numbered positions on the background of stars in order, 1 to 6.

10. Describe the apparent motion of Mars, as viewed from Earth, relative to the background of stars.

_____

_____

11. At any time, did Mars move backward in its orbit?

_____

12. Why does Mars exhibit retrograde motion?

_____

_____

13. Assume, as Ptolemy erroneously did in A.D. 141, that our planet is stationary and the planets and Sun orbit Earth. In this model of the solar system, how could the apparent backward motion of Mars be explained?

_____

_____

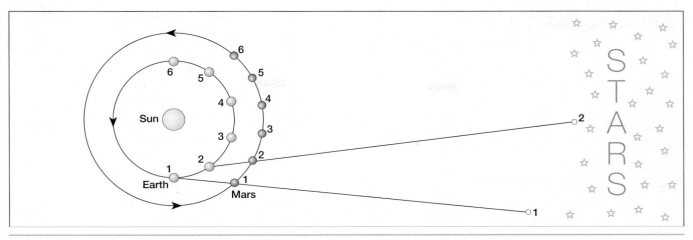

Figure 19.3 Diagram of the Sun, Earth, and Mars for illustrating retrograde motion. (Diagram not drawn to scale.)

Viewing a Planet from Earth

Observing a planet requires a dark sky. In addition, the observer must be on the half of the Earth facing the planet. Noon is experienced at the place on Earth facing directly toward the Sun, while midnight occurs at the location on the opposite side of Earth. Places experiencing sunrise and sunset are located between the noon and midnight positions.

Use Figure 19.2 and the following steps to determine the best times for viewing a planet from Earth.

Step 1: On Figure 19.2, indicate the position of the planet to be observed and Earth's position on the desired date.

Step 2: At Earth's position, draw a circle the size of a dime and mark the locations of noon, sunset, midnight, and sunrise.

Step 3: Draw a line on the circle that separates the half of Earth facing the planet to be observed from the half facing away.

Step 4: Keeping in mind that an earthbound observer must be on the half of Earth facing the planet to see it, note the two places where the line from Step 3 intersects Earth's surface. These will be the times the planet is first visible (*rises*) and when it is no longer visible (*sets*).

Use the May 1, 2012, positions of the planets plotted on Figure 19.2 to answer the following questions.

14. Using the steps outlined above, estimate the May 1, 2012, approximate rising and setting times for each of the following planets:

PLANET	RISE	SET
Venus	_____	_____
Mars	_____	_____

15. On May 1, 2012, what are the best times to observe the following planets with the unaided eye? (*Note:* Times must be between sunset, 6 P.M., and sunrise, 6 A.M.)

Venus: _____

Mars: _____

16. On May 1, 2012, Mercury is visible (in the evening, in the morning, *or* neither)?

Mercury is visible _____.

The Inner Solar System Today

17. Use Table 19.1 and what you have learned in this exercise to construct a model of the inner solar system for today's date on Figure 19.2. Show the positions of the four inner planets and label them with today's date.

18. Based on your model, will tonight—between 6 P.M. and midnight—be a (good *or* poor) time to view Mars?

Tonight will be a _____ time to view Mars.

Companion Website

The companion website provides numerous opportunities to explore and reinforce the topics of this lab exercise. To access this useful tool, follow these steps:

1. Go to www.mygeoscienceplace.com.
2. Click on "Books Available" at the top of the page.
3. Click on the cover of *Applications and Investigations in Earth Science, 7e.*
4. Select the chapter you want to access. Options are listed in the left column ("Introduction," "Web-based Activities," "Related Websites," and "Field Trips").

Notes and calculations.

Locating the Planets

Name _____   **Course/Section** _____

Date _____   **Due Date** _____

1. Label the orbits of Mercury, Venus, Earth, and Mars on Figure 19.4.

2. On Figure 19.4, mark the positions of Mercury, Venus, Earth, and Mars for December 1, 2013.

3. Why does Mars periodically exhibit retrograde motion?

   _____

   _____

   _____

4. Does Venus exhibit retrograde motion? Explain.

   _____

   _____

5. Describe what is meant when a planet is said to be "in" a particular constellation.

   _____

   _____

6. When viewed from Earth, Mercury's position, relative to the background stars, changes dramatically each year. Explain.

   _____

   _____

7. Describe the relative positions of Earth, Mars, and the Sun at the time when Mars is easiest to see from Earth.

   _____

   _____

   _____

   _____

   _____

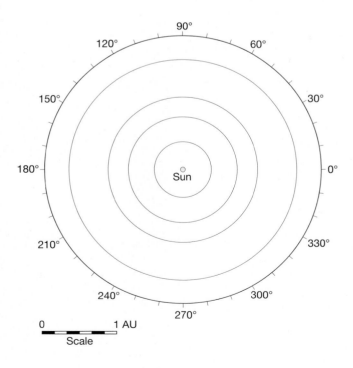

Figure 19.4 Planet locations on December 1, 2013.

Examining the Terrestrial Planets

Objectives

Completion of this exercise will prepare you to:

A. Describe the major geologic processes that have shaped the landforms of the terrestrial planets.

B. Identify landforms on each of the terrestrial planets and the Moon.

C. Describe the procedure for determining the relative ages of a planet's surface features.

D. List the factors that influence the size of impact craters.

Materials

ruler
colored pencils
calculator
metric ruler

sandbox
meterstick
small projectiles

Major Geologic Processes

The forces that drove Earth's evolution and shaped its topography have also operated elsewhere in the solar system. The four main geologic processes that act on Earth's surface are *volcanism, tectonic activity, gradation,* and *impact cratering*. Since each of these processes produces distinctive landforms, identifying and analyzing the surface features they generate makes it possible to unravel the geologic history of Earth and other planetary surfaces.

Volcanism and tectonic activity are driven by a planet's internal forces and tend to increase the relief and enhance the topography of a planet. During **volcanism**, molten rock (magma) and its entrapped gases erupt onto a planet's surface. Volcanic cones with summit *craters*, or large *calderas*, and vast plains covered by lava flows are examples of volcanic landforms.

Mount Shasta consists of overlapping cones, including the prominent secondary cone Shastina. Figure 20.1 provides two views of Mount Shasta, the most massive volcano in the Cascade Range.

1. On the aerial views in Figure 20.1A and B, label the summit (background) and Shastina (foreground).

2. Measure the width of Mount Shasta in Figure 20.1A along the line A–A'.

   _____ kilometers

3. Lava flows and debris flows extend far beyond the base of most volcanoes. Outline the margin of the large, dark-colored lava flow in Figure 20.1B.

4. Using the road that runs along the outer margin of this lava flow in Figure 20.1B for scale, what is the approximate thickness of this flow at its outermost margin: (10 feet, 100 feet, *or* 500 feet)?

   The thickness of the flow is about _____.

Tectonic activity involves the deformation of crustal rock by fracturing, faulting, or folding. This process is responsible for the major structural features of a planet, such as elevated plateaus, rift valleys, and folded terrains. Landforms produced by tectonic activity include *mountains, ridges,* and *fault-generated scarps* (cliffs).

Figure 20.2 shows a portion of the Valley and Ridge Province of the Appalachian Mountains, which is composed of folded sedimentary strata.

5. Label two of the ridges and the intervening valley between them on the image.

6. Measure the width of the area shown.

   _____ kilometers

7. What geologic process *deformed* the rock exposed at the surface: (volcanism, tectonic activity, gradation, *or* impact cratering)?

   _____

Figure 20.1 Mount Shasta, California. (**A.** Courtesy of U.S. Geological Survey. **B.** Photo by John S. Shelton)

Gradation involves the external processes that act to level landforms. Rock is broken down (weathered), moved to lower elevations by gravity (mass wasting), and carried away by erosional agents such as running water, wind, and ice. Earth's atmosphere plays a key role in these processes.

8. Figure 20.3 is an image of a landscape on Earth that has been dissected by running water. In your own words, describe the drainage pattern in this area.

_____

_____

9. Is the erosional agent shown in Figure 20.4 (running water, wind, *or* ice)?

Erosional agent: _____

10. In what direction did the erosional agent in Figure 20.4 move: (left to right *or* right to left)?

The erosional agent moved from _____.

The fourth geologic process acting on a planet's surface, **impact cratering**, is the consequence of rapidly moving *meteorites* striking a planetary surface. Large craters often have *central peaks* and slump blocks around the rim. In addition, they are usually surrounded by a thick blanket of material called *ejecta*. By

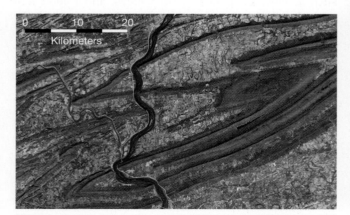

Figure 20.2 Valley and Ridge Province, central Pennsylvania. (Courtesy of NASA)

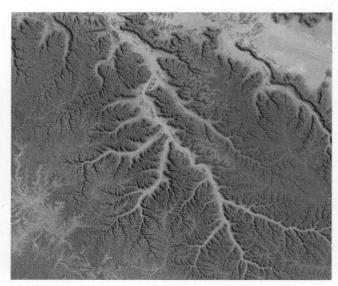

Figure 20.3 Satellite image of dendritic drainage pattern. (Courtesy of NASA)

Figure 20.4 Sand dunes. (Photo by Michael Collier)

contrast, small craters are simple structures with well-defined rims and bowl-shaped depressions. Although impact cratering was important in Earth's evolution, it has played a minor role in producing our modern landscape.

Figure 20.5 is an image of Meteor Crater, an impact crater in Arizona that is about 50,000 years old. Meteor Crater has a diameter of approximately 1,200 meters. Scientists estimate that it was produced by an object about 25 meters across.

11. How many times larger is the diameter of Meteor Crater than the object that produced it?

_____ times larger

12. Suggest a reason why such a comparatively small object can form such a large hole.

_____

Figure 20.5 Meteor Crater. (Courtesy of NASA)

13. Despite being one of the best-preserved and youngest impact craters on Earth, Meteor Crater shows signs of erosion. Describe this evidence.

_____

_____

14. It is possible that Meteor Crater once had a central peak. If it existed, what might have happened to this feature?

15. Label at least three large blocks of ejecta on the image.

Landforms on the Moon

The Moon has two major types of terrains that can be easily seen with the unaided eye—bright highlands and dark smooth regions that resemble seas. The **lunar highlands** are elevated several kilometers above the dark areas called **maria basins**. The maria basins are large impact basins that were subsequently filled with numerous layers of fluid basaltic lava.

16. On the central image of Figure 20.6, label the lunar highlands *A*, label the maria basins *B*, label the large craters called *rayed craters* that have bright rays of ejecta extending far beyond their crater rims *C*, and label the small basins (craters) that have been flooded with basaltic lava *D*. Letters may be used more than once.

17. Which of the areas of the Moon (the bright highlands *or* the dark maria) are more heavily cratered?

_____ are more heavily cratered.

18. Examine Figure 20.6, which shows the side of the Moon that faces Earth. Approximately what percentage of this side of the Moon consists of lunar maria: (20, 40, *or* 70)?

_____ percent

19. Figure 20.6A is an image of Euler crater, located in the western part of Mare Imbrium. Label its central peak, the ejecta surrounding the crater, and the crater rim.

20. Measure the diameter of Euler crater from rim to rim.

_____ kilometers

21. How many times (larger *or* smaller) is Euler crater compared to Meteor Crater?

The Euler crater is about _____

times _____ than Meteor Crater.

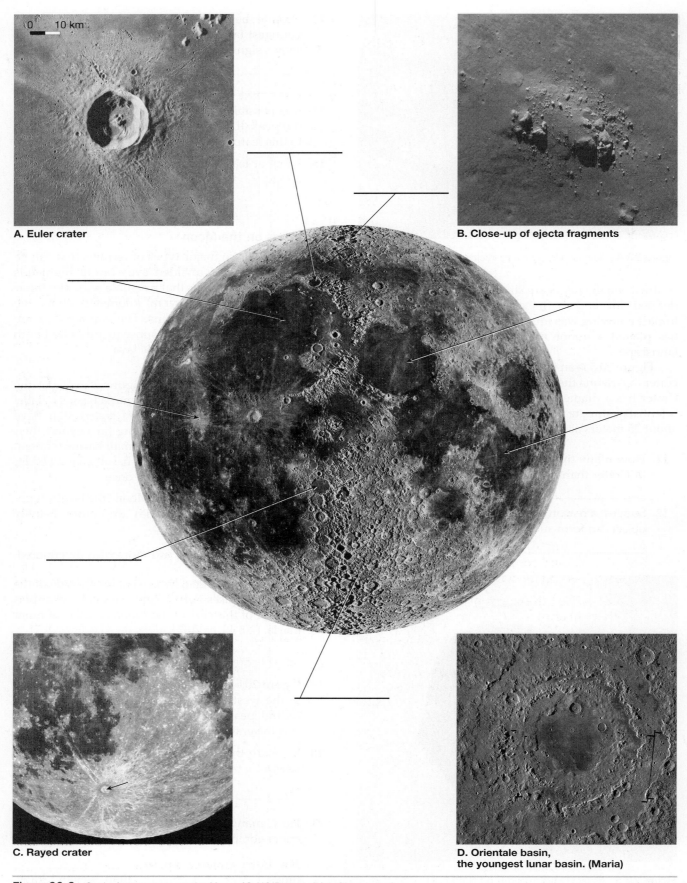

0   10 km

A. Euler crater

B. Close-up of ejecta fragments

C. Rayed crater

D. Orientale basin,
the youngest lunar basin. (Maria)

Figure 20.6 Geologic processes on the Moon. (© UC Regents/Lick Observatory)

Landforms on Mercury

Mercury, the planet closest to the Sun, is only slightly larger than Earth's Moon. Because Mercury lacks an atmosphere, it experiences noontime temperatures approaching 800°F (427°C)—one of the most extreme environments in the solar system. Use the images in Figure 20.7 to answer the following questions.

22. How does the surface of Mercury compare to that of Earth's Moon? List at least one similarity and one difference.

_____

_____

_____

23. Figure 20.7A shows a close-up view of a portion of Mercury's surface. Figure 20.7A resembles what landform on the Moon?

_____

24. Figure 20.7B resembles what feature on the Moon? What is the most likely rock type that covers the central part of this structure?

_____

_____

25. The arrows on Figure 20.7C point to the trace of a cliff-like structure that has an offset of more than 1.6 kilometers (1 mile). Which of the four major geologic processes created this structure?

_____

26. Label a central peak on two of the craters in Figure 20.7D.

27. The ejecta from the large crater at the top left of Figure 20.7D has formed three long chains of very small secondary craters.

    a. Label each of the chains of secondary craters with small arrows.

    b. Label a fault scarp. (*Hint:* Look in the lower-left part of the image.)

28. Of the four major processes that alter a planet's surface, which appears to be the least effective on Mercury?

_____

29. Why does Mercury show little evidence of erosion due to running water, wind, or ice?

_____

Landforms on Venus

Venus, second only to the Moon in brilliance in the night sky, is similar to Earth in size, density, mass, and location in the solar system. However, the similarities end there.

The Venusian atmosphere consists of 97 percent carbon dioxide, a greenhouse gas, which is responsible for surface temperatures that reach 475°C (900°F). Although strong winds occur, surface winds average only a few meters per second, and surface pressures are 90 times greater than those on Earth.

Venus is shrouded in thick, opaque clouds that completely hide the surface from view of traditional cameras. Nevertheless, using radar, which can penetrate clouds, the Magellan spacecraft has provided thousands of radar images of the Venusian surface that reveal its varied topography (Figure 20.8, page 273). In general, radar images show rough topography (fractured rock, rough lava flows, and ejecta) as bright regions, while areas covered by smooth lava flows appear dark. About 80 percent of the surface of Venus consists of low-lying plains covered in lava flows.

Use the Venusian images in Figure 20.8 to answer the following questions.

30. Figure 20.8A shows extensive lava flows on Venus that originated from the volcano Ammavaru, which lies about 300 kilometers west of the image. Outline the margin of the lava flow, which appears bright in this radar image. Recall that rough lava flows appear bright in radar images, and smooth flows appear dark.

Figure 20.8B shows prominent circular volcanic structures referred to as *pancake domes*.

31. Measure the diameter of one of the two largest volcanic domes.

_____ kilometers

32. How does the diameter of the dome compare to the diameter of Earth's volcanic cone you measured in question 2?

Diameter is _____.

33. Do the volcanic domes in this image have (steep conical *or* broad flat) tops?

The volcanic domes have _____. tops.

34. Are the outer margins of these volcanic structures (steep *or* gently sloping)?

The outer margins are _____.

35. Based on your answer to question 34, were these structures produced by (fluid *or* viscous) lavas?

These structures were produced by _____ lavas.

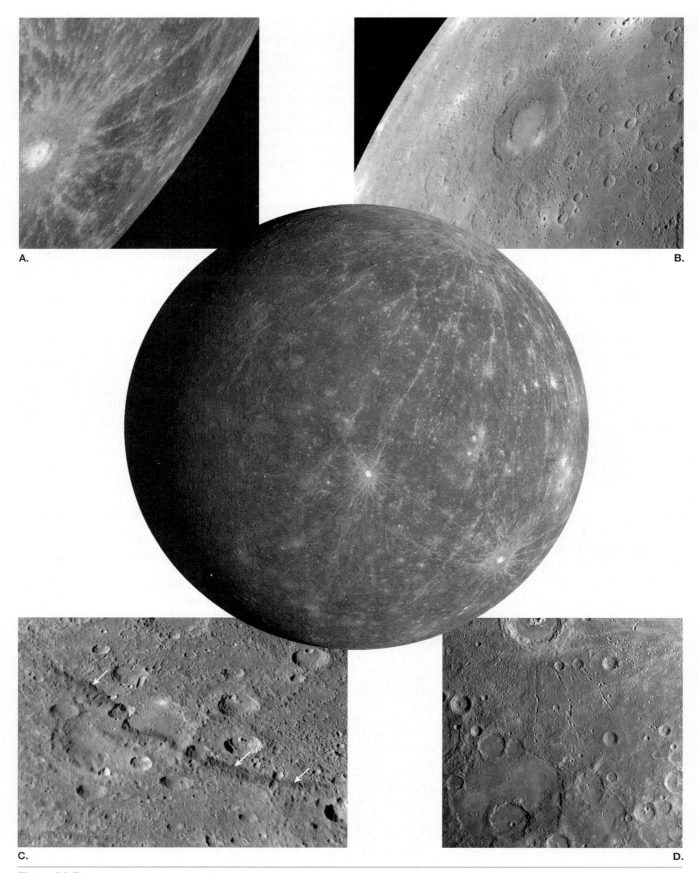

A.

B.

C.

D.

Figure 20.7 Geologic processes on Mercury. (Courtesy of NASA)

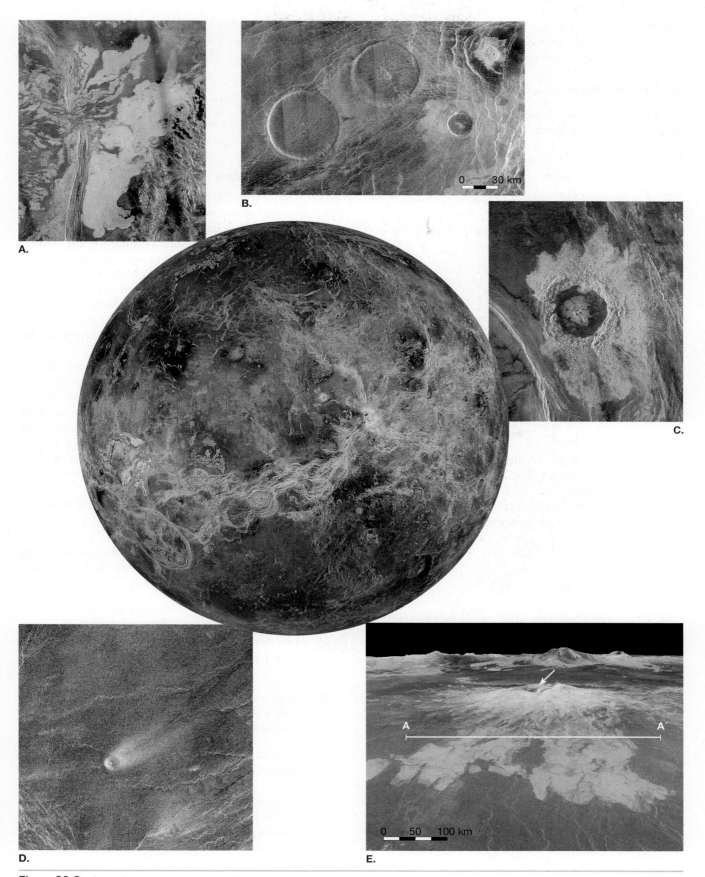

Figure 20.8 Geologic processes on Venus. (Courtesy of NASA)

36. Is the bright feature in the upper-right corner of Figure 20.8B (a volcanic cone *or* an impact crater)? (*Hint:* Compare it to Figure 20.8C.)

   The feature is _____.

   Figure 20.8C is a radar image of Dickinson crater, an impact crater 69 kilometers (42 miles) wide.

37. Compare the size of Dickinson crater to Meteor Crater in Arizona.

38. Dickinson crater is _____ times larger than Meteor Crater.

39. Label the crater's central peak, crater floor, and ejecta on Figure 20.8C.

40. Why are parts of the image in Figure 20.8C bright white in color?

   _____

41. The area surrounding the central peak is dark. What does this tell you about this feature?

   _____

42. Examine Figure 20.8D. Which erosional agent causes the white area extending from this small volcanic cone: (running water, wind, *or* ice)?

   _____

43. The bright areas on the central global view of Venus show highly fractured ridges and canyons of the Aphrodite highlands. What geologic process produced these features: (volcanism, tectonic activity, gradation, *or* impact cratering)?

   _____

   Figure 20.8E shows a false-color satellite image of the volcano Sapas Mons.

44. Using the scale provided, measure the width of the base of Sapas Mons along line A–A'.

   _____ kilometers

45. Approximately how many times wider is Sapas Mons than Mount Shasta, which you examined in question 2?

   Sapas Mons is _____ times wider than Mount Shasta.

46. From its base to its summit, Mount Shasta is about 3.5 kilometers high, whereas Sapas Mons is 1.5 kilometers high. Briefly describe the shape of Sapas Mons.

   _____

47. Approximately 1000 impact craters have been identified on Venus. Is this (fewer *or* more) than on Mercury? Is this (fewer *or* more) than on Earth?

   _____ than on Mercury

   _____ than on Earth

Landforms on Mars

Mars is similar to Earth, with a thin atmosphere, polar ice caps, and seasons. Numerous large shield volcanoes and vast lava plains suggest that Mars has had an extensive volcanic history. Some of the largest volcanoes in the solar system are found on Mars. However, all of the volcanoes on Mars are apparently extinct. Unlike Earth, Mars does not experience plate tectonics, but it does exhibit many tectonic features caused by compressional and extensional stresses. Furthermore, impact craters are very evident on the Martian surface.

Erosion is the dominant geologic process altering the present-day surface of Mars. Mass wasting and wind are agents responsible for landslides and numerous sand dunes, respectively. Moreover, because of the planet's extremely low surface temperatures, the water on or immediately below the surface of Mars is frozen. However, many Martian landforms suggest that, in the past, running water was an effective erosional agent.

After you examine the images of Mars in Figure 20.9, complete the following.

48. The long canyon system near the center of the full-disk image of Mars is Valles Marineris. Measure the length of this feature.

   _____ kilometers

49. How does the length of Valles Marineris compare to the width of the United States?

   _____

50. Based on your answer to question 49, do you think Valles Marineris is (larger *or* smaller) than the Grand Canyon?

   Valles Marineris is _____ than the Grand Canyon.

51. The image shown in Figure 20.9A is a close-up of a valley wall of Valles Marineris. Name the erosional agent responsible for producing the gullies in this image.

Figure 20.9 Geologic processes on Mars. (Courtesy of NASA)

Referring to the impact crater in Figure 20.9B, complete the following.

52. Label the crater rim and the central peak on the image.

53. Because the subsurface of some regions on Mars contains abundant ice, impacts often generate ejecta with a mud-like consistency. Outline the ejecta surrounding this crater.

54. Compare the appearance of the ejecta blanket that surrounds this crater with the ejecta of Dickinson crater shown in Figure 20.8C.

    _____

    _____

Use Figure 20.9C, which shows Olympus Mons, one of four huge volcanoes in a region called Tharsis, to complete the following.

55. What is the approximate width of the Olympus Mons summit caldera? (Note: See white arrow.)

    _____ kilometers

56. How does the width of the caldera atop Olympus Mons compare to the width of Mount Shasta, shown in Figure 20.1A?

    _____

Use Figure 20.9D to answer the following.

57. What are the features at the bottom of the crater? (*Hint:* Refer to the close-up image of similar features in Figure 20.9E.)

    _____

58. What erosional agent produced these features: (wind, water, ice, *or* gravity)?

    _____

59. Draw arrows on Figure 20.9E to show the prevailing wind direction.

Determining the Relative Ages of Landforms

Although the Moon and planets formed at the same time, approximately 4.6 billion years ago, due to variations in the rates of geologic activity, their current surfaces differ significantly in age. When comparing planetary surfaces, impact craters provide a means for determining relative ages. In general, older surfaces show more impact craters per unit area. In addition, old craters lack ejecta blankets and rays because these features are obliterated by subsequent bombardment, and some old craters become partially buried by younger lava flows.

Use Figure 20.6 (page 270) to answer the following questions.

60. Based on crater density, which of these two regions (the lunar highlands *or* the maria basins) is older?

    The _____ are older.

61. Rocks brought back from the lunar maria during the *Apollo* landings are about 3.2–3.8 billion years old. Therefore, are the lunar highlands (older *or* younger) than 3.2–3.8 billion years?

    The lunar highlands are _____ than 3.2–3.8 billion years.

62. Are rayed craters that have bright ejecta deposits around them (older *or* younger) than the regions that surround them? How did you arrive at your answer?

    Rayed craters are _____.

    _____

63. List these lunar features in order, from oldest to youngest: maria, highlands, rayed craters.

    Oldest: _____

    Middle: _____

    Youngest: _____

64. Based on crater density, is the surface of Mercury (Figure 20.7D) (older, younger, *or* about the same age) compared to the maria (Figure 20.6D) on the Moon?

    The surface of Mercury is _____ compared to the maria on the Moon.

65. Does the surface of Mars (Figure 20.9) appear to be (older than, younger than, *or* about the same age as) the Moon's highlands? Does it appear to be (older than, younger than, *or* about the same age as) the surface of Mercury?

    The surface of Mars appears to be _____ the lunar highlands.

    The surface of Mars appears to be _____ the surface of Mercury.

66. Which of the four planetary surfaces that you have investigated has the most active geologic history? Which has the least active geologic history?

    Most active: _____

    Least active: _____

67. Based on crater density, list the surfaces of the four terrestrial planets, from oldest to youngest.

Oldest: _____

Second oldest: _____

Third oldest: _____

Youngest: _____

68. Describe how the surface of Mercury is likely to change during the next billion years.

_____

_____

Impact Cratering Experiment

Impact cratering is one of the most common processes responsible for altering the surfaces of the terrestrial planets. Observe the equipment shown in Figure 20.10 and conduct the following experiment:

Step 1: Gather the equipment necessary to conduct the impact cratering experiment.

Step 2: Add sand to the sandbox. Use a ruler to flatten the surface of the sand.

Step 3: Write a hypothesis describing the suspected relationship between an impact crater's diameter and the mass and velocity of the object that produces it.

_____

_____

_____

Step 4: One at a time, drop each of the balls from heights of 0.5 meter, 1.0 meter, and 1.5 meters on the sand in the box and measure the diameter, in millimeters, of the crater produced each time. Make sure to flatten the surface of the sand between trials. Keeping the variables constant, repeat the experiment three times and average your results. Record the average for each of the trials on the data sheet in Table 20.1.

69. Examine your data closely and state your conclusions concerning the general relationships between

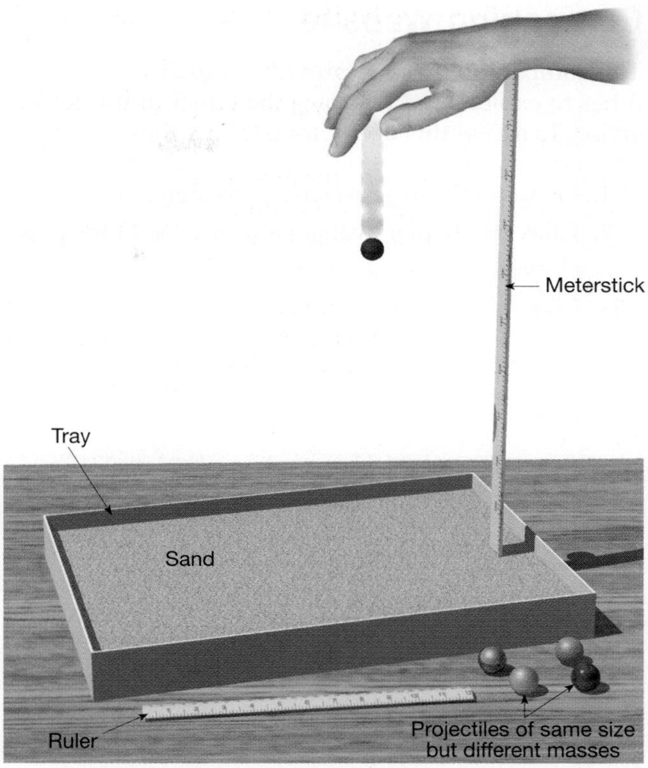

Figure 20.10 Impact cratering lab equipment.

crater size and both the mass and the velocity of the object that produced the crater.

_____

_____

_____

_____

_____

_____

_____

70. Write a general statement that evaluates your impact-cratering hypothesis with reference to your conclusions.

_____

_____

_____

_____

Table 20.1 **Impact Crater Data Sheet**

BALL TYPE		CRATER DIAMETER		
DESCRIPTION	MASS	0.5 M HEIGHT CRATER DIAMETER MM	0.1 M HEIGHT CRATER DIAMETER MM	1.5 M HEIGHT CRATER DIAMETER MM

Companion Website

The companion website provides numerous opportunities to explore and reinforce the topics of this lab exercise. To access this useful tool, follow these steps:

1. Go to www.mygeoscienceplace.com.
2. Click on "Books Available" at the top of the page.

3. Click on the cover of *Applications and Investigations in Earth Science, 7e.*
4. Select the chapter you want to access. Options are listed in the left column ("Introduction," "Web-based Activities," "Related Websites," and "Field Trips").

Examining the Terrestrial Planets

Name _____

Course/Section _____

Date _____

Due Date _____

1. What are the four major geologic processes that have shaped the surfaces of the terrestrial planets and the Moon?

   _____

   _____

   _____

   _____

2. Does the surface of the Moon most resemble the surface of (Mercury, Venus, Earth, *or* Mars)?

   _____

3. Which planet's surface has been mapped exclusively by radar? Why is radar mapping used?

   _____

   _____

4. Briefly describe how the relative ages of various landforms can be determined.

   _____

   _____

   _____

   _____

5. On which planet other than Earth is it most likely that water-deposited sedimentary rocks occur? Explain your choice.

   _____

   _____

6. In Figure 20.11, identify landforms A and B and the planet on which each is found.

   Landform A: _____

   Landform B: _____

   Planet shown: _____

7. Compare the southern half of Figure 20.12 to the northern half. Which surface is older? How did you arrive at your conclusion?

   The _____ half is older.

   _____

   _____

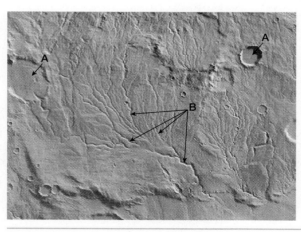

Figure 20.11 Image to accompany question 6.

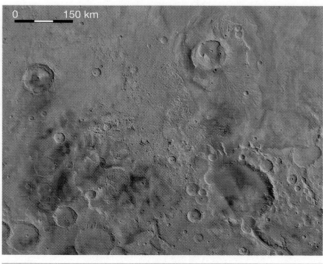

Figure 20.12 Apollinaris Patera and surrounding region. (Courtesy of NASA)

Motions of the Earth–Moon System

Objectives

Completion of this exercise will prepare you to:

- **A.** Recognize and name each of the phases of the Moon.
- **B.** Diagram the position of the Earth, Moon, and Sun for each of the phases of the Moon.
- **C.** Explain the difference between the synodic and sidereal cycles of the Moon.
- **D.** Diagram the positions of the Earth, Moon, and Sun during solar and lunar eclipses.

Materials

metric ruler   hand lens
calculator

Introduction

Earth has one natural satellite—the Moon. In addition to accompanying Earth in its annual trek around the Sun, our Moon orbits Earth about once each month. When viewed from a Northern Hemisphere perspective, the Moon moves counterclockwise (eastward) around Earth. The Moon's orbit is elliptical, causing the Earth–Moon distance to vary by about 6 percent, averaging 384,401 kilometers (238,329 miles).

The motions of the Earth–Moon system constantly change the relative positions of the Sun, Earth, and Moon. The results are some of the most noticeable astronomical phenomena—the *phases of the Moon* and the occasional *eclipses of the Sun and Moon*.

Phases of the Moon

The changing phases of the Moon were among the first astronomical phenomena to be understood. The lunar phases result from the motion of the Moon and sunlight that is reflected from the Moon's surface.

Half of the Moon is always illuminated by the Sun, but the portion that is visible to an observer on Earth changes daily. When the Moon is located between the Sun and Earth, none of the bright side can be seen from Earth—a phase called *new-Moon* ("no-moon"). Conversely, when the Moon lies on the side of Earth opposite the Sun, all of its bright side is visible, producing the *full-Moon* phase. At any position between these extremes, only a fraction of the Moon's illuminated half is visible.

Figure 21.1 is a view of the Earth–Moon system from above the Northern Hemisphere of Earth. Eight positions of the Moon, during its monthly journey around Earth, are illustrated. Complete the following questions using Figure 21.1.

1. For each of the eight numbered positions of the Moon, draw a line through the Moon that separates the half of the Moon that is *visible from Earth* (the half facing Earth) from the half that cannot be seen.

2. With a pencil, darken the half of the Moon that is *not* visible from Earth in each of the eight numbered positions.

3. As the Moon journeys from position 1 to position 5, does the portion of its *illuminated side visible from Earth* (increase or decrease)?

   The portion of the Moon's illuminated side visible from Earth _____.

4. As the Moon journeys from position 5 to position 8, does the portion of its *illuminated side visible from Earth* (increase or decrease)?

   The portion of the Moon's illuminated side visible from Earth _____.

5. Match each of the eight images (A–H) of the lunar phases as seen from Earth with the Moon's corresponding position (1–8) in the space provided below Figure 21.1.

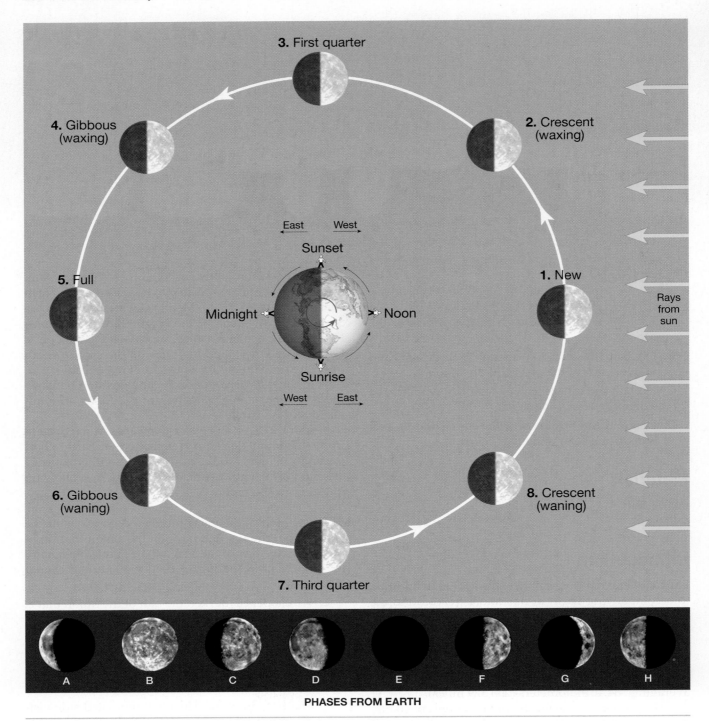

Figure 21.1 The lunar cycle, as viewed from above the Northern Hemisphere of Earth.

POSITION NUMBER	LETTER OF PHASE SEEN FROM EARTH	NAME OF PHASE
1.	_____	_____
2.	_____	_____
3.	_____	_____
4.	_____	_____
5.	_____	_____
6.	_____	_____
7.	_____	_____
8.	_____	_____

Observe the time of day (noon, sunset, and so on) represented by the four positions of an observer on Earth in Figure 21.1. Find the observer in the "noon" position. Because of Earth's counterclockwise rotation, about six hours later, that observer will be in the "sunset" position and will see the Sun setting in the west.

Use the eight positions of the Moon in its orbit and the times represented on Earth to answer the following questions. (*Note:* Except for the new-Moon phase, the Moon is visible from Earth during daylight hours.)

6. Is the full Moon highest in the sky to an Earth-bound observer at (noon, sunset, midnight, *or* sunrise)?

   _____

7. Can a full Moon be observed from Earth at noon? Explain the reason for your answer.

   _____

   _____

8. Does the third-quarter Moon appear highest in the sky at (noon, sunset, midnight, *or* sunrise)?

   The third-quarter is highest at _____.

9. Does a full Moon set—that is, become no longer visible—at (noon, sunset, midnight, *or* sunrise)? Which direction must an observer look (eastward *or* westward) to see the setting full Moon? (*Hint:* Earth's rotation causes the Moon to rise and set.)

   A full Moon sets at _____.

   Observer must look _____ to see the setting full Moon.

10. Can the first- and third-quarter lunar phases be observed during daylight hours? Explain the reason for your answer.

    _____

    _____

11. At approximately what times will the first-quarter Moon rise and set?

    Rise: _____ Set: _____

12. At approximately what times will the third-quarter Moon rise and set?

    Rise: _____ Set: _____

13. Throughout the lunar cycle, the Moon moves further eastward in the sky. Therefore, to an observer on Earth, does the time of day when the Moon is highest in the sky become progressively (earlier *or* later)?

    The time of day when the Moon is highest in the sky become progressively _____.

14. Assume that a crescent-phase Moon is observed in the early evening in the western sky. During the next few days, will the Moon be rising (earlier *or* later)? Will the visible portion of the Moon become progressively (larger *or* smaller)?

    The Moon will be rising _____.

    The visible portion of the Moon will become progressively _____.

Synodic and Sidereal Months

The time interval required for the Moon to complete a full cycle of phases is 29.5 days, a period of time called the **synodic month**. This complete cycle of the phases

of the Moon (that is, one new-Moon to the next new-Moon) is the basis of the word "month" (or "moonth"). Although the cycle of phases requires 29.5 days, the true period of the Moon's 360° revolution around Earth takes only 27.3 days and is known as the **sidereal month**. The difference of approximately 2 days results from the fact that as the Moon revolves around Earth, the Earth–Moon system is also orbiting the Sun.

Figure 21.2 illustrates the month-long motions of the Earth–Moon system around the Sun.

15. On Month 1 in Figure 21.2, indicate the dark half of the Moon on each of the eight lunar positions by shading the area with a pencil.

16. Select from the eight lunar position (labeled 1–8) in Month 1 and indicate which of them represents the following lunar phases:

PHASE	LUNAR POSITION (MONTH 1)
New moon	_____
Third-quarter moon	_____
Full moon	_____
First-quarter moon	_____

17. On Month 1, label the position of the new-Moon phase with the words "new-Moon."

Begin with position 1 (new-Moon phase) in Month 1 and imagine that the Moon is revolving 360° around Earth. At the same time, the Moon is following Earth's orbit toward Month 2.

18. After the 360° revolution is complete, the Moon would return to position 1, and it would be one month later (Month 2). Circle position 1 on the Month 2 diagram.

19. What is the name for one complete 360° revolution of the Moon around Earth? How many days does this take?

    The moon will have completed a _____ month.

    This takes _____ days.

20. Is position 1 in Month 2 the new-Moon phase?

    _____

21. In what position does the new-Moon phase occur in Month 2?

    Position: _____

22. When the Moon moves in its orbit from position 1 to position 2 and is once again in the new-Moon phase, what type of month will it have completed? How many days does this take?

    The Moon will have completed a _____ month.

    This takes _____ days.

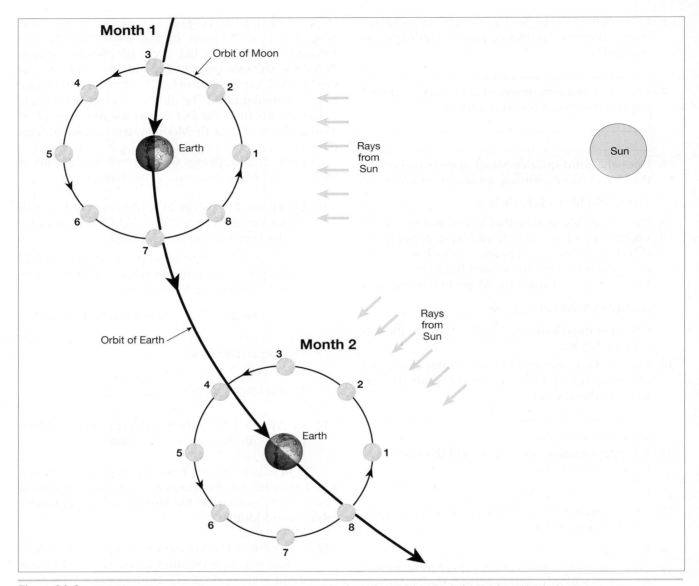

Figure 21.2 Monthly motion of the Earth–Moon system around the Sun, viewed from above the Northern Hemisphere.

23. In your own words, explain the difference between a sidereal month and a synodic month.

_____

_____

Eclipses

Eclipses occur when the Sun, Moon, and Earth are aligned along the same plane. An eclipse can be either a **solar eclipse** (when the Moon moves directly between Earth and the Sun) or a **lunar eclipse** (when the Moon moves within Earth's shadow).

24. In Figure 21.3, sketch and label the positions of the Sun, Earth, and Moon during a solar eclipse.

Figure 21.3 Solar eclipse diagram.

25. Does a solar eclipse occur during the (new-Moon, first-quarter, _or_ full-Moon) phase of the Moon?

_____

Figure 21.4 Lunar eclipse diagram.

26. In Figure 21.4, sketch and label the positions of the Sun, Earth, and Moon during a lunar eclipse.

27. Does a lunar eclipse occur during the (new-Moon, third quarter, *or* full-Moon) phase of the Moon?

_____

28. Suggest a reason why Earth does not experience a solar and lunar eclipse during each month.

_____

_____

_____

Companion Website

The companion website provides numerous opportunities to explore and reinforce the topics of this lab exercise. To access this useful tool, follow these steps:

1. Go to www.mygeoscienceplace.com.
2. Click on "Books Available" at the top of the page.
3. Click on the cover of *Applications and Investigations in Earth Science, 7e.*
4. Select the chapter you want to access. Options are listed in the left column ("Introduction," "Web-based Activities," "Related Websites," and "Field Trips").

Notes and calculations.

Motions of the Earth–Moon System

Name _____   **Course/Section** _____

Date _____   **Due Date** _____

1. Write a description of how the phases of the Moon change during a full lunar cycle. Begin with the new-Moon phase.

   _____

   _____

   _____

   _____

3. Explain the difference between a sidereal month and a synodic month.

   _____

   _____

2. Each of the four photographs in Figure 21.5 were taken when the Moon was at its highest position in the sky. In the space provided below each photo, write the name of the phase represented and the time of day when the picture was taken.

Phase: _____   Phase: _____   Phase: _____   Phase: _____

Time: _____   Time: _____   Time: _____   Time: _____

Figure 21.5 Photos to accompany summary question 2. (Photos © UC Regents/Lick Observatory)

4. Figure 21.6 illustrates the Earth–Moon system viewed from above the Northern Hemisphere. Sketch and label the positions of the Moon that represent the new-Moon, first-quarter, third-quarter, and full-Moon phases.

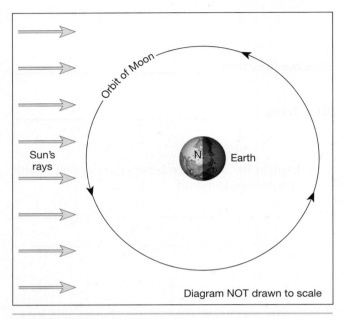

Figure 21.6 Diagram to accompany summary question 4.

PART 5
Earth Science Skills

Photo courtesy of Peter Giovannini/Photolibrary

Location and Distance on Earth

Objectives

Completion of this exercise will prepare you to:

A. Explain the most common system used for locating places and features on Earth.

B. Use Earth's grid system to accurately locate a place or feature.

C. Explain the relationship between latitude and the angle of the North Star (Polaris) above the horizon.

D. Explain the relationship between longitude and solar time.

E. Determine the shortest route and distance between any two places on Earth's surface.

Materials

ruler	calculator
protractor	globe
50- to 80-cm length of string	world wall map
atlas	

Introduction

Globes and maps each have a system of north–south and east–west lines, called **Earth's grid**, that forms the basis for locating points on Earth (Figure 22.1). The grid is, in effect, much like a large sheet of graph paper that has been laid over the surface of Earth. Using the system is very similar to using a graph; that is, the position of a point is determined by the intersection of two lines.

The **latitude** of a location on Earth is the distance of that location north or south of the **equator** (Figure 22.1). The lines (circles) of the grid that extend around Earth in an east–west direction are called **parallels of latitude**. As the name implies, these circles are parallel to one another. Two places on Earth, the **North Pole**

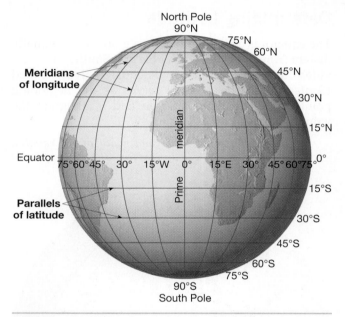

Figure 22.1 Earth's grid system. Parallels of latitude are shown in red, while meridians of longitude are shown in blue.

and **South Pole**, are exceptions; they are points of latitude rather than circles.

The **longitude** of a location on Earth is the distance of that location east or west of the **prime meridian** (Figure 22.1). **Meridians of longitude** are lines (half circles) that extend from the North Pole to the South Pole on one side of Earth. Adjacent meridians are farthest apart on the equator and converge (come together) toward the poles.

The intersection of a parallel of latitude with a meridian of longitude determines the location of a point on Earth's surface.

Earth's shape is nearly spherical. Since parallels and meridians mark distances on a sphere, their designation, like distance around a circle, is given in *degrees* (°). When more precision is necessary, a degree can be subdivided into 60 equal parts, called *minutes* ('), and a minute of angle can be divided into 60 parts, called *seconds* ("). Thus, 31°10'20" means 31 degrees, 10 minutes, 20 seconds.

The scale of a map or globe determines how precisely a place may be located. On detailed maps, it is often possible to estimate latitude and longitude to the nearest degree, minute, and second. On the other hand, when using a world map or globe, it may only be possible to estimate latitude and longitude to the nearest whole degree or two.

In addition to showing location on Earth, latitude and longitude can be used to determine distance. Knowing the shape and size of Earth, the distance in miles and kilometers covered by a degree of latitude or longitude has been calculated.

Determining Latitude

The equator is a circle drawn on a globe that is equally distant from both the North Pole and South Pole. It divides the globe into two equal halves, called **hemispheres**. The equator serves as the beginning point for determining latitude and is assigned the value 0°00'00" latitude.

Latitude is distance north and south of the equator, measured as an angle in degrees from the center of Earth (Figure 22.2).

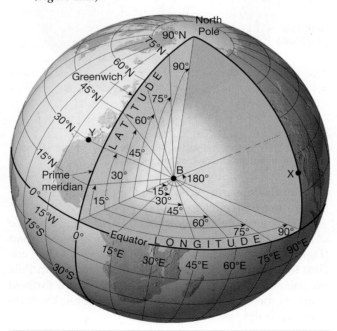

Figure 22.2 Measuring latitude and longitude. The angle measured from the equator to the center of Earth (B) and then northward to the parallel where Point Y is located is 30°. Therefore, the latitude of Point Y is 30°N. All points on the same parallel as Y are designated 30°N latitude.

The angle measured from the prime meridian where it crosses the equator to the center of Earth (B) and then eastward to the meridian where Point X is located is 90°. Therefore, the longitude of Point X is 90°E. All points on the same meridian as X are designated 90°E longitude.

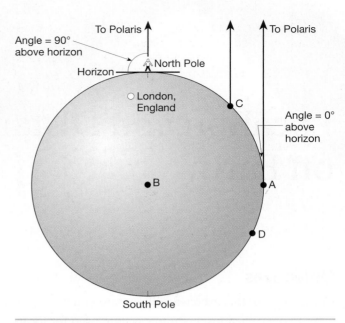

Figure 22.3 Hypothetical Earth.

Latitude begins at the equator, extends north to the North Pole, designated 90°00'00"N latitude (a 90° angle measured north from the equator), and also extends south to the South Pole, designated 90°00'00"S latitude. *The poles and all parallels of latitude, with the exception of the equator, are designated either N (if they are north of the equator) or S (if they are south of the equator).*

1. Locate the equator on a globe. Figure 22.3 represents Earth, with Point B its center. Sketch and label the equator on the diagram in Figure 22.3. Also label the Northern Hemisphere and Southern Hemisphere on the diagram.

2. On Figure 22.3, make an angle by drawing a line from Point A on the equator to Point B (the center of Earth). Then extend the line from Point B to Point C in the Northern Hemisphere. The angle you have drawn (∠ABC) is 45°. Therefore, by definition of latitude, Point C is at 45°N latitude.

3. Draw a line on Figure 22.3 parallel to the equator that intersects Point C. All points on this line are 45°N latitude.

4. Using a protractor, measure ∠ABD on Figure 22.3. Then draw a line parallel to the equator that also goes through Point D. Label the line with its proper latitude.

On a map or globe, parallels may be drawn at any interval.

5. How many degrees of latitude separate the parallels on the globe you are using?

The globe has _____ degrees of latitude between each parallel.

6. Keep in mind that the lines (circles) of latitude are parallel to the equator and to each other. Locate some other parallels on the globe. Sketch and label a few of these on Figure 22.3.

7. Use the diagram that illustrates parallels of latitude in Figure 22.4 to answer questions 7a and 7b.

   a. Draw and label the following additional parallels of latitude on the figure:

      5°N latitude

      10°S latitude

      25°N latitude

   b. Refer to Figure 22.4. Write out the latitude for each designated point as was done for Points A and B. Remember to indicate whether the point is north or south of the equator by writing an *N* or *S* and include the word *latitude*:

      Point A: <u>30°N latitude</u>   Point D: _____

      Point B: <u>5°S latitude</u>   Point E: _____

      Point C: _____   Point F: _____

8. Use a globe or an atlas to locate the cities listed below and give their latitude to the nearest degree. Indicate *N* or *S* and include the word *latitude*:

   Moscow, Russia: _____

   Durban, South Africa: _____

   Your college campus city: _____

9. By using a globe or an atlas, give the name of a city or feature that is equally as far south of the equator as your college campus city is north.

   _____

10. What is the farthest one can be from the equator in degrees of latitude?

    _____ degrees

11. What are the names of the two places on Earth that are farthest from the equator to the north and to the south?

    _____

    _____

There are five special parallels of latitude marked and named on most globes.

12. Use a globe or an atlas to locate the following special parallels and indicate the name given to each:

    NAME OF PARALLEL

    66°30′N latitude: _____

    23°30′N latitude: _____

    0°00′ latitude: _____

    23°30′S latitude: _____

    66°30′S latitude: _____

Latitude and the North Star

Today, most ships use Global Positioning System (GPS) navigational satellites to determine their location. (For information about the GPS, visit the website listed at the end of this exercise.) However, early explorers used the angle of the North Star (a star named Polaris) above the horizon to determine their north–south position in the Northern Hemisphere. As shown on Figure 22.3, someone standing at the North Pole would look overhead (at a 90° angle above the horizon) to see Polaris. Their latitude would be 90°00′00″N. In contrast, someone standing on the equator, 0°00′00″ latitude, would observe Polaris on the horizon (at a 0° angle above the horizon).

Use Figure 22.3 to answer the following questions.

13. Is the angle of Polaris above the horizon for someone standing at Point C (45°, 90°, *or* 180°)?

    _____ degrees

14. Describe the relationship between a particular latitude and the angle of Polaris above the horizon at that latitude.

    _____

    _____

15. What is the angle of Polaris above the horizon at the following cities?

    Fairbanks, Alaska: _____ degrees

    St. Paul, Minnesota: _____ degrees

    New Orleans, Louisiana: _____ degrees

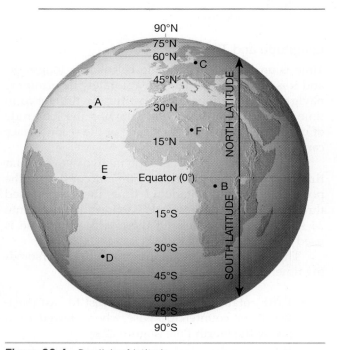

Figure 22.4 Parallels of latitude.

Determining Longitude

Meridians are the north–south lines (half circles) on the globe that converge at the poles and are farthest apart along the equator. Notice on the globe that all meridians are alike. The choice of a zero, or beginning, meridian was arbitrary. The meridian that was chosen by international agreement in 1884 to be 0°00'00" longitude passes through the Royal Astronomical Observatory at Greenwich, England, located near London. This internationally accepted reference for longitude is named the **prime meridian**.

> *Longitude is distance, measured as an angle in degrees east and west of the prime meridian (Figure 22.2).*

Longitude begins at the prime meridian (0°00'00" longitude) and extends to the east and to the west, halfway around Earth, to the 180°00'00" meridian, which is directly opposite the prime meridian. *All meridians, with the exception of the prime meridian and the 180° meridian, are designated either E (if they are east of the prime meridian) or W (if they are west of the prime meridian).*

16. Locate the prime meridian on a globe. Sketch and label it on the diagram of Earth shown in Figure 22.3.

17. Label the Eastern Hemisphere and the Western Hemisphere on Figure 22.3.

On a map or globe, meridians can be drawn at *any* interval.

18. How many degrees of longitude separate each of the meridians on the globe or map you are using?

    The globe or map has _____ degrees of longitude between each meridian.

19. Keep in mind that meridians are farthest apart at the equator and converge at the poles. Sketch and label several meridians on Figure 22.3.

20. Refer to Figure 22.5. Write out the longitude for each designated point as was done for Points A and B below. Remember to indicate east (*E*) or west (*W*) of the prime meridian and include the word *longitude*.

    Point A: <u>30°E longitude</u>   Point D: _____

    Point B: <u>20°W longitude</u>   Point E: _____

    Point C: _____   Point F: _____

21. Use a globe or an atlas to locate the cities listed below and give their longitude to the nearest degree. Indicate *E* or *W* and include the word *longitude*:

    Wellington, New Zealand: _____

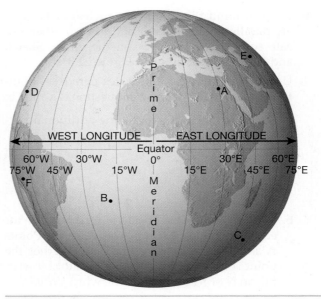

Figure 22.5 Meridians of longitude.

Honolulu, Hawaii: _____

Your college campus city: _____

22. Using a globe or an atlas, give the name of a city, feature, or country that is at the same latitude as your college campus city but equally distant from the prime meridian in the opposite hemisphere.

    _____

23. The farthest a place can be directly east or west of the prime meridian is (45, 90, *or* 180) degrees of longitude.

    _____ degrees

Longitude and Time

Time is an important factor in calculating longitude, and it is especially valuable in accurately determining navigational locations. By knowing the difference in time between two places and knowing the longitude of one of the places, the longitude of the second place can be determined.

Time on Earth can be kept in two ways. **Solar time**, or **Sun time**, determines time according to the position of the Sun. **Standard time**, the system used throughout most of the world, divides the globe into 24 standard time zones. Of these two, solar time is used to determine longitude.

The following facts are important to understanding time:

- Earth rotates on its axis from west to east (eastward), or counterclockwise when viewed from above the North Pole (Figure 22.6).
- It is noon, Sun time, on the meridian that is directly facing the Sun (the Sun has reached its

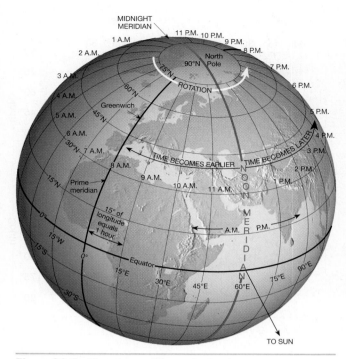

Figure 22.6 The noon meridian and solar time.

highest position in the sky, called the *zenith*) and midnight on the meridian on the opposite side of Earth.

- The time interval from one noon by the Sun to the next noon averages 24 hours and is known as the *mean solar day.*

- Earth turns through 360° of longitude in one mean solar day, which is equivalent to 15° of longitude per hour, or 1° of longitude every 4 minutes of time.

- Places that are east or west of each other, regardless of the distance, have different solar times. For example, people located to the east of the noon meridian have already experienced noon; their time is afternoon (P.M.—*post* ["after"] *meridiem* ["noon meridian"]). People living west of the noon meridian have yet to reach noon; their time is before noon (A.M.—*ante* ["before"] *meridiem* ["noon meridian"]). *Time becomes later going eastward and earlier going westward.*

Use the facts of time to answer the following questions. Remember to indicate A.M. or P.M. with your answers.

24. What would be the solar time of a person living 1° of longitude west of the noon meridian?

Solar time: _____ (A.M., P.M.)

25. What would be the solar time of a person located 4° of longitude east of the noon meridian?

Solar time: _____ (A.M., P.M.)

26. If it is noon, solar time, at 70°W longitude, what is the solar time at each of the following locations?

72°W longitude: _____

65°W longitude: _____

90°W longitude: _____

110°E longitude: _____

Until the invention of specialized and very accurate clocks, called *chronometers*, early navigators could not determine longitude. Today, most navigation is done using satellites, but ships still carry chronometers as a backup system.

A shipboard chronometer is set to keep the time at a known place on Earth, such as the prime meridian. If it is noon by the Sun where the ship is located, and at that same instant the chronometer indicates that it is 8 A.M. on the prime meridian, the ship must be 60° of longitude (4 hours difference × 15° per hour) east (the ship's time is later) of the prime meridian (Figure 22.6). The difference in time need not be in whole hours. A 30-minute difference in time between two places would be equivalent to 7.5° of longitude, 20 minutes would equal 5°, and so forth.

27. Imagine that you are on a ship, and based on the position of the Sun, you determine that it is exactly noon on board. What is the ship's longitude if, at that instant, the time on the prime meridian is the following? (*Note:* Drawing a diagram showing the prime meridian, the ship's location east or west of the prime meridian, and the difference in hours may be helpful.)

6:00 P.M.: _____

1:00 A.M.: _____

2:30 P.M.: _____

Using Earth's Grid System

Using both parallels of latitude and meridians of longitude, you can accurately locate any point on Earth's surface.

28. Using Figure 22.7, determine and record the latitude and longitude of each of the following points. Point A has been completed for reference. Remember to indicate the direction (*N, S, E,* or *W*) and latitude or longitude. Standard practice dictates that *latitude* is listed first, followed by *longitude*.

Point A: ___30°N___ latitude, ___60°E___ longitude

Point B: _____ latitude, _____ longitude

Point C: _____ latitude, _____ longitude

Point D: _____ latitude, _____ longitude

Point E: _____ latitude, _____ longitude

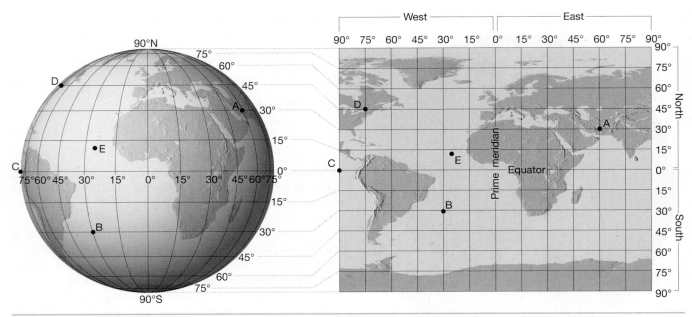

Figure 22.7 Locating places using Earth's grid system.

29. Locate the following points on Figure 22.7. Place a dot on the figure at the proper location and label each point with the designated letter.

Point F: 15°S latitude, 75°W longitude

Point G: 45°N latitude, 0° longitude

Point H: 30°S latitude, 60°E longitude

Point I: 0° latitude, 30°E longitude

30. Use a globe, a map, or an atlas to determine the latitude and longitude of the following cities:

Kansas City, Missouri: _____

Miami, Florida: _____

Oslo, Norway: _____

Auckland, New Zealand: _____

Quito, Ecuador: _____

Baghdad, Iraq: _____

31. Beginning with a world map, and then proceeding to a regional map, determine the city or feature at each of the following locations:

19°28′N latitude, 99°09′W longitude:

_____

41°52′N latitude, 12°37′E longitude:

_____

1°30′S latitude, 33°00′E longitude:

_____

Great Circles, Small Circles, and Distance

Great Circles

A **great circle** is the largest possible circle that can be drawn on a globe (Figure 22.8). Some of the characteristics of a great circle are:

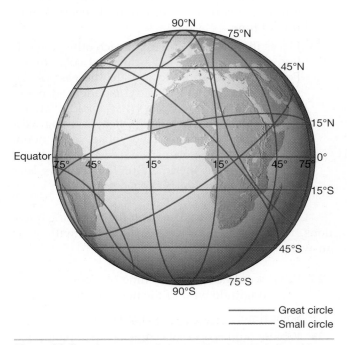

Figure 22.8 A few of the infinite number of great circles and small circles that can be drawn on the globe.

- A great circle divides the globe into two equal parts, called *hemispheres*.
- An infinite number of great circles can be drawn on a globe. Therefore, a great circle can be drawn that passes through any two places on Earth's surface.
- The shortest distance between two places on Earth is along the great circle that passes through those two places.
- If Earth were a perfect sphere, 1° of angle along a great circle would cover identical distances everywhere. But Earth is slightly *flattened* at the poles and *bulges* at the equator, which results in small differences in lengths of degrees. Generally, however, *1° of angle along a great circle equals approximately 111 kilometers, or 69 miles.*

Referring to Figure 22.8 and keeping in mind the characteristics of great circles, examine a globe and answer the following questions.

32. Remember that great circles do not necessarily follow parallels or meridians. Estimate several great circles on the globe by wrapping a piece of string anywhere around the globe such that it divides the globe into *two equal halves*. You should be able to see that there are an infinite number of great circles that can be marked on the globe.

33. Which parallel(s) of latitude is/are a great circle(s)?

_____

Each meridian of longitude is a half circle. If each meridian is paired with the meridian on the opposite side of the globe, a circle is formed.

34. Which meridians, paired with their opposite meridians, are great circles?

_____

Small Circles

Any circle on the globe that does not meet the characteristics of a great circle is considered a **small circle** (Figure 22.8). Therefore, a small circle *does not* divide the globe into two equal parts and *is not* the shortest distance between two places on Earth. Referring to Figure 22.8 and keeping the characteristics of small circles in mind, examine a globe and answer questions 35–37.

35. In general, which parallels of latitude are small circles?

_____

36. Which two latitudes are actually points, rather than circles?

_____

37. In general, which meridians, paired with their opposite meridians, are small circles?

_____

38. Refer to the characteristics of great and small circles to complete the following statements.

   a. All meridians are halves of (great *or* small) circles.

   All meridians are halves of _____ circles.

   b. With the exception of the equator, all parallels are (great *or* small) circles.

   All parallels are _____ circles.

   c. The equator is a (great *or* small) circle.

   The equator is a _____ circle.

   d. The poles are (points *or* lines) of latitude, rather than circles.

   The poles are _____ of latitude.

Determining Distance Along a Great Circle

Determining the distance between two places on Earth when both are on the equator or the same meridian requires two steps:

Step 1: Determine the number of degrees between the two places (degrees of longitude on the equator or degrees of latitude on a meridian).

Step 2: Multiply the number of degrees by 69 miles (or 111 kilometers), the approximate number of kilometers or miles per degree for any great circle.

Use a globe and these steps to answer questions 39 and 40.

39. Approximately how many miles would you travel, going from 10°W longitude to 40°E longitude along the equator?

_____ miles

40. Approximately how many kilometers is London, England, directly north of the equator?

_____ kilometers

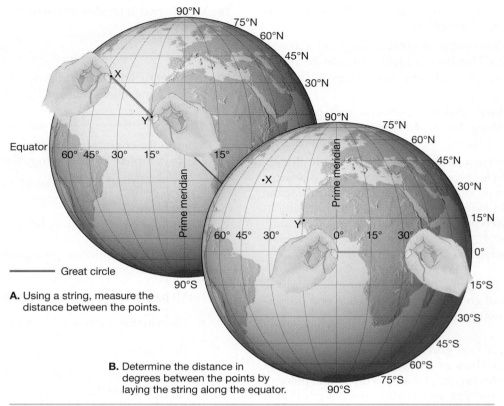

Figure 22.9 Determining the distance between two places on Earth along a great circle other than the equator or meridian. In the example illustrated, the distance between X and Y along the great circle is 30°, which is approximately equivalent to 2070 miles (3330 kilometers).

Determining the shortest distance between two places on Earth that are *not* both on the equator or the same meridian requires these four steps:

Step 1: On a globe, determine the great circle that intersects both places.

Step 2: Stretch a piece of string along the great circle between the two places on the globe and mark the distance between them on the string with your fingers (Figure 22.9A).

Step 3: While still marking the distance with your fingers, place the string on the equator, with one end on the prime meridian. Determine the number of degrees along the great circle between the two places by measuring the marked string's length in degrees of longitude along the equator, which is also a great circle (Figure 22.9B).

Step 4: Multiply the number of degrees along the great circle by 69 miles (or 111 kilometers) to arrive at the approximate distance. (For example, the great circle distance between X and Y in Figure 22.9 would be approximately 2070 miles (30° × 69 miles/degree), or 3330 kilometers (30° × 111 kilometers/degree).

41. Use a globe, a piece of string, and the four steps to determine the approximate great circle distance in degrees, miles, and kilometers from Memphis, Tennessee, to Tokyo, Japan.

    Degrees along the great circle between Memphis and Tokyo = _____°

    Distance along the great circle between Memphis and Tokyo = _____ miles _____ km

Determining Distance Along a Parallel

Since all parallels except the equator are small circles, the length of 1° of longitude along a parallel, other than the equator, will always be less than 111 kilometers (69 miles). Table 22.1 shows the length of 1° of longitude at various latitudes on Earth.

42. Examine a globe. What do you observe about the distance around Earth along each parallel as you get farther away from the equator?

    _____

Table 22.1   **Longitude as Distance**

°LAT.	LENGTH OF 1° LONG. km	miles	°LAT.	LENGTH OF 1° LONG. km	miles	°LAT.	LENGTH OF 1° LONG. km	miles
0	111.367	69.172	30	96.528	59.955	60	55.825	34.674
1	111.349	69.161	31	95.545	59.345	61	54.131	33.622
2	111.298	69.129	32	94.533	58.716	62	52.422	32.560
3	111.214	69.077	33	93.493	58.070	63	50.696	31.488
4	111.096	69.004	34	92.425	57.407	64	48.954	30.406
5	110.945	68.910	35	91.327	56.725	65	47.196	29.314
6	110.760	68.795	36	90.203	56.027	66	45.426	28.215
7	110.543	68.660	37	89.051	55.311	67	43.639	27.105
8	110.290	68.503	38	87.871	54.578	68	41.841	25.988
9	110.003	68.325	39	86.665	53.829	69	40.028	24.862
10	109.686	68.128	40	85.431	53.063	70	38.204	23.729
11	109.333	67.909	41	84.171	52.280	71	36.368	22.589
12	108.949	67.670	42	82.886	51.482	72	34.520	21.441
13	108.530	67.410	43	81.575	50.668	73	32.662	20.287
14	108.079	67.130	44	80.241	49.839	74	30.793	19.126
15	107.596	66.830	45	78.880	48.994	75	28.914	17.959
16	107.079	66.509	46	77.497	48.135	76	27.029	16.788
17	106.530	66.168	47	76.089	47.260	77	25.134	15.611
18	105.949	65.807	48	74.659	46.372	78	23.229	14.428
19	105.337	65.427	49	73.203	45.468	79	21.320	13.242
20	104.692	65.026	50	71.727	44.551	80	19.402	12.051
21	104.014	64.605	51	70.228	43.620	81	17.480	10.857
22	103.306	64.165	52	68.708	42.676	82	15.551	9.659
23	102.565	63.705	53	67.168	41.719	83	13.617	8.458
24	101.795	63.227	54	65.604	40.748	84	11.681	7.255
25	100.994	62.729	55	64.022	39.765	85	9.739	6.049
26	100.160	62.211	56	62.420	38.770	86	7.796	4.842
27	99.297	61.675	57	60.798	37.763	87	5.849	3.633
28	98.405	61.121	58	59.159	36.745	88	3.899	2.422
29	97.481	60.547	59	57.501	35.715	89	1.950	1.211
30	96.528	59.955	60	55.825	34.674	90	0.000	0.000

43. Use Table 22.1 to determine the length of 1° of longitude at each of the following parallels:

LENGTH OF 1° OF LONGITUDE

15° latitude: _____ km, _____ miles

30° latitude: _____ km, _____ miles

45° latitude: _____ km, _____ miles

80° latitude: _____ km, _____ miles

44. Use the Earth's grid illustrated in Figure 22.7 to determine the distances between the following points:

Distance between Points D and G:

_____ degrees × _____ miles/degree

= _____ miles

Distance between Points B and H:

_____ degrees × _____ km/degree

= _____ km

Memphis, Tennessee, and Tokyo, Japan, are both located at about 35°N latitude.

45. Use a globe or world map to determine how many degrees of longitude separate Memphis, Tennessee, from Tokyo, Japan.

_____ degrees of longitude separate Memphis, Tennessee, and Tokyo, Japan.

46. From Table 22.1, what is the length of 1° of longitude at latitude 35°N?

_____ miles

47. How many miles is Tokyo, Japan, *directly* west of Memphis, Tennessee?

_____ miles

48. In question 41, you determined the great circle distance between Memphis, Tennessee, and Tokyo, Japan. How many miles shorter is the great circle route between these cities than the east–west distance along a parallel (see question 47)?

The great circle route is _____ miles shorter.

Companion Website

The companion website provides numerous opportunities to explore and reinforce the topics of this lab exercise. To access this useful tool, follow these steps:

1. Go to www.mygeoscienceplace.com.
2. Click on "Books Available" at the top of the page.
3. Click on the cover of *Applications and Investigations in Earth Science, 7e.*
4. Select the chapter you want to access. Options are listed in the left column ("Introduction," "Web-based Activities," "Related Websites," and "Field Trips").

Location and Distance on Earth

Name _____   **Course/Section** _____

Date _____   **Due Date** _____

1. In Figure 22.10, prepare a diagram illustrating Earth's grid system. Include and label the equator and prime meridian. Explain the system used for locating points on the surface of Earth, referring to the diagram if necessary.

   _____

   _____

   _____

2. Define the following terms:

   Parallel of latitude: _____

   _____

   Meridian of longitude: _____

   _____

   Great circle: _____

   _____

3. Determine whether the following statements are true or false. If a statement is false, reword it to make it a true statement.

   T F　a. The distance measured north or south of the prime meridian is called latitude.

   T F　b. Pairing any meridian with its opposite meridian on Earth forms a great circle.

   T F　c. The equator is the only meridian that is a great circle.

4. What is the relationship between the latitude of a place in the Northern Hemisphere and the angle of Polaris above the horizon at that place?

   _____

5. Approximately how many miles does 1° equal along a great circle?

   1° along a great circle equals _____ miles.

6. What is the latitude and longitude of your home city?

   Latitude: _____

   Longitude: _____

7. Use a globe or map to determine, as accurately as possible, the latitude and longitude of Athens, Greece.

   Latitude: _____

   Longitude: _____

8. Write a brief paragraph describing how to determine the shortest distance between two places on Earth's surface.

   _____

   _____

   _____

Figure 22.10　Diagram of Earth's grid system.

9. From question 48 of the exercise, how many miles shorter is the great circle route between Memphis, Tennessee, and Tokyo, Japan, than the straight east–west distance along a parallel?

    _____ miles shorter

10. Approximately how many miles is it from London, England, to the South Pole? (Show your calculation.)

    _____

    _____ miles

11. Using Figure 22.11, determine and record the latitude and longitude of each of the following points.

    Point A: _____ latitude, _____ longitude

    Point B: _____ latitude, _____ longitude

    Point C: _____ latitude, _____ longitude

    Point D: _____ latitude, _____ longitude

    Point E: _____ latitude, _____ longitude

12. You are shipwrecked and floating in the Atlantic Ocean, somewhere between London, England, and New York, New York. Fortunately, you managed to save your globe. You were most recently in London, so your watch is still set for London time. It is noon, by the Sun, at your location. Your

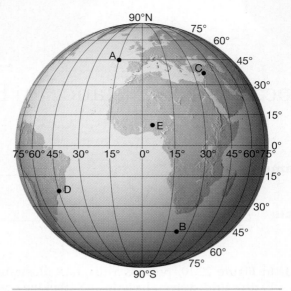

Figure 22.11 Locating places using Earth's grid.

watch indicates that it is 4 P.M. in London. Are you closer to the United States or to England? Explain how you arrived at your answer.

_____

_____

_____

The Metric System, Measurements, and Scientific Inquiry

Objectives

Completion of this exercise will prepare you to:

A. List the units for length, mass, and volume that are used in the metric system.

B. Use the metric system for measurements.

C. Convert units within the metric system.

D. Understand and use micrometers and nanometers for measuring very small distances as well as astronomical units and light-years for measuring large distances.

E. Determine the approximate density and specific gravity of a solid substance.

F. Conduct a scientific experiment, using accepted methods of scientific inquiry.

Materials

metric ruler
metric tape measure or
  meterstick
nickel coin
small rock
thread
large graduated cylinder
  (marked in milliliters)

calculator
paper clip
metric balance
"bathroom" scale
  (metric)

Introduction

Earth science involves studying phenomena on many different scales—from atoms to galaxies. Almost every scientific investigation requires accurate measurements. To describe objects, Earth scientists use units of measurement that are appropriate for the particular feature or phenomenon being studied. For example, they would use centimeters or inches, instead of kilometers or miles, to measure the width of this page; they would use kilometers or miles, rather than centimeters or inches, to measure the distance from New York to London, England.

Most areas of science have developed units of measurement that meet their particular needs. However, regardless of the unit used, all scientific measurements are defined within a broader system so that they can be understood and compared. In science, the fundamental units have been established by the *International System of Units* (SI, Système International d'Unités) (Table 23.1).

The Metric System

The **metric system** is a decimal system (based on fractions or multiples of 10) that uses only one basic unit for each type of measurement: the **meter** (m) as the unit of length (Figure 23.1), the **liter** (l) as the unit of volume (1 liter is equal to the volume of 1 kilogram of pure water at 4°C [39.2°F], about 1.06 quarts.), and the **gram** (g) as the unit of mass (Figure 23.2). In the English system, the units used to express the same relations are feet, quarts, and ounces.

Working with the Metric System

In the metric system, the basic units of weights and measures are related by multiples of ten. It is similar to the U.S. monetary system, where 10 pennies = 1 dime and

Table 23.1 Base Units of the SI[1]

UNIT	QUANTITY MEASURED	SYMBOL
Meter	Length	m
Kilogram	Mass	kg
Second	Time	s
Kelvin	Thermodynamic temperature	K
Ampere	Electric current	A
Mole	Quantity of a substance	mol
Candela	Luminous intensity	cd

[1]From these base units, other units are derived to express quantities such as power (watt, W), force (newton, N), energy (joule, J), and pressure (pascal, Pa).

METER (m)

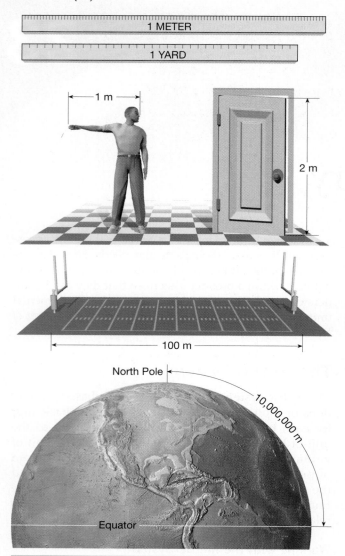

Figure 23.1 The SI unit of length is the meter (m), which is slightly longer than a yard. Originally described as one ten-millionth of the distance from the equator to the North Pole, it is currently defined as the distance traveled by light in a vacuum in 0.0000000033 second.

10 dimes = 1 dollar. In the English system of weights and measures, no such regularity exists; for example, 12 inches = 1 foot and 5280 feet = 1 statute mile. Thus, the advantage of the metric system is *consistency*.

Table 23.2 lists the prefixes that are used in the metric system to indicate how many times more (in multiples of 10) or what fraction (in fractions of 10) of the basic unit is present. Therefore, from the information in the table, you see that 1 *kilo*gram (kg) = 1000 grams, while 1 *milli*gram (mg) = 1/1000 gram.

To familiarize yourself with metric units, determine the following measurements, using the equipment provided in the laboratory.

GRAM (g)

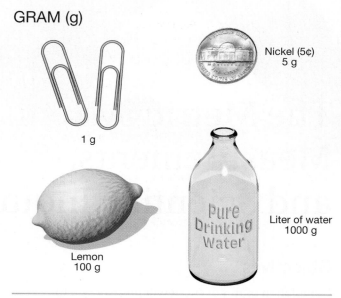

Figure 23.2 The basic unit of mass in the metric system is the gram (g), approximately equal to the mass of 1 cubic centimeter of pure water at 4°C (39.2°F). A gram is about the weight of 2 paper clips, and an ounce is about the weight of 40 paper clips.

Measuring length:

1. Use a metric measuring tape (or meterstick) to measure your height as accurately as possible to the nearest hundredth of a meter (called a centimeter).

   _____ centimeters (cm)

2. Use a metric ruler to measure the length of this page as accurately as possible to the nearest tenth of a centimeter (called a millimeter).

   _____ millimeter (mm)

3. Accurately measure the length of your shoe to the nearest millimeter.

   _____ millimeters (mm)

Measuring volume:

4. Use a graduated cylinder to measure the volume of the paper cup to the nearest milliliter.

   _____ milliliters (ml)

Measuring mass:

5. Weigh the following and record your results. (Follow the directions of your instructor for using a metric balance.)

   Sample of rock: _____ grams (g)

   Paper clip: _____ grams (g)

   Nickel coin: _____ grams (g)

Table 23.2   Metric Prefixes and Symbols

PREFIX[1]	SYMBOL[2]	MEANING
giga-	G	1 billion times base unit (1,000,000,000 × base)
mega-	M	1 million times base unit (1,000,000 × base)
kilo-	k	1 thousand times base unit (1000 × base)
hecto-	h	1 hundred times base unit (100 × base)
deka-	da	10 times base unit (10 × base)
BASE UNIT	m (meter)–base unit of length l (liter)–base unit of volume g (gram)–base unit of mass	
deci-	d	one-tenth the base unit (.1 × base)
centi-	c	one-hundredth the base unit (.01 × base)
milli-	m	one-thousandth the base unit (.001 × base)
micro-	μ	one-millionth the base unit (.000001 × base)
nano-	n	one-billionth the base unit (.000000001 × base)

[1]A prefix is added to the base unit to indicate how many times more or what fraction of the base unit is present. For example, 1 kilometer (km) = 1000 meters, and 1 millimeter (mm) = 1/1000 meter.

[2]When writing in the SI system, periods are not used after the unit symbols, and symbols are not made plural. For example, a length of 50 centimeters would be written as "50 cm," not "50 cm." or "50 cms."

(*Note:* Two terms that are often confused are *mass* and *weight*. Mass is a measure of the amount of matter an object contains. Weight is a measure of the force of gravity on an object. For example, the mass of an object would be the same on both Earth and the Moon. However, because the gravitational force of the Moon is less than that of Earth, the object would weigh less on the Moon. On Earth, mass and weight are directly related, and often the same units are used to express the two.)

6. Use the metric "bathroom" scale. Weigh yourself as accurately as possible, to the nearest tenth of a kilogram. (*Note:* If a metric scale is not available, convert your weight in pounds to kilograms by multiplying your weight in pounds by 0.45.)

   _____ kilograms (kg)

Metric Conversions

As stated earlier, one important advantage of the metric system is that it is based on multiples of ten. As shown on the metric conversion diagram in Figure 23.3, conversion from one unit to another can be accomplished simply by *moving the decimal point* to the left if going to larger units or by moving the decimal point to the right if going to smaller units.

For example, if you measure the length of a piece of string and find that it is 1.43 decimeters long, in order to convert its length to millimeters, start with 1.43 on the "deci-" step of the diagram. Then move the

decimal two places (steps) to the right (the "milli-" step). The length, in millimeters, is 143.0 millimeters.

7. Use the metric conversion diagram in Figure 23.3 to convert the following:

   a. 2.05 meters (m) = _____ centimeters (cm)

   b. 1.50 meters (m) = _____ millimeters (mm)

   c. 9.81 liters (l) = _____ deciliters (dl)

   d. 5.4 grams (g) = _____ milligrams (mg)

   e. 6.8 meters (m) = _____ kilometer (km)

   f. 4,214.6 centimeter (cm) = _____ meters (m)

   g. 321.50 gram (g) = _____ kilogram (kg)

   h. 70.73 hectoliter (hl) = _____ dekaliters (dal)

8. Use a metric tape measure (or meterstick) to determine the length of your laboratory table as accurately as possible, to the nearest hundredth of a meter. Then convert the length to each of the units in question 8b.

   a. Length of table: _____ meters

   b. Length of table equals:

      _____ millimeters (mm)

      _____ centimeters (cm)

      _____ kilometers (km)

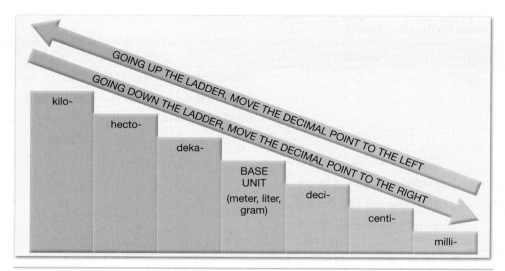

Figure 23.3 Metric conversion diagram. Beginning at the appropriate step, if going to larger units, move the decimal to the left for each step crossed. When going to smaller units, move the decimal to the right for each step crossed. For example, 1.253 meters (base unit step) would be equivalent to 1253.0 millimeters (decimal moved three steps to the right, the milli- step).

Metric–English Conversions

Because the change to the metric system will occur gradually over the next few decades, we will be forced to use both the metric and English systems simultaneously. If we are not able to convert from one system to the other, we will occasionally be inconvenienced or mildly frustrated.

Use the conversion tables on the inside back cover of this manual to figure out the metric equivalent for each of the following units.

Length conversion:

9. 1 inch = _____ centimeters

10. 1 meter = _____ feet

11. 1 mile = _____ kilometers

Volume conversion:

12. 1 gallon = _____ liters

13. 1 cubic inch = _____ cubic centimeters

Mass conversion:

14. 1 ounce = _____ gram

15. 1 pound = _____ kilogram

Temperature

Temperature represents one relatively common example of using different systems of measurement. On the Fahrenheit temperature scale, 32°F is the melting point of ice, and 212°F marks the boiling point of water (at standard atmospheric pressure). On the **Celsius scale**, ice melts at 0°C and water boils at 100°C. On the **Kelvin scale**, ice melts at 273 K.

Conversion from one temperature scale to another can be accomplished using either an equation or a graphic comparison scale. To convert Celsius degrees to Fahrenheit degrees, the equation is °F = (1.8)°C + 32°. To convert Fahrenheit degrees to Celsius degrees, the equation is °C = (°F − 32°)/1.8. To convert Kelvins (K) to Celsius degrees, subtract 273 and add the degree symbol.

16. Convert the following temperatures to their equivalents. Do the first four conversions using the appropriate equation, and do the others using the temperature comparison scale on the inside back cover of this manual.

  a. On a cold day, it was 8°F = _____ °C.

  b. Ice melts at 0°C = _____ °F.

  c. Room temperature is 72°F = _____ °C.

  d. A hot summer day was 35°C = _____ °F.

  e. Normal body temperature is 98.6°F = _____ °C.

  f. A warm shower is 27°C = _____ °F.

  g. Hot soup is 72°C = _____ °F.

  h. Water boils at 212°F = _____ K.

17. Use the temperature comparison scale on the inside back cover to answer the following:

  a. The thermometer reads 28°C. Will you need your winter coat?

  _____

b. The thermometer reads 10°C. Will the outdoor swimming pool be open today?

_____

c. If your body temperature is 40°C, do you have a fever?

_____

d. The temperature of a cup of cocoa is 90°C. Will it burn your tongue?

_____

e. Your bath water is 15°C. Will you have a scalding, warm, or chilly bath?

_____

f. "Who's been monkeying with the thermostat? It's 37°C in this room." Are you shivering or perspiring?

_____

Metric Review

Use what you have learned about the metric system to determine whether the following statements are _reasonable_. Write "yes" or "no" in the blanks. _Do not_ convert these units to English equivalents; just _estimate_ the value of each.

18. A man weighs 90 kilograms.

_____

19. A fire hydrant is 1 meter tall.

_____

20. A college student drank 3 kiloliters of coffee last night.

_____

21. The room temperature is 295 K.

_____

22. A dime is 1 millimeter thick.

_____

23. Sugar will be sold by the milligram.

_____

24. The temperature in Paris today is 80°C.

_____

25. The bathtub has 80 liters of water in it.

_____

26. You will need a coat if the outside temperature is 30°C.

_____

27. A pork roast weighs 18 grams.

_____

Special Units of Measurement

Scientists often use special units to measure various phenomena. Most of them are defined using SI units.

Very Small Distances

Two units commonly used to measure very small distances are the **micrometer** (symbol μm), also known as the **micron**, and the **nanometer** (symbol nm).

By definition, 1 micrometer equals 0.000001 m (one-millionth of a meter). There are 1 million micrometers in 1 meter and 10,000 micrometers in 1 centimeter. Also by definition, 1 nanometer equals 0.000000001 m (one-billionth of a meter).

28. Are there (10, 100, _or_ 1000) nanometers in a micrometer?

There are _____ nanometers in a micrometer.

29. What is the length of a 2.5-centimeter line, expressed in micrometers and nanometers?

The line is _____ micrometers long.

The line is _____ nanometers long.

30. Some forms of radiation (for example, light) travel in very small waves, with distances from crest to crest of about 500 nanometers (0.5 μm) How many of these waves would it take to equal 1 centimeter?

_____ waves in 1 centimeter

Very Large Distances

Astronomers must measure very large distances, such as the distances between planets or the distances to the stars and beyond. To simplify their measurements, they have developed special units, including **astronomical units** (symbol AU) and **light-years** (symbol LY).

The astronomical unit is a unit for measuring distance within the solar system. One astronomical unit is equal to the average distance of Earth from the Sun. This average distance is 150 million kilometers, which is approximately equal to 93 million miles.

31. The planet Saturn is 1427 million kilometers from the Sun. How many AUs is Saturn from the Sun?

_____ AUs from the Sun

The light-year is used for measuring distances to the stars and beyond. One light-year is defined as the distance that light travels in a vacuum in one year. This distance is about 6 trillion miles (that is, 6,000,000,000,000 miles).

32. Approximately how many kilometers will light travel in one year?

_____ kilometers per year

33. The nearest star to Earth, excluding our Sun, is named Proxima Centauri. It is about 4.27 light-years away. What is the distance of Proxima Centauri from Earth in both miles and kilometers?

_____ miles

_____ kilometers

Density and Specific Gravity

Two important properties of a material are its **density** and **specific gravity**. Density is the mass of a substance per unit volume, usually expressed in grams per cubic centimeter (g/cm^3) in the metric system. The specific gravity of a solid is the *ratio* of the mass of a given volume of a substance to the mass of an equal volume of some other substance taken as a standard (usually water at 4°C). Because specific gravity is a ratio, it is expressed as a pure number and has no units. For example, a specific gravity of 6 means that the substance has six times more mass than an equal volume of water. Because the density of pure water at 4°C is $1g/cm^3$, the specific gravity of a substance will be numerically equal to its density.

The approximate density and specific gravity of a rock or another solid can be determined by using the following steps:

Step 1: Determine the mass of the rock, using a metric balance.

Step 2: Fill a graduated cylinder that has its divisions marked in milliliters approximately two-thirds full with water. Note the level of the water in the cylinder, in milliliters.

Step 3: Tie a thread to the rock and immerse the rock in the water in the graduated cylinder. Note the new level of the water in the cylinder.

Step 4: Determine the difference between the beginning level and the after-immersion level of the water in the cylinder.

Step 5: Calculate the density and specific gravity by using the following information and appropriate equations.

A milliliter of water has a volume approximately equal to 1 cubic centimeter (cm³). Therefore, the difference between the beginning water level and the after-immersion water level in the cylinder equals the volume of the rock in cubic centimeters. Furthermore, *1 cubic centimeter (1 milliliter) of water has a mass of approximately 1 gram.* Therefore, the difference between the beginning water level and the after-immersion water level in the cylinder is the mass of a volume of water equal to the volume of the rock.

Using the steps listed above for determining density and specific gravity, complete questions 34 and 35.

34. Determine the density and specific gravity of a small rock sample by completing questions 34a–34f.

a. Mass of rock sample: _____ grams

b. After-immersion level of water: _____ ml

Beginning level of water in cylinder: _____ ml

Difference: _____ ml

c. Volume of rock sample: _____ cm^3

d. Mass of a volume of water equal to the volume of the rock: _____ g

e. Density of rock:

$$\text{Density} = \frac{\text{Mass of rock (g)}}{\text{Volume of rock (cm}^3)}$$

$$= \underline{\qquad} g/cm^3$$

f. Specific gravity of rock:

Specific gravity

$$= \frac{\text{Mass of rock (g)}}{\text{Mass of an equal volume of water (g)}}$$

$$= \underline{\qquad}$$

35. As a means of comparison, your instructor may ask that you determine the density and/or specific gravity of other objects. If so, record your results in the following spaces.

a. Object: _____

Density: _____ g/cm^3

Specific gravity: _____

b. Object: _____

Density: _____ g/cm^3

Specific gravity: _____

Methods of Scientific Inquiry

Scientists use many methods in attempting to understand natural phenomena. Some scientific discoveries represent purely theoretical ideas, and others occur by chance. However, scientific knowledge is often gained by following this sequence of steps:

Step 1: Establish a **hypothesis**—a tentative, or untested, explanation.

Step 2: Gather data and conduct experiments to test the hypothesis.

Step 3: Accept, modify, or reject the hypothesis, on the basis of extensive data gathering or experimentation.

The following simple inquiry should help you understand the process.

Step 1—Establishing a Hypothesis

Observe all the people in the laboratory and pay particular attention to each individual's height and shoe length.

36. Based on your observations, write a hypothesis that relates a person's height to his or her shoe length.

    Hypothesis: _____

    _____

    _____

Step 2—Gathering Data

Previously, in questions 1 and 3 of the exercise, each person in the laboratory measured his or her height and shoe length.

37. Gather your data by asking 10 or more people in the lab for their height and shoe length measurements. Enter your data in Table 23.3, recording height to the nearest hundredth of a meter and shoe length to the nearest millimeter.

Step 3—Evaluating the Hypothesis Based Upon the Data

Plot all your data from Table 23.3 on the height versus shoe length graph in Figure 23.4, by locating a person's height on the vertical axis and his or her shoe length on the horizontal axis. Then place a dot on the graph where the two intersect.

38. Describe the pattern of the data points (dots) on the height versus shoe length graph in Figure 23.4.

For example, are the points scattered all over the graph, or do they appear to follow a line or curve?

_____

_____

39. Draw a single line on the graph that appears to average, or best fit, the pattern of the data points.

40. Describe the relationship of height to shoe length that is illustrated by the line on your graph.

_____

_____

_____

41. Ask several people, whose height and shoe length *are not* part of your data set, for their height. Then see how accurately your line predicts what their shoe length should be. Do this by marking each person's height on the vertical axis and then follow a line straight across to the right until you intersect the line on the graph. Read the predicted shoe length from the axis directly below the point of intersection.

42. Summarize how accurately your graph predicts a person's shoe length, given only his or her height.

_____

_____

_____

43. Using your graph's ability to make predictions as a guide, do you think you should accept, reject,

Table 23.3   Data Table for Recording Height and Shoe Length Measurements of People in the Lab

PERSON	HEIGHT (NEAREST HUNDREDTH OF A METER)	SHOE LENGTH (NEAREST MILLIMETER)
1	_____ . _____ m	_____ mm
2	_____ . _____ m	_____ mm
3	_____ . _____ m	_____ mm
4	_____ . _____ m	_____ mm
5	_____ . _____ m	_____ mm
6	_____ . _____ m	_____ mm
7	_____ . _____ m	_____ mm
8	_____ . _____ m	_____ mm
9	_____ . _____ m	_____ mm
10	_____ . _____ m	_____ mm
11	_____ . _____ m	_____ mm
12	_____ . _____ m	_____ mm
13	_____ . _____ m	_____ mm
14	_____ . _____ m	_____ mm
15	_____ . _____ m	_____ mm

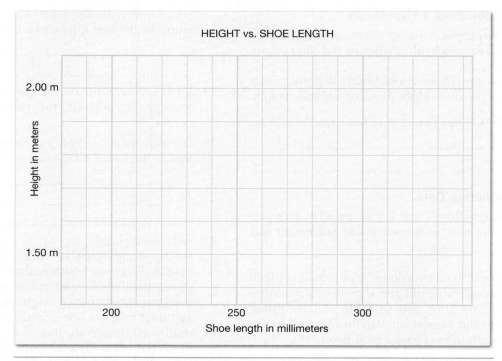

Figure 23.4 Height versus shoe length graph.

or modify your original hypothesis? Give the reason(s) for your choice.

_____

_____

_____

Your study has been restricted to people in your laboratory.

44. Why would your ability to make predictions have been more accurate if you had used the heights and shoe lengths of 10,000 people to construct your graph?

_____

_____

_____

Drawing hasty conclusions with limited data can often cause problems. In science, you can never have too much data. Experiments are repeated many times by many different people before results are accepted by the scientific community.

Companion Website

The companion website provides numerous opportunities to explore and reinforce the topics of this lab exercise. To access this useful tool, follow these steps:

1. Go to www.mygeoscienceplace.com.
2. Click on "Books Available" at the top of the page.
3. Click on the cover of *Applications and Investigations in Earth Science, 7e.*
4. Select the chapter you want to access. Options are listed in the left column ("Introduction," "Web-based Activities," "Related Websites," and "Field Trips").

The Metric System, Measurements, and Scientific Inquiry

Name _____   **Course/Section** _____

Date _____   **Due Date** _____

1. List the basic metric unit and symbol used for each of these measurements:

   Length: _____

   Mass: _____

   Volume: _____

2. Convert the following units:
   a. 2 liters = _____ deciliters
   b. 600 millimeters = _____ meter
   c. 72° F = _____ °C
   d. 0.32 kilogram = _____ grams
   e. 12 grams = _____ milligrams

3. Indicate by answering "yes" or "no" whether the following statements are reasonable:
   a. A person is 600 centimeters tall.

   _____

   b. A bag of groceries weighs 5 kilograms.

   _____

   c. It took 52 liters of gasoline to fill the car's empty gasoline tank.

   _____

4. List your height and shoe length, using the metric system.
   a. Height: _____ meters
   b. Shoe length: _____ millimeters

5. How many micrometers are there in 3.0 centimeters?

   _____ micrometers = 3.0 centimeters

6. How many waves, each 500 nanometers long, would fit along a 2-centimeter line?

   _____ waves would fit along a 2-centimeter line.

7. What is the distance, in kilometers, to a star that is 6.5 light-years from Earth?

   _____ kilometers from Earth

8. Uranus, one of the most distant planets, is 2870 million kilometers from the Sun. What is its distance from the Sun in astronomical units?

   Uranus is _____ astronomical units from the Sun.

9. Explain the difference between the terms *density* and *specific gravity*.

   _____

   _____

   _____

10. At the conclusion of the height vs. shoe length experiment, did you accept, reject, or modify your original hypothesis? Give a reason for this decision.

   _____

   _____

   _____

Appendices

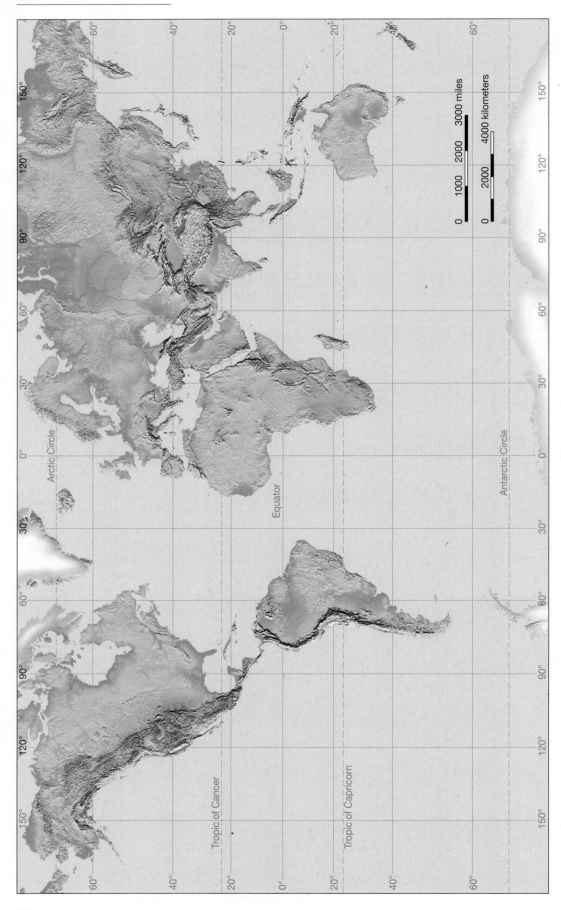

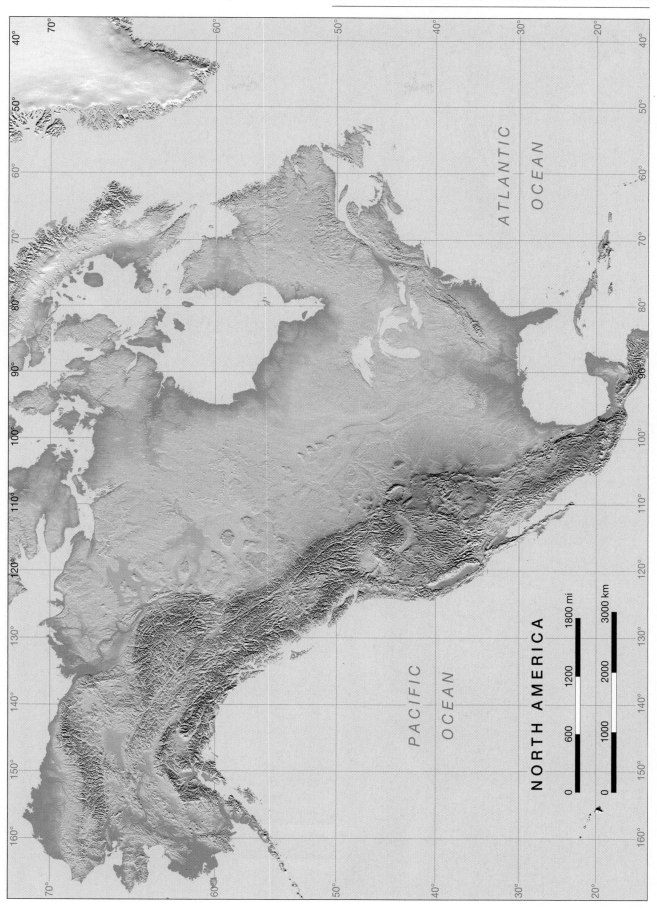

ATLANTIC
OCEAN

PACIFIC
OCEAN

NORTH AMERICA

0	600	1200	1800 mi

0	1000	2000	3000 km

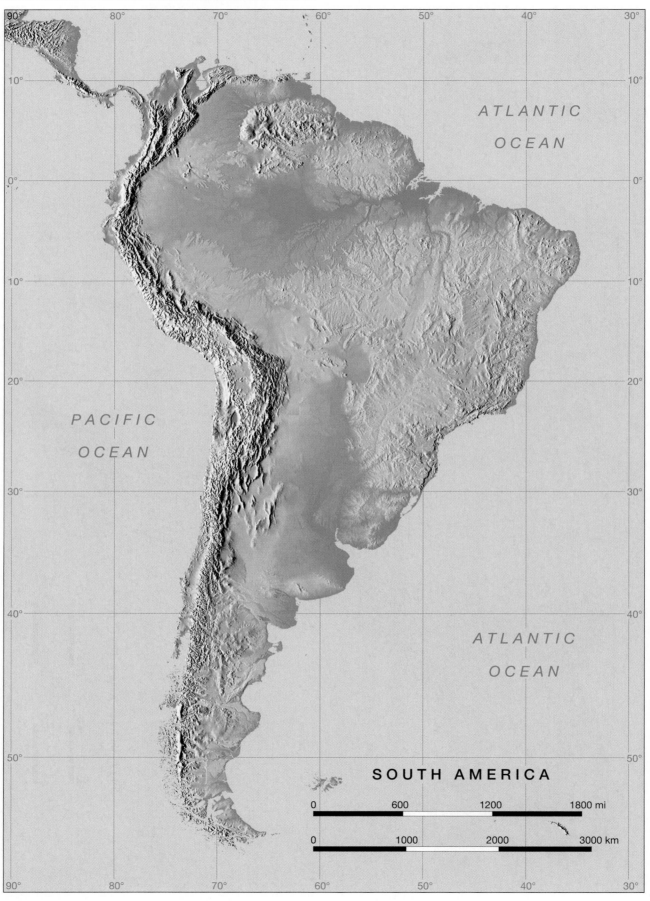

ATLANTIC
OCEAN

PACIFIC
OCEAN

ATLANTIC
OCEAN

SOUTH AMERICA

0	600	1200	1800 mi

0	1000	2000	3000 km

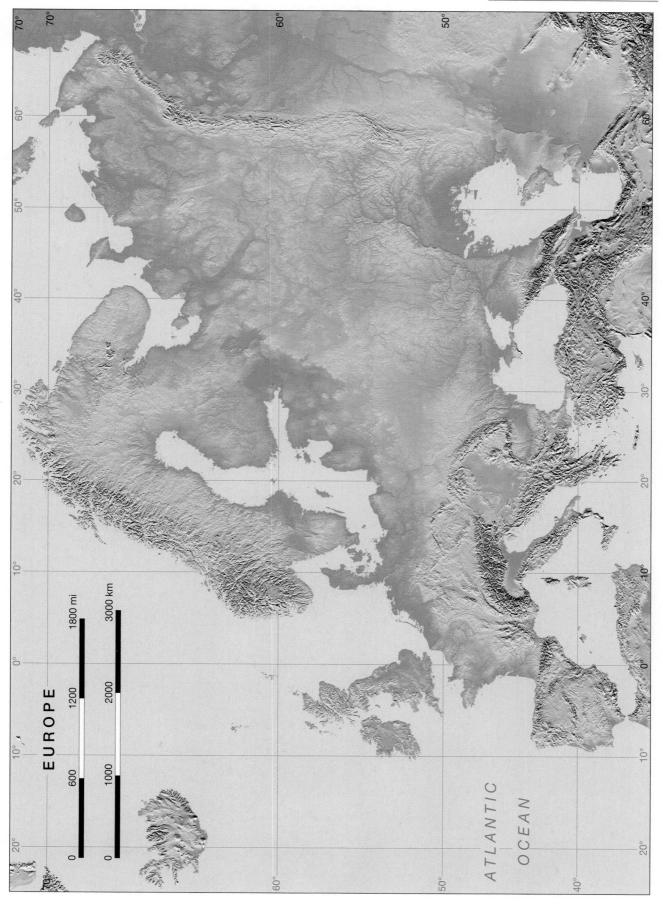

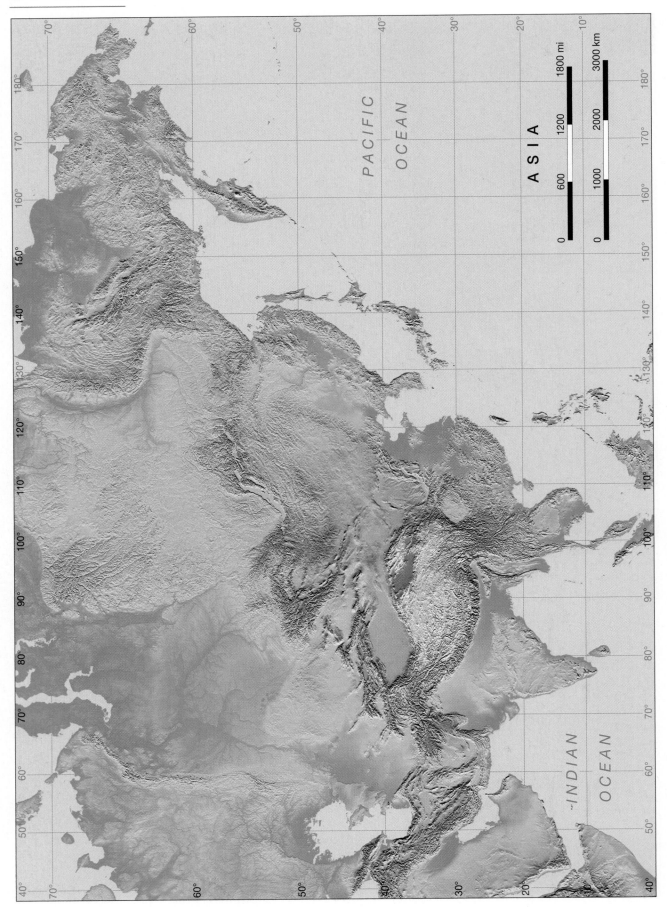

PACIFIC

OCEAN

ASIA

INDIAN

OCEAN

1800 mi

3000 km

1200

2000

600

1000

0

0

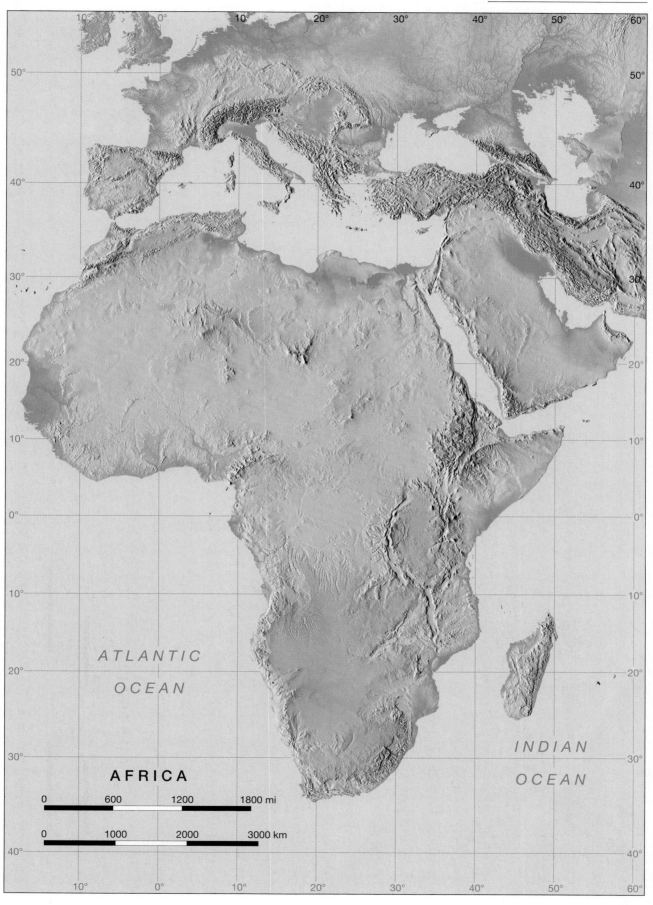

10°   0°   10°   20°   30°   40°   50°   60°

50°

40°

30°

20°

10°

0°

10°

ATLANTIC

OCEAN

20°

INDIAN

OCEAN

30°

AFRICA

| 0 | 600 | 1200 | 1800 mi |

| 0 | 1000 | 2000 | 3000 km |

40°

10°   0°   10°   20°   30°   40°   50°   60°

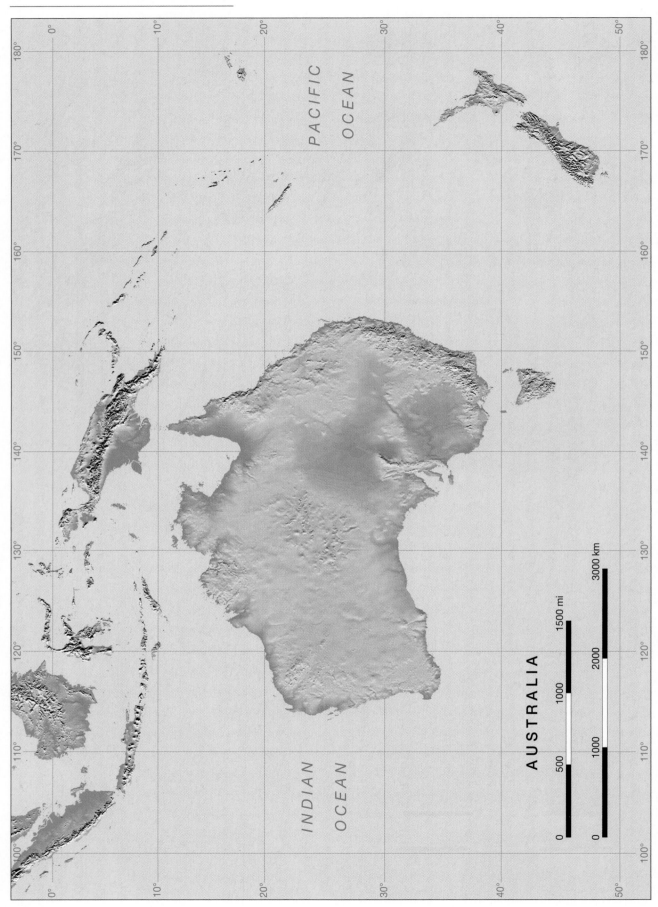

PACIFIC OCEAN

180°
170°
160°
150°
140°
130°
120°
110°
100°

0°
10°
20°
30°
40°
50°

INDIAN OCEAN

AUSTRALIA

0   500   1000   1500 mi

0   1000   2000   3000 km